APPLICATIONS OF PRESSURE-SENSITIVE PRODUCTS

T0175727

Handbook of Pressure-Sensitive Adhesives and Products

Fundamentals of Pressure Sensitivity

Technology of Pressure-Sensitive Adhesives and Products

Applications of Pressure-Sensitive Products

APPLICATIONS OF PRESSURE-SENSITIVE PRODUCTS

EDITED BY

ISTVÁN BENEDEK

MIKHAIL M. FELDSTEIN

CRC Press
Taylor & Francis Group
Boca Raton London New York

CRC Press is an imprint of the
Taylor & Francis Group, an **informa** business

CRC Press
Taylor & Francis Group
6000 Broken Sound Parkway NW, Suite 300
Boca Raton, FL 33487-2742

First issued in paperback 2019

ISBN-13: 978-1-4200-5935-9 (hbk)(vol 3)
ISBN-13: 978-0-367-38654-2 (pbk)(vol 3)
ISBN-13: 9781420059342 (set)

Library of Congress Cataloging-in-Publication Data

Applications of pressure-sensitive products / editors, Istvan Benedek and Mikhail M. Feldstein.
    p. cm.
Includes bibliographical references and index.
ISBN 978-1-4200-5935-9 (alk. paper)
    1. Pressure-sensitive adhesives. I. Benedek, Istvan, 1941- II. Feldstein, Mikhail M. III. Title.

TP971.A66 2009
668'.3--dc22
                                           2008012199

Visit the Taylor & Francis Web site at
http://www.taylorandfrancis.com

and the CRC Press Web site at
http://www.crcpress.com

Contents

8 Test of Pressure-Sensitive Adhesives and Products

Preface

Extensive progress in the use of pressure-sensitive adhesives (PSAs) and pressure-sensitive products (PSPs) in past years imposed a detailed and critical examination of their science and technology. Based on our experience in both scientific activity and industrial areas, as well as the special knowledge of outstanding scientists and technologists as contributors, we have addressed all aspects of PSAs in the form of a handbook. The huge volume of data accumulated in this field over the past decade presents a delicate problem due to the gap between the fundamentals of PSAs and their practice. The volume and diversification of data as well as the boundary between theory and application imposed the need to impart our treatise in three books.

The destination of this handbook is twofold. On the one hand, it is addressed to scientists focusing on the fundamental processes underlying the complex phenomenon of pressure-sensitive adhesion; on the other hand, it is intended for industrial researchers who are involved in the practical application of these fundamentals for the development of various products and specialists working in various end-use domains of PSPs. *Fundamentals of Pressure Sensitivity* contains detailed characterizations of the processes occurring in PSAs at all stages of the life of an adhesive joint: its formation under compressive force, under service as the bonding force is removed, and under adhesive bond fracture when the main type of deformation is extension.

Technology of Pressure-Sensitive Adhesives and Products describes particular features of the behavior of different classes of PSAs, for example, rubber–resin-based adhesives, acrylics, and silicones, and discusses the synthesis of pressure-sensitive raw materials, their formulation, and the manufacture of PSAs and PSPs. The PSPs manufactured using various raw materials and technologies (described in *Technology of Pressure-Sensitive Adhesives and Products*) are applied in different domains.

This book, *Applications of Pressure-Sensitive Adhesives and Products*, describes the main classes and representatives of PSPs, as well as their competitors, end use, application domains, application technology, and tests.

Chapter 1 presents the "Construction and Classes of Pressure-Sensitive Products." Various PSPs were developed and requirements for new applications facilitate their continuous diversification. Chapter 1 describes the build-up, classes, and main representatives of classic PSPs. Examination based on their construction (components and build-up) allows further detailed discussion of some special products developed recently that have increased importance (e.g., electroconducting, electro-optical, and medical PSPs). The PSPs are described according to their build-up (i.e., adhesive-coated, adhesiveless, carrierless, and linerless construction), the status of the adhesive (i.e., solvent-, water-borne, and hot-melt-based adhesives), their removability (permanent/removable), and their environmental (temperature and chemical) resistance. Labels, tapes, protective films, and forms are presented as the main representative product classes.

"Electrically Conductive Adhesives in Medical Applications" are the subject of Chapter 2. Preparation, electroconducting, and adhesive properties of such adhesives are discussed. Application of electrically conducting adhesives in iontophoretic systems for controlled transdermal drug delivery is also described.

"Pressure-Sensitive Adhesives for Electro-Optical Applications" are described in Chapter 3. Two main types of PSAs for electro-optical applications are discussed, a protective coating laminate for the color filter in the liquid crystal display, which requires a UV-curable, optically clear PSA to protect and flatten the color layer for improved image quality with easy handleability without involving solvent, and low-refractive-index PSAs as supportive materials to help control light distribution with low reflectance and low interference in light-transmitting devices. The chemistry, the physics, and the processing of these PSAs are examined.

"End-Use Domains and Application Technology of Pressure-Sensitive Adhesives and Products" are the subject of Chapter 4. Chapter 4 describes the end-use domains of the main PSPs (i.e., labels, tapes, protective films, and forms), the special products developed for each application field, the application technology (methods, devices, and machines), the deapplication technology, the postmodification technology, and the non-pressure-sensitive competitors for such products.

Chapter 5 describes the preparation methods of "Skin Contact Pressure-Sensitive Adhesives," which find increasing application in transdermal drug delivery systems. The first part of Chapter 5 deals with hydrophilization of conventional hydrophobic PSAs via mixing with hygroscopic absorbents of moisture (see also Chapter 7 in *Technology of Pressure-Sensitive Adhesives and Products*). In the second part of Chapter 5, the so-called hydrocolloid formulations are described, based on emulsions.

"Factors Governing Long-Term Wear of Skin-Contact Adhesives" are examined in Chapter 6. Until recently, the problems of PSA wear on human skin comprised an area of empirical qualitative expertise of researchers dealing with the development of adhesive dressings and patches for transdermal drug delivery. Chapter 6 is the first publication to set the basis for new fundamental science that renders the quantitative prediction of wearing time from the results of a simple *in vivo* peel test on volunteers. Mechanisms of PSA interaction with the epidermis of human skin, which have great practical and theoretical importance, are also the subject of investigation in Chapter 6.

"Competitors for Pressure-Sensitive Adhesives and Products" are discussed in Chapter 7. Advances in macromolecular chemistry led to the development of adhesive-free tacky films used *per se* or by means of a chemical or physical treatment. Chapter 7 describes the manufacturing competitors of adhesive-coated PSPs. Developments in macromolecular chemistry and polymer processing facilitated the manufacture of self-adhesive plastic films. Such films are tacky due to their special chemical composition or macromolecular build-up, having a synthesis-based or a formulation-given tack and peel, or they are

physically surface-treated. A broader range of products with pressure-sensitive properties exists now compared with some decades ago, and some are not coated with PSAs.
Rather, they have an inherent pressure sensitivity. Other products are adhesive-free.
PSPs include PSA-coated products, but the two product groups (PSAs and PSPs) may
differ significantly. PSPs are more than PSAs because product construction can contribute to self-adhesive behavior. In some domains of PSP application, a technical solution may be given by engineering of plastics or adhesives. The self-adherence can be
controlled by application conditions (e.g., temperature and pressure). Such products are
removable and, thus, they compete with PSA-based removable self-adhesive products
with a low tack/peel resistance (e.g., protective films and masking tapes). Their manufacture, performance characteristics, and end use are presented in comparison with classic
adhesive-coated self-adhesive products.

Chapter 8, "Test of Pressure-Sensitive Adhesives and Products," describes the test
methods used for PSAs and PSPs. Fundamental research focused on PSAs and PSPs, as
well as the manufacture technology of pressure-sensitives, requires test methods that
allow control and characterization of the raw materials used for PSAs, tests of the processability of the adhesive, tests of the adhesive performances of the PSAs and PSPs,
tests of the convertibility of the pressure-sensitive web, and tests of the application performance of the products. Such tests are tools for industrial quality assurance and serve
as a control modality of technological discipline or end-use applicability. Various in-
house and standard test methods were developed and are still in use. Some (as noted in
Fundamentals of Pressure Sensitivity) allow the correlation of macromolecular science
with practical requirements. A critical view of test methodology (science versus industry) is presented.

This book includes the scientific basis of suitability of raw materials for specific applications (i.e., chemical and physical, rheology) and the manufacture (formulation) of the
adhesive and the PSP (converting the adhesive). We have selected self-adhesive labels,
tape, and protection films as the main laminates; whenever possible, a comparison with
and extension to other products and applications is included. To illustrate the different
topics and issues discussed, we have referred to a number of commercially available
products. These products are only mentioned to clarify the discussion and in no way
constitute any judgment about inherent performance characteristics or their suitability
for specific applications or end uses.

We were pleased to see the participation of scientists and industrial experts working
in very different areas of the field on this book. We thank our contributors for their
efforts.

The Editors

Editors

István Benedek is an industrial consultant based in Wuppertal, Germany. After exploring his initial interest in macromolecular science, he transferred to the plastics processing and adhesive converting industry as research and development manager, where he has worked for three decades. He is the author, coauthor, or editor of several books on polymers, including *Pressure-Sensitive Adhesives Technology* (Dekker, New York, 1996), *Development and Manufacture of Pressure-Sensitive Products* (Dekker, New York, 1999), *Pressure-Sensitive Formulation* (VSP, Utrecht, the Netherlands, 2000), *Pressure-Sensitive Adhesives and Applications* (Dekker, New York, 2004), *Development in Pressure-Sensitive Products* (CRC, Boca Raton, FL, 2006), *Pressure-Sensitive Design, Theoretical Aspects* (VSP, Leiden, the Netherlands, 2006), and *Pressure-Sensitive Design and Formulation, Applications* (VSP, Leiden, the Netherlands, 2006), as well as more than 100 scientific research and technical reports, patents, and international conference papers on polymers, plastics, paper/film converting, and web finishing. He is a member of the Editorial Advisory Board of the *Journal of Adhesion Science and Technology*. Dr. Benedek received his PhD (1972) in polymer chemistry and engineering technology from Polytechnic University of Temeswar.

Mikhail M. Feldstein, one of the world's leading experts in the development of new polymeric composites with tailored performance properties that span pressure-sensitive adhesives and other materials designed for medical and pharmaceutical applications, was born in 1946 in Moscow. In 1969 he graduated with honors from M.V. Lomonosov Moscow State University, Faculty of Chemistry, and in 1972 he earned his PhD in polymer science from the same university for the investigation of polyelectrolyte complexes with ionic surfactants and lipids. His early research interests were associated with the mechanisms of the formation and molecular structure of interpolymer complexes. Since 1972 he has worked in the industry of polymers for medical usage as a developer of hydrophilic pressure-sensitive adhesives for skin application in transdermal therapeutic systems and wound dressings. He received international recognition comparatively late: his earliest contacts with colleagues beyond the borders of former Soviet Union date to 1994 only. In 1999, a famous scientist and vice president of the Russian Academy of Sciences, academician Nicolai A. Platė, invited him to join A.V. Topchiev Institute of Petrochemical Synthesis of the Russian Academy of Sciences, one of the most well-known academic institutes in polymer science. Later that year, Feldstein established long-term and large-scale research cooperation with a leading pharmaceutical company, Corium International, Inc. (CA). In 2005, Feldstein earned his DrSc in polymer science from the A.V. Topchiev Institute of the Russian Academy of Sciences.

Since the second half of the 1990s, Feldstein has focused on the molecular origins of pressure-sensitive adhesion and the interrelationship between adhesion and other properties of polymer blends. Based on gained insight into the phenomenon of adhesion at a molecular level, he has developed the first-ever technology for obtaining numerous novel pressure-sensitive adhesives of controlled hydrophilicity and performance properties by the simple mixing of nonadhesive polymer components in certain ratios. Feldstein is the author of nearly 200 research papers, 7 book chapters, and 25 patents. He is a member of Adhesion Society and Controlled Release Society. Feldstein is also an associate editor of the *Journal of Adhesion*.

Contributors

Sergey V. Antonov
A.V. Topchiev Institute of Petrochemical
  Synthesis
Russian Academy of Sciences
Moscow, Russia

István Benedek
Pressure-Sensitive Consulting
Wuppertal, Germany

Anatoly E. Chalykh
A.N. Frumkin Institute of Physical Chemistry
  and Electrochemistry
Russian Academy of Sciences
Moscow, Russia

E.-P. Chang
Avery Research Center
Pasadena, California

Daniel L. Holguin
Avery Research Center
Pasadena, California

Valery G. Kulichikhin
A.V. Topchiev Institute of Petrochemical
  Synthesis
Russian Academy of Sciences
Moscow, Russia

Alexey V. Shapagin
A.N. Frumkin Institute of Physical Chemistry
  and Electrochemistry
Russian Academy of Sciences
Moscow, Russia

Anna A. Shcherbina
A.N. Frumkin Institute of Physical Chemistry
  and Electrochemistry
Russian Academy of Sciences
Moscow, Russia

Parminder Singh
Corium International Inc.
Menlo Park, California

J. Anand Subramony
Dr. Reddy's Laboratories Ltd.
Bachupalli, Qutubullapur, India

Natalya M. Zadymova
Chemical Department
M.V. Lomonosov Moscow State University
Moscow, Russia

1

Construction and Classes of Pressure-Sensitive Products

István Benedek
*Pressure-Sensitive
Consulting*

1.1 Build-Up of Pressure-Sensitive Products

Applications of Pressure-Sensitive Products correlates the fundamentals of pressure sensitivity described in *Fundamentals of Pressure Sensitivity* and the manufacturing technology of pressure-sensitive adhesives (PSAs) and pressure-sensitive products (PSPs) presented in *Technology of Pressure-Sensitive Adhesives and Products* with their application technology, end-use domains, and test methods. This first chapter serves as an introduction, which, together with Chapter 4, constitutes a basis for systematic discussion of various special aspects of PSP application in the next chapters. This chapter will inform the reader about the latest developments in the application and end use of PSAs and PSPs.

PSPs may have a simple or sophisticated construction, depending on their end use. They can be classified according to their fundamental build-up; however, the main classes of PSPs (labels, tapes, and protective webs) cover a wide range of application fields. Thus, they can have various specific constructions as well. On the other hand, more or less expensive products can be used in the same application field. The build-up of PSPs was described in detail by Benedek in a previous book [1]. This chapter presents only the principles of build-up of the main PSPs. Their classification and a detailed discussion of their application will be presented in Chapter 4, as well as a discussion of special products by experts from various domains of pressure-sensitive application (e.g., medical or electronic fields). Generally, PSPs are web- or sheet-like constructions that exhibit self-adhesion. In principle, such products include the component(s) that ensures the required mechanical properties and the component(s) that provides adhesion. In special cases, a sole component (e.g., the PSA for carrierless PSPs or the carrier for self-adhesive films) can play the role of both components. Development of the control of debonding resistance, that is, advances in removable products, was decisive for the development of adhesiveless self-adhesive products. Generally, removability requires peel reduction. Peel reduction is correlated to tack reduction. On the other hand, the application of PSPs requires a minimum amount of tack. Thus, certain removable low-peel and low-tack products must undergo forced lamination (at room temperature [RT] or at elevated temperature) during their application with a higher application pressure, that is, they need the same application technology as adhesive-free self-adhesive plastic films. That means that for certain end uses self-adhesive products without PSA can also be suggested (see Chapter 7 in this book).

1.1.1 Adhesive-Coated Pressure-Sensitive Products

In the first stage of PSP development, PSAs were used to impart pressure sensitivity. In a classic, pressure-sensitive adhesive-coated product, adhesivity was provided by the PSA. Therefore, the adhesive was the most important component of PSPs. According to the sheet-like, that is, the discrete form of the PSP (e.g., labels and forms), or the web-like form (e.g., tapes and protective films), the coated PSA layer must be protected by a separate release liner; otherwise, this function is given by the release properties of the carrier. Thus, monowebs (i.e., monolayer constructions or multiwebs, such as pressure-sensitive laminates) are manufactured (see also *Technology of Pressure-Sensitive Adhesives and Products*, Chapter 10). The construction of PSPs was discussed in detail by Benedek in Ref. [1]. Monoweb constructions (uncoated and coated) and multiweb constructions were described, uncoated and coated monoweb constructions were investigated in comparison with coated ones, and various coated monoweb constructions were also examined comparatively. The terminus technicus "monoweb" is employed in this case for a PSP that is used as produced (i.e., as a continuous one-component, film-like material). This product can have a sophisticated multilayer build-up, which is a permanent construction, in comparison with temporary multilayer constructions that can be disassembled during their use (i.e., multiwebs). Figure 1.1 summarizes the main constructions of monoweb PSPs.

1.1.1.1 Monoweb Products

The complex, multilayer structure of labels containing a solid-state carrier and a solid-state release liner, temporarily bonded together by an adhesive, can be simplified for

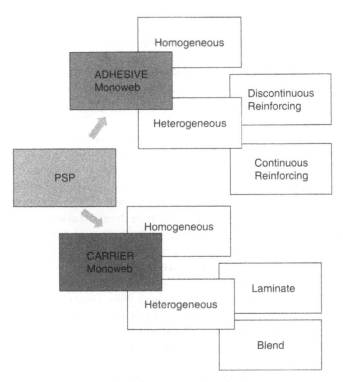

FIGURE 1.1   The main monoweb pressure-sensitive constructions.

tapes and adhesive-coated protective films, which possess only one solid-state carrier component. Principally, this is allowed by their web-like manufacture and use, in which the carrier material also plays the role of backing. Classic monoweb constructions (e.g., tapes and protective films) are coated monowebs (i.e., they possess a PSA adhesive layer coated on a nonadhesive carrier material). As discussed previously (see Chapter 10 in *Technology of Pressure-Sensitive Adhesives*), the adhesive layer can be applied with various (coating) techniques such as those used in the converting (coating or printing) industry or in plastics manufacturing (extrusion and calendering). As discussed in *Technology of Pressure-Sensitive Adhesives and Products*, Chapter 8, because of the broad range of raw materials available for PSAs and the sophisticated adhesive coating technologies now in use (see *Technology of Pressure-Sensitive Adhesives and Products*, Chapter 10), the adhesive properties of a PSP can be easily regulated by coating a solid-state carrier material with a low-viscosity PSA with controllable chemistry-related adhesive characteristics that can also be regulated by coating geometry.

The choice of mono- or multiweb PSP construction depends on its adhesivity. Principally, PSPs coated with a contact adhesive (e.g., envelopes) with low tack, which adheres only to itself, can be considered monowebs (see also *Technology of Pressure-Sensitive Adhesives and Products*, Chapter 2, and Chapters 4 and 8 of this book).

Generally, in such products the existence of a solid-state carrier material is imposed by its application technology. The carrier acts as a face stock material as well as a release liner. Both functions are illustrated by tapes. Tapes must be more aggressive; therefore, they require a thick PSA layer coated on a nonadhesive carrier material. As known from the manufacture of carrier material for tapes [2], generally such carrier material functions as a release liner also; therefore, it possesses release performance or is transformed into an adhesive-repellent material by coating with or embedding of release substances or modeling its surface (e.g., embossing; see *Technology of Pressure-Sensitive Adhesives and Products*, Chapter 10).

Depending on the end use of the PSP, the solid-state carrier material and the PSA coated on this carrier can be quite different. The geometry of both components (solid-state carrier material and PSA) and the construction of the PSP can also differ. The particular build-up of a PSP is a function of required performance characteristics and of the manufacturing procedure. The product build-up-related formulation was described by Benedek in Ref. [3]. PSP manufacture may require the use of other supplemental non-adhesive, coated layers (e.g., primer, lacquer, printing ink, and release agents; see also *Technology of Pressure-Sensitive Adhesives and Products,* Chapters 8 and 10, and Chapter 4 of this book). Thus, from the point of view of their time-independent construction, the main parts of so-called monoweb PSPs are actually multiwebs.

Tapes are the most important class of monoweb PSPs. Their construction as a function of the product components are described in Section 1.2.4.2; their construction as a function of their special application domains are discussed in Chapter 4 and their test methods are presented in Chapter 8. Transfer PSAs (so-called adhesives from the reel), coated temporarily on a siliconized release liner (double-side siliconized) can be considered adhesive-coated monowebs. In this case, the solid-state carrier material is the release liner. The build-up of adhesive-coated PSP webs as a function of their components is described in Section 1.2.4.3. From the point of view of their construction, protective films can be considered large surface tapes (or vice versa), namely, special removable tapes.

Only a few monoweb labels were developed (see Chapter 4). Because of the balanced adhesive performance of labels (see *Technology of Pressure-Sensitive Adhesives and Products*, Chapter 11) and their low-pressure, high-speed application technology (automatic labeling), it is very difficult to manufacture labels without a separate release liner (i.e., labels with a monoweb structure). However, special monoweb labels (roll labels) have also been developed. Linerless labels are supplied as a continuous tape-like monoweb material. A special coating on the top surface of the label prevents blocking of the adhesive layer. John Waddington in the U.K. launched monoweb labels in the 1980s, but their application equipment is more expensive and their printing is more complex than that of other label types.

The number of coated layers and their nature in a monoweb PSP depends on its end-use requirements. For instance, a laminated PSA tape with good chemical, heat, and water resistance that is to be used for soldering portions of a printed circuit board is manufactured using a craft paper carrier material, impregnated with latex, treated with a primer or corona discharge, coated on one side with a rubber–resin adhesive and on the other side with a polyethylene film and release. Such a multilayer construction (i.e., multiweb) is used as a monoweb. Table 1.1 illustrates the main product constructions used as monowebs.

TABLE 1.1   Build-Up of Main PSPs Used as Monowebs

Product	Construction	Carrier Role During		PSP Grade
		Manufacture	Application	
	One-Component Monoweb			
Carrierless tape	Reinforced	Temporary	Temporary	Transfer tape
Self-adhesive film	Tackified	Permanent	Permanent	Protective film Separation film
	Multicomponent Monoweb			
Common tape	Two-component PSP: PSA–carrier	Permanent	Permanent	Packaging and special tapes
	Three-component PSP: PSA–carrier–release layer	Permanent	Permanent	Packaging and special tapes Monoweb label Overlaminating film
	Multicomponent PSP: PSA–primer–carrier–release layer	Permanent	Permanent	Packaging and special tapes
	Multicomponent PSP: PSA–primer–carrier–printing ink–release layer	Permanent	Permanent	Packaging and special tapes
	Multicomponent PSP: PSA–primer–multilayer carrier–printing ink–release layer	Permanent	Permanent	Packaging and special tapes
	Multicomponent PSP: PSA–primer–impregnated carrier–printing ink–release layer	Permanent	Permanent	Packaging and special tapes
Common protective webs	Two-component PSP: PSA–carrier	Permanent	Permanent	Paper- and film-based protective webs
	Three-component PSP: PSA–carrier–release	Permanent	Permanent	Paper- and film-based protective webs
	Multicomponent PSP: PSA–primer–multilayer carrier–printing ink–release printing ink–release	Permanent	Permanent	Film-based protective webs

The range of common monoweb products is large. Although the old, theoretical definition of PSAs accentuates their instantaneous adhesion (tack), in practice products with very low tack are also used more and more frequently, mostly for removable applications. As mentioned previously, for such products the so-called application tack is tested, which supposes bonding under pressure. PSA-based PSPs were developed as monowebs also (e.g., contact adhesive-based and cold-seal adhesive-based products in which bonding occurs to the adhesive itself without or under pressure; see Table 1.2); they work like common PSPs due to their viscoelasticity. Therefore, from the practical point of view, it would be more correct to define PSPs as products that bond under very different conditions due to their special rheology.

TABLE 1.2   Special Adhesive-Coated Monoweb Products

Product Class	Construction	Build-Up			Application Conditions		Commercial Product
		Base Polymer	Release		Application Pressure	Substrate	
Contact adhesivated	Adhesive-coated carrier	NRL	—		Low	Itself	Envelops, bags
Cold sealable	Adhesive- and release-coated carrier	Acrylic	PA		High	Itself	Bags, various packaging materials
RT/HT laminable	Adhesive- and release-coated carrier	NR, acrylic	Carbamate		Medium	Various materials	Protection film

1.1.1.2   Laminates

As illustrated in Table 1.1, certain PSPs that are considered monoweb construction from the functional point of view possess a multilayer build-up based on multilayer carrier film or numerous "soft" coated layers. They can be considered laminates.

In principle, laminate constructions are not new. They were developed in the packaging material industry. The introduction of new face stock materials, especially film-like materials, required the improvement of certain characteristics such as dimensional stability, cuttability, and temperature resistance. This was possible partially because of the development of laminated face stock materials. Such multiweb constructions have been produced by bonding the solid-state components through adhesive coating or by extrusion. Theoretically, the common adhesive-coated laminates (e.g., various, mainly heterogeneous packaging materials) differ from PSPs with respect to the nature of the laminating adhesive and the character of the adhesive bond only. They are manufactured using a permanent adhesive, that is, they are permanent laminates (except for some special film laminates used for so-called easy-peel applications, which, although permanently laminated, can be debonded). Generally, in the construction of PSPs, such permanent laminates can be built up in a temporary laminate. Permanent laminates serve as carrier materials; temporary laminates are PSPs. Some PSPs, (e.g., security labels and forms) are constructed with a carrier that includes both types of laminates, temporary and permanent.

The components of the pressure-sensitive laminate depend on its construction, end use, application technology, and final form of PSP. Principally, there are sheet laminates and roll laminates.

1.1.1.3   Multiweb Products

In principle, a pressure-sensitive laminate is built up from a carrier material, an adhesive, and a release liner [1,4]. Because of the balanced adhesive performance characteristics (i.e., high tack, high peel resistance, and adequate cohesion) and their discontinuous (non-web-like) character, which imposes their high-speed handling and automatic application, labels require a continuous, supplemental carrier material, that is, a release liner. The liner allows labels to be processed (i.e., printed and confectioned) as a continuous web (see also Chapter 4) and protects their adhesive layer. The liner makes the web-like application of sheet-like (e.g., die-cut) products possible. Almost all label stocks and double-side pressure-sensitive tapes use a separate release web. Such products with temporary multiple solid-state components are laminates (see also *Technology of Pressure-Sensitive Adhesives and Products,* Chapter 10).

Classic manufacture of the finished product requires build-up of a PSP from its components (i.e., carrier material[s] and PSAs). In practice, other solid-state product components such as the release liner and cover film, as well as liquid-state components such as the primer and antistatic agents, can be built in one or more layers, yielding a laminate with a complex structure.

For instance, certain tapes must be primed to ensure good anchorage of the adhesive on the carrier (see also *Technology of Pressure-Sensitive Adhesives and Products,* Chapter 10). For some tapes a release layer should be coated on the back side of the carrier. Such a layer is not necessary if the material of the carrier exhibits abhesive properties, like certain nonpolar plastic films. For special tapes a separate release film should be interlaminated. Such tapes include double-side products and transfer tapes, which need a separate solid-state release or carrier layer. In this way, various multilayer products are manufactured with multilayers of soft components (e.g., adhesive, primer, or ink) or multilayers of solid-state components (e.g., carrier, liner, or overlaminating film).

Pressure-sensitive laminates manufactured to protect the adhesive-coated surface of a tape or label are an example of temporary construction. In such products the release liner may protect the adhesive or other components incorporated in the adhesive layer (see also Chapter 4). Another role of the separate release liner is that of a continuum passing through the coating, converting, and labeling machines carrying the discontinuous label. Decals or labels applied by hand do not need this function. In some cases, such as the manufacture of transfer tapes, the use of a solid-state component is a technological modality used only to build up and apply the product. A patent [5] describes the use of (ultraviolet [UV]-light-induced) photopolymerization of acrylic monomers directly on the carrier to manufacture tape. This patent uses a temporary carrier, an endless belt, which does not become incorporated in the final product. The tapes have PSA on both sides of the carrier. In this case the laminate is an auxiliary construction built up to allow the construction of the final laminate from the label or tape and the substrate. Labels and certain tapes are built up during manufacturing and application as temporary laminates and are used as permanent laminates.

Multiweb constructions can include solid-state components and adhesives that differ in their chemical nature and build-up (e.g., paper, plastic films, fabrics, nonwovens, and foam). For instance, double-side-coated tapes can have a PSA on one side and a reaction mixture of uncross-linked polyurethane (PU) on the other side. A polyvinyl chloride (PVC) tape applied on a PU foam reacts with the adherent and adheres to it chemically. Double-side mounting tapes may have different adhesives on each side of the carrier (see also Chapter 4). Labels may also have a very complex construction. For instance, label construction may contain seven layers as follows: release liner, release layer, adhesive, primer, face stock, ink, and top coat [6]. Such multilayer self-adhesive labels comprise a carrier label containing silicones, an adhesive layer, a printed message, a carrier layer (e.g., polyester), a release layer (e.g., silicones), a second adhesive layer, a second carrier layer (e.g., polyethylene [PE]), and a printed message. Labels may have more than two adhesive–carrier layer–message layer units.

1.1.1.3.1 *Pressure-Sensitive Adhesives with Multilayer*

As discussed previously (see *Technology of Pressure-Sensitive Adhesives and Products*, Chapter 10), certain PSPs are manufactured as a continuous laminate from a primed-carrier material and a release-coated liner. Postfinishing of such a web during its conversion (see *Technology of Pressure-Sensitive Adhesives and Products*, Chapter 10) can lead to its coating with other supplemental soft layers (e.g., printing ink and lacquer). Labels are variances of such construction. According to their active (adhesive) surfaces, tapes can be classified as single- or double-side coated. Double-side-coated tapes are used with or without a carrier. The adhesive characteristics as well as the chemistry of the adhesive layers may be quite different. In Chapter 3, Holguin and Chang describe a special, *in situ*-made pressure-sensitive composite based on a curable label construction comprising a curable PSA layer (1–2 µm thick) and curable epoxy layer (0.5–1.5 µm thick) coated onto a transparent carrier film. Common industrial practice uses multilayer adhesives constructed from a rigid primer and soft adhesive to control energy dissipation, that is, removability (see also *Technology of Pressure-Sensitive Adhesives and Products*, Chapter 1).

1.1.1.3.2 *Pressure-Sensitive Products with a Multilayer Carrier*

As discussed previously (see *Technology of Pressure-Sensitive Adhesives and Products*, Chapter 10), for certain products the carrier material is a laminate. In other cases, the product itself has a sophisticated multilayer structure including various solid-state components. The main representatives of such products are business forms (see Chapter 3), in which various solid-state and soft components are "assembled" together ensure a multiple-detachable construction, with some parts that display permanent adhesion and other parts that are only temporarily bonded. Such products are constructed composites (see also *Technology of Pressure-Sensitive Adhesives and Products*, Chapter 1). They can also include soft components with a composite structure (e.g., adhesives). The release liner itself may also have a multilayer construction. For instance, different degrees of release are achieved for a double-side liner and a release paper, coated on one side with PE and using the same release component.

1.1.1.4 Products with Special Adhesive Build-Up

Products with special adhesive construction contain a composite-like adhesive or a PSA that is coated with special geometry. Continuous or discontinuous carrier materials as well as adhesive layers with a special geometry and multilayer composite build-up can be used. The discontinuity of the carrier and the adhesive may serve to fulfill special mass- or heat-transfer requirements to allow regulation of adhesive performance (e.g., removability [7]) or converting and end-use characteristics (e.g., cuttability [2,8] or easy tear [9]). Some monoweb tapes include a partially discontinuous carrier material, a discontinuous adhesive, or both. A porous carrier may be required as well. The carrier can also have an asymmetric construction (geometry, surface quality, etc.). A patterned PSA transfer tape may possess an asymmetric carrier material, which has one surface with a series of recesses and another that is smooth. In some cases the cross-section of the carrier differs from a usual parallelepiped. Masking tapes can have special carrier constructions to allow conformability. For instance, a masking tape carrier can have a stiffened, wedge-shape, adhesiveless, longitudinal section (the thickest away from the tape centerline) extending from one edge, with a pleated structure to conform to small radii. A packaging tape can have a central tear section. The central tearaway portion is built up with two parallel lines of perforations extending the entire length of the flexible part. Adhesive tapes with the adhesive-coated area narrower than the substrate and one edge with an adhesive-free strip have an easy-untie property and are useful for bundling electronic parts, building materials, vegetables, and other items.

1.1.1.4.1 Filled Adhesive

As discussed previously (see *Technology of Pressure-Sensitive Adhesives and Products*, Chapter 8), the PSA can be filled during formulation. As described in detail in Refs. [1,2], the carrier material can also have a filled structure. Filling reduces the costs and ensures special properties such as improved cohesion (and converting properties) and control of thermal or electrical conductivity (see also Chapter 4). Glass microbubble-filled mounting tapes imitate the conformability of cellular (air-filled) structures. Electrical conductivity or nonconductivity may be required for insulating, packaging, and medical tapes and is achieved with special fillers. Filled medical PSAs were described in Ref. [10]. Pressure-sensitive hydrogels are a special domain of bioadhesives that contain water, as described by Feldstein *et al.* in Refs. [11,12]. Filled hydrocolloids were discussed by Kulichikhin *et al.* in Ref. [13]. Filled adhesives play an important role in the manufacture of carrierless PSPs as well (see Section 1.1.3). Developments in "water-filled" PSPs are discussed in detail by Feldstein, Singh, and Cleary in *Technology of Pressure-Sensitive Adhesives and Products*, Chapter 7.

1.1.1.4.2 Discrete Adhesive

The adhesive (if coated on a carrier) can be discontinuous also. Coating of the adhesive as a noncontinuous layer with discrete areas allows control of the bonding–debonding properties [14]. Producers are looking to manufacture their own base label stock or special stripe-, patch-, or spot-coated adhesive construction and labels with adhesive-free zones. For instance, a pressure-sensitive removable adhesive tape that is useful for paper products comprises two outer discontinuous layers (based on tackified styrene–isoprene–styrene block copolymer) and a discontinuous middle release layer. The tape can be manufactured with less adhesive than conventional tapes and its top side can be printed and perforated. Adhesive tapes with a narrower adhesive-coated width than that of the substrate and one edge with an adhesive-free strip have easy-untie properties and are useful for bundling electronic parts, building materials, vegetables, and other commodities.

The adhesive layer can have a rough surface as well, with channels to enclose air flow between the adherent and the sheet. The surface roughness of the sheet has a

50- to 1000-μm width and a 10- to 1000-μm height of convexes [15]. Adhesive tape may have adhesive only along the lengthwise edges (to allow easy perforation) [16]. A removable display poster is manufactured by coating distinct adhesive and nonadhesive strips on the carrier. The adhesive strips are situated on the same plane, with the plane elevated with respect to the product surface. For a pressure-sensitive label with a wrinkle-resistant, lustrous, opaque-facing layer for application to collapsible-wall containers (squeeze bottles), the carrier has a thermoplastic core layer with upper and lower surfaces and voids, a void-free thermoplastic skin layer fixed to the upper surface and optionally to the lower surface of the core layer, and discrete areas of PSA. For the core layer, a blend of isotactic polypropylene (PP)/polybutene terephthalate with PP as the skin layer and circular dots of hot-melt PSA (HMPSA) as the adhesive is used.

Multilayer adhesive construction can also be manufactured. Here the PSA-coated carrier material functions as a transfer base for a heat-activatable adhesive. After application of the heat-sensitive adhesive, the PSA-coated tape can be peeled off [17].

1.1.2 Adhesiveless Pressure-Sensitive Products

Self-adhesive products generally have either a carrier and a PSA layer or a pressure-sensitive carrier. PSPs that are built up, such as an uncoated monoweb, are composed of a carrier material only. This carrier must have a special chemical nature or undergo special physical treatment to allow self-adhesion under special application conditions (pressure, temperature, and previous surface treatment). The development of macromolecular chemistry and extrusion technology allows the manufacture of carrier materials with built-in pressure sensitivity, that is, PSPs constructed like an uncoated monoweb but behaving, when applied, like a coated web. Carrier materials with adhesivity were discussed in detail in Refs. [2,18]. Generally, such conformable, autoadhesive monowebs are plastic films (see Chapter 7). Uncoated monowebs used as PSPs can have a homogeneous or heterogeneous structure. The whole carrier can be autoadhesive (e.g., ethylene-vinyl acetate copolymers [EVAc] or very-low-density polyolefin-based films) or it may possess an adhesive layer defined by manufacture (coextrusion) or by diffusion of a self-adhesive, built-in (embedded) component (see Chapter 7).

The manufacture of a self-supporting adhesive material is a complex procedure. The adhesive bond is the result of chemical attraction as well as physical anchorage. Both require contact surface and interpenetration, that is, flow. Such a cold flow depends on chemical basis, the product's geometry, and application conditions. Unfortunately, mechanically resistant, self-supporting products exhibit only a limited flow, with the exception of PVC, which is (due to its plasticizing ability) an ideal material to achieve self-adhesive performance without loss of mechanical strength. Depending on its formulation and softness, PVC can be used as a self-adhesive medical tape (carrier) or as an adhesive for such products. For instance, decorative decals (e.g., adhesive films based on PVC with a very high plasticizer level) possess a "softened" monoweb construction. Such monoweb tapes are mostly special or experimental products. Only the production of superficially soft plastic film carrier materials and the development of thermoplasts with elastic (rubber-like) and viscoelastic properties allow the manufacture of noncoated PSPs.

The new generation of protective films is built up from a (adhesive) carrier without a PSA layer. Theoretically, such protective films are the simplest PSPs, built up as a one-component, pressure-sensitive (self-adhesive) carrier material. In practice, the manufacture of an uncoated, self-adhesive adhesive material is a complex procedure, described in Ref. [2] (see also Chapter 7).

1.1.3 Carrierless Pressure-Sensitive Products

Generally, the carrier plays the role of a face stock material. It is the most important solid-state part of the product, functioning as a packaging-, protecting-, and information-carrying component. This is a common role of the carrier material in labels, tapes, and protective webs, and only a few web-like PSPs do not have a solid-state face stock material (e.g., transferable letters, decalcomania, and transfer tapes; see Chapter 4). Some PSAs may be used as PSP per se, without a carrier material. For some of these products (e.g., decalcomania and transferable letters), the PSA layer also plays the role of information carrier. However, because of its discontinuity and to ensure the required mechanical resistance during manufacture and application, such PSPs also require a release liner. Transfer-printing materials like Letraset can be considered carrierless or temporary laminates with a monoweb character (after application).

The design and formulation of carrierless products was discussed by Benedek in Ref. [3]. Various technical possibilities for manufacturing a self-sustaining, carrierless, self-adhesive product exist. Principally, the carrier can be softened or the adhesive can be reinforced. Strengthening (stiffening) of the macromolecular structure of the adhesive is possible using several methods. A method is based on partial stiffening, using macromers; the self-sustaining structure of the adhesive can macroscopically also be reinforced, where the (fibrillar or globular) solid-state fillers or gaseous fillers ensure the stiffening of the adhesive. For such construction, the level of the filler and the rheology of the polymer matrix play a special role [19]. Coating is used to manufacture the web-like self-supporting adhesive, which can be postfoamed and postcured. The chemical basis and some manufacturing methods for carrierless tapes are summarized by Benedek in Ref. [2]. Tackified acrylic emulsions can be mechanically foamed, coated, and impregnated with acrylic adhesive; silica-filled acrylics can be coated and cured; acrylic filled with expandable microbubbles can be coated, cured, and blown; and a blend of chlorinated PE, acrylonitrile rubber, and liquid chlorinated paraffin can be extruded.

For instance, sealing tapes (without carrier), based on tackified butyl rubber, are applied with an extruder. Sheet-like hot melts are used for thermal lamination of various web-like materials. For such applications, EVA and ethylene propylene copolymers, copolyesters, copolyamides, vinyl chloride copolymers, and thermoplastic PURs have been developed. For instance, self-adhesive films (tapes) used to bond roofing insulation and laminate dissimilar material formulations contain EVAc–polyolefin blends, rosin, waxes, and antioxidants. They are processed as hot melt and cast as 1- to 4-mm films (application temperature 180°C). A carrierless, self-sustaining pressure-sensitive film can also be manufactured by laminating together an (acrylic) adhesive-based film and a rubber-based film.

"Hardening" of an adhesive can also lead to a self-adhesive, carrier-like product. Special tapes exist that have an adhesive with a carrier-like character. Such behavior can be achieved by cross-linking, foaming, filling, or reinforcing the adhesive layer. The adhesive can contain a metallic network to ensure electrical conductivity as well. As discussed in Ref. [1], a thick, triple-layer adhesive tape (with filler material in the center layer) can be manufactured by photopolymerization. This is really a carrierless tape that has the same chemical composition (except filler) for the "carrier" (i.e., self-supporting and adhesive) middle layer and the outer adhesive layers.

Generally, constructions with porous carrier materials in which the adhesive can penetrate into the carrier (i.e., it can impregnate) can always be considered partially carrierless and partially adhesiveless tapes. Such PSPs have layers with both carrier and adhesive characteristics. In reality, transfer tapes with a carrier need the carrier only as technological aid. During manufacture they are carrier-based; during application they are carrierless. Monoweb labels without a built-in carrier have also been manufactured [20].

In some cases the tape is an adhesive monoweb, but behaves before or after use like a nonadhesive product. Such behavior is due to its superficial cross-linking. According to Meinel [21], in this way a carrierless PSP with better conformability can be obtained (the adhesive layer possesses adequate strength to permit it to be used without a carrier material). The product exhibits both adhesive and adhesion-free surfaces. The tack-free surface is achieved by superficial cross-linking of the adhesive. This cross-linked portion of the adhesive has greater tensile strength and less extensibility. In some cases the polymer can be cross-linked only along the edges of the adhesive layer. The tack-free edge prevents oozing and dirtiness. The superficially cross-linked adhesive layer may be stretched to fracture the cross-linked portions, exposing the tacky core to bond it. The surface of the adhesive becomes virtually nontacky under light pressure, but becomes tacky when the product is pressed against the adherent surface.

1.1.4 Linerless Pressure-Sensitive Products

As discussed previously (see Section 1.1.1.1), in web-like PSPs the release-coated/or -uncoated carrier works as a release liner. Plastic-based carriers work as release liners due to their low adhesivity or reduced contact surface. Low adhesivity is given by the nature of the film raw material (nonpolar polymers), by the supplemental release coating of the film, or by reduced contact surface based on the geometry of the film. Certain plastic films do not exhibit chemical affinity or contain antiblocking agents [2]. In such cases the uncoated (nonpolar) plastic films can work as a release liner. Antiblocking agents are materials added to an adhesive (or carrier) formulation to prevent the adhesive coating from adhering to its backing when the adhesive-coated carrier material is rolled or stacked at ambient or elevated temperatures and relative humidities (see also *Technology of Pressure-Sensitive Adhesives and Products*, Chapter 8). Certain technological additives (e.g., surface active agents in PVC or fillers) can work as imbedded release agents as well (see also *Technology of Pressure-Sensitive Adhesives and Products*, Chapter 8).

Reduced contact area is ensured by a special folded or embossed surface (see also *Technology of Pressure-Sensitive Adhesives and Products*, Chapter 10) of the plastic film. Using such surfaces, polar plastics may also function as an abhesive component.

For instance, a contoured special PVC film displays only 1/10 of its full surface as contact area. The shape of the embossing plays an important role also (see also *Technology of Pressure-Sensitive Adhesives and Products*, Chapter 10). Release liners with folded, contoured film are used for tapes that are applied on round items, that is, where conformability is very important.

1.2 Classes of Pressure-Sensitive Products

Classification of PSPs was discussed by Benedek in Ref. [1]. PSPs can be classified according to their adhesive and their use. The PSAs coated on PSPs differ according to their physical status, their base formulation, which ensures permanent or temporary bonding, and their environmental (i.e., chemical and temperature) resistance. The product class determines the carrier type and geometry, the adhesive type and geometry, the release type and geometry, and the build-up of PSP. For instance, very different carrier materials, release materials, adhesive, coating weight, and laminate construction are required for labels, tapes, or protective films.

1.2.1 Pressure-Sensitive Product Classes According to Pressure-Sensitive Adhesive-Status

As discussed in *Technology of Pressure-Sensitive Adhesives and Products*, Chapter 8, as related to their raw materials, PSAs can be formulated with dispersed/solved adhesives or with 100% solids, which lead to PSPs with different performance characteristics. Generally, each adhesive type (i.e., solvent-based, water-based, or hot-melt adhesives) can be used for the whole range of PSPs (e.g., labels, tapes, or protective webs); however, preferences in their use exist according to their end-use performance characteristics (as a result of their various raw materials and manufacturing technology) and economic considerations. As discussed in detail in Ref. [1], the physical state of the PSA strongly influences its chemical composition, the adhesive properties [7], which affect the converting properties [8], and the end-use performance characteristics [9]. As discussed in Ref. [22], which describes the practice of adhesive design and formulation for PSP classes, very different formulations are needed for the same end use (e.g., removability, water solubility) for solvent-based, water-based, or hot-melt adhesives. Formulation of dispersed/solvent adhesives is the subject of *Technology of Pressure-Sensitive Adhesives and Products*, Chapter 8.

1.2.1.1 Solvent-Based Pressure-Sensitive Products

Because of the large range of raw materials, which includes tackified elastomers as well as viscoleastomers, their multiple possibilities of formulation, which include tackification, cross-linking, and filling, and their low content in nonadhesive formulation components (see also *Technology of Pressure-Sensitive Adhesives and Products*, Chapter 8), solvent-based adhesives possess almost unlimited application possibilities for various PSPs (see also Chapter 4) that are hindered by environmental and economic considerations only. PSPs with solvent-based adhesives were described by Benedek in Ref. [22].

The adhesive properties of solvent-based adhesives were examined in comparison to water-based adhesives in Ref. [7]. Formulation of solvent-based adhesives for labels (paper, film, special, and permanent/removable labels); tapes (rubber–resin), and protective films was also discussed. Special solvent-based acrylic formulations for removable products and for water-soluble PSPs were discussed by Czech in Refs. [10,23]. The performance characteristics (e.g., adhesive properties, converting properties, and aging resistance [thermal stability and UV light stability]) of solvent-based rubber–resin formulations and acrylics were comparatively examined by Benedek in Ref. [7]. Solvent-based acrylic PSAs have some special characteristics compared with classic rubber–resin PSAs. First, a higher solids content (more than 20–30%) is possible, and the solid content/viscosity ratio can be adjusted exactly. Special solvents are generally used, and both solutions and dispersions can be produced. In general, fewer or no plasticizers and antioxidants are used; finally, cross-linking is possible using various methods.

1.2.1.2 Water-Based Pressure-Sensitive Products

Mainly forced by economic considerations, PSPs with water-based PSAs compete with solvent-based PSPs principally for common products, for example, labels and tapes for general use (see also Chapter 4). Special features of water-based acrylic PSAs compared with other water-based PSAs were examined by Benedek in Ref. [7]. Generally, water-based acrylic PSAs possess a higher solids content and a higher solids content/viscosity ratio than common carboxylated butadiene rubber (CSBR)/EVAc dispersions; they also respond better to dilution. Water-based acrylic PSAs exhibit a lower surface tension than common EVAc or CSBR dispersions. Current surface tension values range from 34 to 40 mN/m, whereas EVAc dispersions express values above 45 mN/m. Rubber-based dispersions have a surface tension between 45 and 60 mN/m. Water-based acrylic PSAs demonstrate higher coagulum (grit) than CSBR. Water-based acrylic PSAs generate less foam formation than CSBR and EVAc. Finally, water-based acrylic PSAs demonstrate better wet-out properties than EVAc- and CSBR-based PSAs and can be converted at higher speeds than CSBR and EVAc dispersions. Water-based label and tape formulations were discussed in Ref. [22]. Formulation of water-based PSAs is described in *Technology of Pressure-Sensitive Adhesives and Products*, Chapter 11; their manufacture is discussed in Chapter 10 of the same book.

1.2.1.3 100% Solids-Based Pressure-Sensitive Products

Hot-melt PSAs are used for common PSPs, for which low-cost manufacturing equipment and high processing speed (no drying required) are important, and the formulation-related adhesion–cohesion balance can be easily shifted using a high coating weight. Such products mainly include common tapes and some common labels (see also Chapter 4). Hot-melt label and tape formulations were described in Ref. [22]. Hot-melt PSAs based on styrenic polymers were discussed by Park in Ref. [24]. Formulation of HMPSAs was discussed in *Technology of Pressure-Sensitive Adhesives and Products*, Chapter 8; their manufacture was described in Chapter 10 of the same book. Developments in formulation and use of HMPSAs were discussed by Hu and Paul in *Technology of Pressure-Sensitive Adhesives and Products*, Chapter 3.

Radiation-cured 100% solids allow the in-line synthesis of PSAs (see *Technology of Pressure-Sensitive Adhesives and Products,* Chapter 1), that is, the simultaneous manufacture of the PSA and of the pressure-sensitive web. This technology is used mainly for special products. Formulation of radiation-curable adhesives is the subject of *Technology of Pressure-Sensitive Adhesives and Products,* Chapter 8.

1.2.2 Pressure-Sensitive Product Classes According to Bond Character

In some application domains a permanent adhesive joint is needed; in other end-use domains temporary adhesion is required. Both product classes (i.e., permanent and removable PSPs) can be manufactured using PSAs based on various raw materials and with various manufacture technologies. Both product classes include labels as well as tapes (see Chapter 4); protective webs must be removable.

1.2.2.1 Permanent Pressure-Sensitive Products

Permanent PSAs ensure a "permanent" adhesive bond. Their adhesion is not limited in time and generally is higher than the mechanical resistance of the whole PSP, that is, the laminate cannot be dismounted (peeled-off) without destroying it. The main products in this class are tapes, labels, and forms (see Chapter 4). Design and formulation of permanent adhesives was described in Ref. [22]. Generally, the adhesive performance characteristics of PSAs are investigated with permanent adhesives and balanced formulation (see *Technology of Pressure-Sensitive Adhesives and Products,* Chapter 8), and common tests for such characteristics are carried out using permanent adhesives [7] (see also Chapter 8). Formulation of permanent adhesives was discussed in *Technology of Pressure-Sensitive Adhesives and Products,* Chapter 8.

1.2.2.2 Removable Pressure-Sensitive Products

Removability was discussed in detail by Benedek in Refs. [3,7,14,22]. Removable PSAs build up a time- and debonding force-dependent adhesive joint, where the components of the pressure-sensitive laminate can be peeled off with the possibility of time-dependent relamination (repositionable PSPs), time-independent relamination (readherable PSPs), or without this possibility (semipermanent PSAs). The main PSPs of this class are protective webs [9] (see also Chapter 4), special tapes [9] (see Chapter 4), and labels [9] (see also Chapter 4). Their formulation was examined in *Technology of Pressure-Sensitive Adhesives and Products,* Chapter 8.

As discussed in detail in *Technology of Pressure-Sensitive Adhesives and Products,* Chapter 8, the main possibility to achieve removability is based on formulation (see also Table 2.8.3 in *Technology of Pressure-Sensitive Adhesives and Products*). However, contact surface reduction (see also *Technology of Pressure-Sensitive Adhesives and Products,* Chapter 10) allows the regulation of peel resistance. Peel resistance and removability can be controlled by regulating the ratio between the adhesion surface and the application surface, which should be smaller than 1. To achieve such conditions, the adhesive layer should be discontinuous or should have a virtually discontinuous surface. Pattern or strip coating (see *Technology of Pressure-Sensitive Adhesives and Products,* Chapter 10) allows the use of nonremovable raw materials for removable applications. Various formulation (e.g., filling, cross-linking), product build-up (e.g., embossed, profiled carrier), and coating technology-related (e.g., pattern coating) possibilities exist to reduce the contact surface between the adhesive and the substrate. The various removable PSPs and their end use are discussed in Chapter 4. Removability-related tests are described in Chapter 8.

1.2.3   Pressure-Sensitive Product Classes According to Environmental Behavior

PSAs and PSPs display different environmental (chemical and temperature) resistance according to the formulation of the PSA (see *Technology of Pressure-Sensitive Adhesives and Products,* Chapter 1) and build-up of the product. This is required by the various application domains of such products (see Chapter 4). The environmental resistance of PSPs includes their chemical (solvents, water), temperature (low and high temperature), and time–temperature (aging) resistance [9].

1.2.3.1   Water-Resistant and Water-Soluble Pressure-Sensitive Products

Water-resistant and water-soluble PSAs were developed for various applications. Generally, high water resistance is required for common products in outdoor applications (e.g., protective films for construction and plotter films; see also Chapter 4) or special products used in contact with water or in a humid environment (e.g., bottle labels, thermal insulation tapes, and medical tapes; see Chapter 4). Low water resistance, that is, water solubility or dispersibility, is required mainly by environmental considerations related to recycling of PSPs [4,25] or for special applications (e.g., medical, splicing, tapes) [9], see also Chapter 4.

1.2.3.1.1   *Water-Resistant Pressure-Sensitive Products*

The main water-resistant PSPs include labels, tapes, and protective films. Their formulation was discussed in *Technology of Pressure-Sensitive Adhesives and Products,* Chapter 11. Such formulations are based mainly on solvent-based adhesives that do not contain water-soluble additives and allow an advanced degree of cross-linking (see *Technology of Pressure-Sensitive Adhesives and Products,* Chapter 8) or on 100% solids. Water resistance of the PSPs depends on the carrier [2]. For instance, filled high-density PE (HDPE) can be used as a face stock material for water-resistant dimensionally stable labels that can be printed using different methods. Like PE-coated paper, PE laminates can also be used for improved humidity resistance. As discussed in detail in Ref. [26], various criteria and test methods exist to evaluate water resistance (see also Chapter 8).

1.2.3.1.2   *Water-Soluble Pressure-Sensitive Products*

The main water-soluble PSPs include special labels and tapes (e.g., bottle labels, splicing tapes, medical tapes; see Chapter 4). For instance, for splicing tapes that must resist temperatures up to 220–240°C, carrier materials must be water dispersible and the adhesive should be water soluble, but resistant to organic solvents. Their formulation was discussed in *Technology of Pressure-Sensitive Adhesives and Products,* Chapter 8. Such formulations are based mainly on special solvent-based acrylics that contain water-solubilizing monomers and on water-based adhesives formulated with various solubilizers [22] (see *Technology of Pressure-Sensitive Adhesives and Products*, Chapter 8) or on 100% solids. As discussed in detail in Ref. [26], various criteria and test methods exist to evaluate water solubility (see also Chapter 8). Advances in hydrogels are discussed in *Technology of Pressure-Sensitive Adhesives and Products*, Chapter 7.

1.2.3.2 Temperature-Resistant Pressure-Sensitive Products

Various PSPs are used at low or high application temperatures [9] (see Chapter 4). They include mainly special labels and special tapes. Formulation for temperature resistance was discussed in Ref. [22] (see also *Technology of Pressure-Sensitive Adhesives and Products*, Chapter 8).

1.2.3.2.1 *High-Temperature-Resistant Pressure-Sensitive Products*

High-temperature-resistant PSPs include mainly special tapes, for example, insulation or mounting tapes (see Chapter 4). Temperature-resistant adhesives are also required for thermally printed labels and copy labels. Temperature resistance depends on both the adhesive and the carrier material [2]. The manufacture of main carrier materials, their performance characteristics, and preferred use are described in Refs. [2,8]. Formulation of PSAs with increased temperature resistance is discussed in *Technology of Pressure-Sensitive Adhesives and Products*, Chapter 8. Their test is described in *Applications of Pressure-Sensitive Products*.

1.2.3.2.2 *Low-Temperature-Resistant Pressure-Sensitive Products*

Low-temperature-resistant PSPs include special labels and tapes for outdoor use (e.g., insulation tapes, mounting tapes) [9,22] (see Chapter 4) or low-temperature storage (see Chapter 4). Formulation of PSAs with low-temperature resistance is discussed in *Technology of Pressure-Sensitive Adhesives and Products*, Chapter 8; their test is described in Chapter 8 of this book.

1.2.4 Pressure-Sensitive Product Classes According to Their Use

Most PSPs are web-like constructions laminated together during their manufacture or application. Their solid-state components function also as information carriers. Exceptions include some special cases in which the adhesive layer itself carries information or plays an esthetic or protective role. Labels are the main PSP class used as discrete items or continuous web-like materials, which function as carriers for informations. Certain tapes and protective films can also play this role [2].

1.2.4.1 Labels

Generally, labels constitute a part of the packaging. Principally, packaging materials must contain, preserve, and present as a support for advertisements and sales aid. Containing is a mechanical function. Labels (except some special dermal dosage products; see Chapter 4) do not fulfil this requirement. Although some antitheft or closure labels exist (see Chapter 4), the function to preserve is not a general requirement for labels.

Their principal function is presentation. Labels are pressure-sensitive items (supplied as sheet or reel) that carry information and are made to be applied (laminated) on a solid-state adherent surface. The build-up of labels has been described in a detailed manner by Benedek in the companion books [2,4,9]. Historically, the production of the PSPs started with the manufacture of common tapes, that is, with the production of monowebs, coated with permanent PSA (see also Table 1.1). The development of removable PSAs allowed the manufacture of removable tapes, labels, and protective films.

Labels are temporary laminates composed of a face stock material with an adhesive pressure-sensitive surface that is protected by a separate abhesive component, the release liner. Its abhesive properties are (generally) given by a coated abhesive layer. Therefore, the construction of the release liner is similar to that of the (coated) face stock: a solid-state carrier bears a coating layer. Various carrier and coating materials and various coating techniques are used for the manufacture of the release liner [4], but principally, the manufacture of labels is the lamination of two coated carrier materials, that is, of the PSA-coated face stock material and the release-coated liner. Advances in this domain are discussed in detail in *Technology of Pressure-Sensitive Adhesives and Products*, Chapter 10. In practice, transfer coating is common for labels (see also *Technology of Pressure-Sensitive Adhesives and Products,* Chapter 10), that is, the adhesive-coated release liner is laminated onto the face stock material. During conversion, the continuous web-like face stock material is die cut (see also *Technology of Pressure-Sensitive Adhesives and Products,* Chapter 10) and during labeling the discontinuous face stock material is transferred from the release liner to the substrate and the temporary laminate is destroyed (see also Chapter 4). In the case of double-side-coated tapes, first one of the release liners is taken away and then the other (see also Chapter 4).

The need for a continuous, solid-state, separate abhesive laminate component for labels arises from their aggressive adhesivity and their discontinuous character. For labels the abhesive properties of the liner must be exactly controlled; for tapes and protective films such performances play a secondary role only. In the actual stage of development the precise regulation of the release force is possible through the use of a separate, abhesive-coated solid-state laminate component only. For reel labels (see the following) the continuous liner allows application by labeling as well. This is an expensive and environmentally inadequate technical solution. In addition, it does not satisfy some special requirements for tapes and protective materials (e.g., resistance to abrasion, lubrication).

According to the application domains of PSPs, very different performance characteristics are required for the solid-state carrier material used as a face stock or as a release liner. Therefore, the grades of carrier material differ as well. Generally, the carrier material should have adequate mechanical characteristics that satisfy the end-use requirements of the product class. In practice, such requirements may vary considerably. For instance, for labels, dimensional stability of the carrier material is required (i.e., control of its plasticity, elasticity, elongation, and shrinkage; see Chapter 4). If dimensional stability is required, ideally it should not depend on the product geometry. Principally, low- or high-gauge carrier materials must display the same mechanical properties. On the other hand, some labels, for example, tamper-evident labels, require a carrier material with low mechanical resistance, that is, with excellent conformability and destructibility (low internal strength; see also Chapter 4). Generally, labels can be classified as common labels (see Chapter 4) and special labels (see Chapter 4).

The range of possible raw materials for PSAs for labels is very broad (see also *Technology of Pressure-Sensitive Adhesives and Products,* Chapter 8). Their carrier materials were discussed by Benedek in Refs. [2,8] (see also *Technology of Pressure-Sensitive Adhesives and Products,* Chapter 10). For label manufacture, two-component rubber–resin or one-component viscoelastic polymer-based formulations can be coated in the molten, dissolved, or dispersed state. Radiation curing has been recommended for liquid (dissolved) prepolymers or molten, high-polymer-based formulations. First, rubber–resin PSAs based on natural raw materials (natural rubber and rosin derivatives) were used; for labels they possess the advantage of well-balanced adhesive properties, easy regulation of adhesive performance, and low cost. Unfortunately, they have a limited resistance to aging. Later, acrylics and carboxylated styrene–butadiene dispersions and HMPSAs based on styrene–butadiene block copolymer (SBC) were introduced as competitors to acrylics. For special applications a broad range of commercial and experimental low-volume products are available. Acrylics are used as solvent-based, water-based, hot-melt, and radiation-curable macromer-based formulations for labels with various build-up and end use (see also *Technology of Pressure-Sensitive Adhesives and Products,* Chapter 8). Label manufacture is the most important application field of water-based PSA. For HMPSA for labels, first SBCs and later acrylate-based block copolymers were introduced. The manufacture of UV-cross-linkable (100% solids) acrylates is more expensive than that of acrylic dispersions. However, such products are an alternative to solvent-based adhesives in which a common hot-melt coating line can be equipped with UV curing lamps without too much capital investment.

1.2.4.1.1 Common Labels

Common labels include reel labels and sheet labels. Reel labels are supplied as a roll-like material where the label is a die-cut discrete item, which must be labeled by its transfer from the continuous release layer. Reel labels are laminated by labeling (in web form); sheet-labels are applied (generally) by handmade lamination. Generally, reel labels possess small dimensions (see Chapter 4). Reel labels must satisfy the general quality requirements for machining of web-like materials. Their lamination by labeling is a high-speed operation that imposes the use of materials with improved mechanical characteristics and a high-tack adhesive (Figure 1.2; see Chapter 4).

In label construction the carrier is the laminate component with a packaging and protection function, that is, with special esthetic and mechanical characteristics. This performance characteristic of the carrier is more pronounced for sheet-like labels, which generally possess a large area and are used for visualization, that is, they require excellent optical quality (see also Chapter 4).

The main carrier materials used for labels are paper and plastic films. Formulation of PSAs must take into account their different bulk and surface characteristics (see *Technology of Pressure-Sensitive Adhesives and Products,* Chapter 8). The choice of the carrier for labels is influenced by the nature and surface of the substrate, the end-use environment (weathering and applications climate), mechanical requirements, printing methods, processing conditions, and special requirements (e.g., Food and Drug

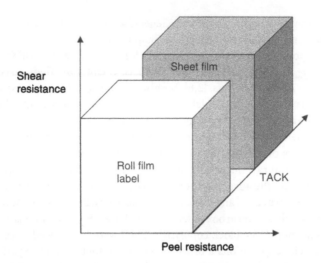

FIGURE 1.2 Schematic presentation of the main adhesive characteristics of roll and sheet labels.

Administration, Bundes Gesundheitsamt (German Sanitary Administration formerly known as the BGA) approval (see also *Technology of Pressure-Sensitive Adhesives and Products,* Chapter 8). The carrier for labels can be classified as uncoated monolayer (e.g., paper, films), coated monolayer (e.g., coated, printed, metallized paper and foils), and composite material. The main carrier materials, their manufacture, and performance characteristics were described by Benedek in Refs. [2,8]. The main criteria for the choice of a carrier material include esthetics, coatability, printability, application, end-use properties, converting properties, converting properties of the laminate, and price [8]. Design and formulation of PSAs for paper and film labels were described by Benedek in Ref. [22]. The influence of the carrier on formulation was discussed in Ref. [3] (see also *Technology of Pressure-Sensitive Adhesives and Products,* Chapter 8).

Paper is the most-used carrier material for labels. Paper used as face stock material for labels was discussed in detail by Benedek in Refs. [2,8]. Quality criteria for paper used as face stock material for labels include weight, thickness, tensile strength, elongation at break, Bekk smoothness, and bursting pressure. For most labels it is used as both face stock and liner material. Generally, low-weight, high-rigidity papers are required for labels. Common label papers have a weight of 70–80 g/m²; paper for common pressure-sensitive labels has a weight of 83–90 g/m². Various grades, such as uncoated, coated, cast-coated, and colored papers, are applied. First, cast-coated papers were introduced; now machine-coated paper qualities are preferred. These include thermopapers and vellum- and clay-coated papers. The proportion of satinated kraft paper is increasing. Fluorescent paper labels were introduced in 1958. Latex-impregnated paper was proposed for weather-resistant labels and calendered (90 g/m²) paper for opaque labels; paper with high absorptivity was suggested for labels printed with so-called cartridge inks; and oil- and fat-resistant paper is used for thermal printing. Latex-impregnated and plastic-coated papers possess an opacity degree of 84–88%. For table labels printed by nonimpact printing (see also *Technology of Pressure-Sensitive Adhesives and Products*, Chapter 10), open porous papers are preferred. Metallized paper and film/paper laminates are used as well. Metallized paper displays low curl in comparison with metallized film. Paper-backed aluminum foils and metallized face material have the highest opacity (100%). Plastic-coated paper attains a gloss degree of 96%.

Paper as a face stock material for labels should display special mechanical and surface characteristics. Adequate mechanical characteristics are required to ensure its dimensional stability (see also Chapter 8) and special behavior during cutting and labeling (see also *Technology of Pressure-Sensitive Adhesives and Products*, Chapter 10). Surface characteristics should allow the paper's coatability (printing and varnishing). The main requirements for a paper carrier for labels are high gloss, high surface strength, good wettability, and good printing ink anchorage. The mechanical and surface characteristics of paper can be improved by impregnating, coating, and laminating (see also *Technology of Pressure-Sensitive Adhesives and Products*, Chapter 10).

Plastic films used as a face stock material for labels are coated as a web and laminated and converted as a web. Therefore, the main requirements for carrier materials used for labels regard their dimensional stability and surface quality (see also Chapter 8). On the other hand, labeling speed is a function of label stiffness. For high-speed labeling, films with a minimum thickness of 50 μm are suggested (see also Chapter 4). The face stock must be able to meet printing and die-cutting specifications as well. The main plastic films used for labels, as well as their chemical and trade names and suppliers, are listed in Ref. [8], in which the performance characteristics of various plastic films used as face stock material for labels such as stiffness, die-cuttability, thermal resistance, ductility, tear resistance, and coatability were also examined. The choice of film face stock material (e.g., PE, PS, PP, and suspension PVC [SPVC]) as a function of the end use of laminate is based on chemical resistance, moisture resistance, shrinkage resistance, flexibility, and clarity. Versatility of cast polypropylene (CPP), oriented polypropylene (OPP), and PE as face stock materials was examined in comparison; PVC was compared with PE; PE was examined in comparison with PP.

Principally, the choice of a plastic carrier material for labels depends on internal or external use, postprintability and writability, temperature resistance, chemical resistance, weatherability, abrasion resistance, labeling ability, and stiffness. Table 1.3 lists the main characteristics of various plastic films used as carrier materials.

For label application the film must be stiff enough for dispensing and flexible enough to conform to various container shapes and resist repeated deformations. The applied label must withstand harsh environments and rough handling. This means that plastic films with a pronounced rigidity and surface quality are recommended as carriers for labels. Such films must allow good register control (in printing; see also *Technology of Pressure-Sensitive Adhesives and Products*, Chapter 10), easy die cutting (in conversion; see also *Technology of Pressure-Sensitive Adhesives and Products*, Chapter 10), and reliable high-speed dispensing (in application; see also Chapter 4). Therefore, the specialist in label design and manufacturing of the labels should know which polymers and plastic film manufacturing methods produce plastic film suitable for label application. For the PSP manufacturer, caliper (thickness and geometry), gloss degree, opacity, and corona treatment are the main performance characteristics of synthetic face stock materials.

TABLE 1.3 Performance Characteristics of Different Plastic Films Used as Face Stock Material

Film Grade	Stiffness	Die-Cuttability	Thermal Resistance	Coatability	Tear Resistance	Thickness	Profile	Transparency	Shrink Resistance	Chem. Res.
				Performance Characteristics of Plastic Films						
LDPE	○	○○	○○	○	••	•••	○○	○	○○	○
HDPE	•	•	•	○	•	•	•	○○	•	•
OPP	••	••	••	○	○○	○	•••	•••	•••	•
CPP	••	•••	•••	•	••	•••	•	••	••	••
PVC-H	•••	•••	••	•••	○	•••	•••	•••	○○	○
PVC-S	○○	○○	○	••	○	••	•••	•	○○	○○
PET	•••	•••	•••	•	○○	••	••	•••	•••	•••
PS	•••	••	••	•	○	•••	••	•••	•••	•

Note: •••, very good; ••, good; •, medium; ○, low; ○○, very low.

The film label market has grown at a rate of more than 30% per year. The main film carrier materials used for labels are polyethylene terephthalate (PET), OPP, low-density polyethylene (LDPE), HDPE, polystyrene, polycarbonate (PC), polyamide (PA), cellulose acetate, PVC, polyacrylate, and laminates.

PVC was the first film-like synthetic material used as a carrier for labels and some years ago it was the most commonly used carrier material. Soft 80- to 120-μm film is used for self-adhesive labels. Films with thickness lower than 60 μm are made from Hand PVC, and for large-surface products calendered PVC is used. For special products (e.g., sheet labels for outdoor application) cast PVC has been used. Thin PVC is used as carrier for tamper-evident labels (see also Chapter 4). PVC exhibits the desired tear and temperature resistance for computer applications. It has excellent UV stability. The main disadvantages of PVC are its caliper variation, shrinkage, and plasticizer migration. In the past decade PE and PP replaced PVC. Comparison of various polymers concerning their flatness, transparency, rigidity, printability, density, cuttability, temperature stability, and price gives different ratings. Thus, PVC offers the best cuttability and printability compared with PP and PET. PP has good density, temperature stability, and price; and PET is the most rigid and transparent film material.

Polyolefin films are the most used plastic carrier materials for labels as face stock and release liners. Some years ago nonoriented blown PE films were used as label carrier material only. The main disadvantage of LDPE as a carrier material for labels is its limited stiffness. Different processing methods, conditions, and material combinations have been suggested to eliminate this disadvantage. Biaxially oriented, cast, and coextruded PE films were developed. Cast LDPE, oriented LDPE, LDPE–polystyrene blends, LDPE–LDPE laminates, and coextrudates and coated products (e.g., clay-coated PE/polystyrene [Pst] film) can also be used. From the range of PEs, HDPE is the most adequate material for labels. Where a combination of high stiffness and puncture resistance is required, that is, a substitute for cardboard and paper, bimodal HDPE is used (as synthetic paper). Cross-linked films have been proposed as well. The high stiffness required for labels can be achieved by orienting the PE film. Films made by machine-direction orientation exhibit outstanding stiffness in the machine direction and flexibility in the cross direction. Cross-direction monooriented films display low cross-direction shrinkage (see also Chapter 8) and are used for labels.

PP films for labels must fulfill the following requirements: usability for transparent and opaque face stock, high mechanical strength (yield and stiffness), moisture and chemical resistance, good weathering characteristics, good printability (matrix printability also), good embossing and die-cutting characteristics, ability to be combined with other films, ability to be metallized, and recyclability. Chemical resistance is related to the end use of the labels and to the environment. For instance, for application on containers, resistance to organic solvents, lubricants, and oils and to humidity, steam, and water are required.

Nonoriented and mono- and bioriented thermofixed PP films are available. For most applications an isotropic material is necessary; therefore, biaxially oriented films should be used. Oriented films give a higher yield at lower gauges. For instance, a PP film with a thickness of 20 μm can be manufactured, yielding 56.2 m²/kg film; a biaxially oriented product with a thickness of only 12.5 μm can be used for a 89 m²/kg film. Coated, pearled, white opaque, and metallized PP films have been manufactured as carriers for labels. Pearled white film may give a plastic-coated paper effect. Such (coextruded [coex]) films can incorporate white pigment in the core or voids in the common "cavitated" films. Transparent films having a thickness of 40 or 50 μm and opaque/white films with a gauge of 28, 35, 40, 50, and 60 μm have been suggested for labels. Coextruded, white, metallized, nonlacquered, and acrylic-lacquered oriented PP films as carrier for labels are manufactured with a thickness of 21–60 μm. PP films provide good machine performance and, when coated, a glossy surface for high-quality print.

Very stiff PP (with a modulus of 2400 N/mm^2) and low-modulus PP (with a modulus of 100 N/mm^2) can be used as carrier materials for labels and tapes [27]. The stiffness of common film materials used for tapes (e.g., oriented Pst, PET, and PVC) is situated at 2500–3500 N/mm (E-modulus). Higher stiffness required for labels can be achieved with special PP homopolymers. Filled PP or heterophase copolymers exhibit a modulus of 3000 N/mm^2. Biaxially oriented Pst replaced PVC, but has since been replaced by PET and PP [28].

Bioriented PP (BOPP) is the fastest growing face stock material with sheet- and reel-fed options, printable in almost any format. It provides the opportunity for significant downgauging from PE. Unfortunately, its dispensability is not the best. Label-Lyte label films (white and transparent two-side materials for roll stock lamination) based on PP may be uncoated or coated; the uncoated film has a thickness of 19–24 μm (mono) or 60 μm (three-layer coex). The coated film as a monolayer possesses a thickness of 50 μm; the coex film is 60 μm thick.

Film/film and film/paper laminates are suggested as carrier material for labels. Humidity-resistant papers are made by PE coating (extrusion coating or laminating). PE (LDPE or HDPE) is coated on paper by extrusion coating (with a coating weight of 20–30 g/m^2). By laminating, LDPE is applied on paper with a coating weight of 10–20 g/m^2 [29]. For extrusion coating a PE grade with a density of 0.915 to 0.925 g/cm^3 and melt flow index of 190/2.16 is recommended [30]. For special end uses, where transparency of a partially delaminated multilayer label is required, paper/film laminates with the PE film as medium layer are suggested. A white opaque PP film overlaminated with PET is used for labeling of soft drinks. Metallized paper and laminated metallic films are also used as face stock for labels [31]. Since 1981, there has been a trend to replace aluminium/paper laminates with metallized papers. Such papers are less humidity sensitive and their demetallizing offers new decoration possibilities.

A large variety of products, for example, common paper, satinated craft papers, clay-coated papers, polymer-coated papers, and plastic films, are used as raw materials for liners (see also Chapter 4). Release liners in weights from 68 to 150 g/m^2, silicone coated on either or both sides, with supercalendered, clay-coated, and polycoated paper, are supplied. High-quality film liners based on PP, PE, and PS are also available [32]. According to Tomuschat [33], glassine paper and PET are currently the standard materials for release liners in the roll label industry. Owing to their high price, these materials are responsible for a major share of the overall costs of label stock material. By replacing paper with 30 μm BOPP and thermally cured silicone with UV-cured silicone (0.8 g/m^2), 135-μm global thickness is achieved, compared with standard label stock, whereas standard paper with 80 g/m^2 is coated with 20 g/m^2 PSA and combined with thermally siliconized (coating weight 1.1 g/m²) glassine liner (62 g/m²). Such construction reduces waste up to 55%.

Reduced thickness is good for more labels (18% more); also, material costs are reduced up to 60%.

Comparison between paper and plastics was presented in Ref. [8]. Stiffness, dimensional stability, and die-cutting properties of common films versus paper were examined. With regard to their stiffness, the different face stock materials can be listed in the following ranking:

$$OPP > PVC > Paper > PS > PE \tag{1.1}$$

With regard to their cost effectiveness, the different face stock materials can be listed in the following ranking:

$$Paper > OPP > PE > PVC > PS > \tag{1.2}$$

1.2.4.1.2 Special Labels

As discussed in detail in Chapter 4, special labels exist that differ in the character of their bond (e.g., permanent and removable labels) or in other end-use properties (e.g., environmental and temperature resistance). The development of removable PSAs allowed the manufacture of removable labels, tapes, and later large-surface PSPs displaying permanent adhesion during processing of the laminate and removability afterward. Such products are protective webs (see Chapters 1 and 4). Labels are also classified as permanent and temporary labels according to the characteristics of the adhesive bond. Repositionable labels are a special class of removable labels that adhere to different surfaces, but remove cleanly and can be reapplied. Generally, their final adhesion builds up over a few hours. Different grades of removability are known according to their time-dependent characteristics of adhesion build-up, such as removable, repositionable, and semipermanent labels. Water-resistant and water-soluble labels are also manufactured (see also Chapter 4). Freezer labels constitute a special class of low-temperature-applicable labels (see also Chapter 4). Formulation of such special labels is discussed in *Technology of Pressure-Sensitive Adhesives and Products*, Chapter 8.

1.2.4.2 Tapes

Tapes were the first PSPs to be produced with a coated monoweb construction (see Section 1.1.1.1). Generally, tapes are PSPs that have a solid-state carrier with a coated or built-in pressure-sensitive layer. Some tapes also have a release layer. Because of the need to be applied as a continuous web and due to their general use as a bonding element of multiple adherents, tapes generally must possess a mechanically resistant carrier material with an aggressive PSA. The mechanical characteristics of the carrier material and the adhesive characteristics of the built-in or coated adhesive differ according to the end use of the product [9]. Unlike labels, tapes do not have a multiweb, laminate construction before use (except for their self-wound character or some special tapes).

Generally, a tape carrier can be any reasonably thin, flat, and flexible material. Thus, it may be woven, nonwoven, metallic, electrically resistant, natural or synthetic, tear resistant or fragile, water resistant, or water soluble. Dimensional stability of the carrier material is required for labels and, to some extent, for certain tapes (e.g., packaging and mounting tapes; see also Chapter 8). For other tapes (e.g., masking or wire-wound tapes or hygienic products; see Chapter 4) and deep-drawn protective films, carrier deformability is needed (see Chapter 4). Such performance characteristics are given mainly by non-paper-carrier materials (see Sections 1.2.2.1 and 1.2.2.2). With the exception of labels and some common packaging tapes, application requires regulation of the dimensions and dimensional stability of the carrier material (i.e., control of its plasticity, elasticity, elongation, shrinkage, etc.).

For tapes and protective films the back side of the self-adhesive (coated or uncoated) material can act as a liner. This function may be ensured by a supplemental release coating (see *Technology of Pressure-Sensitive Adhesives and Products*, Chapter 10) or by the choice of an adequate carrier material with a low level of adhesivity (e.g., nonpolar polyolefins or emulsion PVC [EPVC]). The surface of films based on EPVC is coated with a thin layer of emulsifier. This layer acts as an abhesive substance and allows low-resistance unwinding of the rolls. In practice, for pressure-sensitive tapes with high tack adhesive, a coated release layer is required. This is not necessary if the carrier itself is abhesive, like certain nonpolar plastic films or those containing slip or other abhesive additives. The main plastic films that work also as release liners for tapes are polyester, PP, PE, and PS. The abhesive can be coated on the back side of the carrier material or included in it. For instance, a biaxially oriented multilayer PP film used as carrier or liner for tapes includes a second layer with a thickness of less than one third of the total thickness of the adhesive tape, which contains an antiadhesive substance. Another construction includes the release substance randomly distributed in the carrier material.

There is a trend on the market to provide customers with printed packaging tapes (see also Chapter 4). The text or graphics are imprinted on the nonadhesive side of the carrier material before the tape is made. The problem with HMPSA is that a release coating has already been applied to the backing material before it is imprinted. Such a coating is necessary because hot melts do not release easily from the nonadhesive side of the carrier material. In some cases a separate release film (paper) should be interlaminated. Double-side-coated tapes are products of this type.

Like labels, a primary classification of tapes is based on their general or special characteristics. According to their carrier material, tapes may be divided into paper-based, film-based, and textile-based tapes. The textile carrier used may be woven or nonwoven. Textile carriers ensure nonextensibility and good anchorage. A plastic carrier may be a film or a foam. A film carrier acts as a barrier against chemicals (plasticizers, surfactants, antioxidants, etc.). A foam carrier ensures conformability on uneven surfaces and equalizing of the thermal dilatation coefficients. Combinations of these materials are applied. Some tapes are manufactured with no carrier.

Tapes with carriers are classified as single-side-coated and double-side-coated products. Depending on their adhesive characteristics, tapes may be grouped into permanent and removable products. According to their end use, there is a broad range of possible classifications. Packaging, masking, protecting, marking, closure, fixing, mounting, and insulating tapes are some examples (see also Chapter 4).

1.2.4.2.1 Common Tapes

The use of common tapes is based on their adhesion and permanent bond, which ensures assembly of various items (see Chapter 4). Theoretically, pressure-sensitive tapes have the same construction as wet-adhesive tapes (see also Chapter 2). Such tapes are paper or plastic film based. Paper-based common tapes were the first tapes manufactured; later, film-based tapes were developed. As discussed previously, for pressure-sensitive adhesive tapes based on polymeric film carrier, the release surface is usually the surface of the polymeric film remote from the adhesive, which allows the tape to be conveniently stored in the form of a spirally wound roll. In practice, pressure-sensitive tapes that have a high tack adhesive or high coating weight, for example, double-side tapes or transfer tapes, must have a separate solid-state release component (see Chapter 1).

Common and special tapes are manufactured using paper as a carrier material. They are adhesive-coated PSPs that have various paper qualities and adhesive coatings, depending on their end use. Extensible, pleated, and conformable paper is required as a carrier for special tapes. Extensible, deformable paper is required for medical tapes.

Film-based tapes are manufactured using plastic films as carrier material. These products can have a coated adhesive, a built-in adhesive, or a virtually adhesive-free construction (see also Chapter 4). The self-adhesivity and conformability of certain plastic films and the nonpolar abhesive surface of some plastic carrier films allow the design of adhesive-free and release layer-free tape construction. The first plastic films for tapes were introduced from the range of common packaging films. Their development occurred in parallel with the general development of raw materials, film manufacturing, and transformation methods. The development of other web-like polymer-based products (e.g., fabric, nonwoven, foam) influenced the development of the packaging industry and the carrier materials. Hard and soft PVC, cellulose hydrate, PE, cellulose acetate, and PP have been the most used carrier materials for tapes. Generally, oriented films have been applied.

Permanent tapes give permanent adhesive bonding. The main representatives of this class are packaging tapes (see Chapter 4). High-strength carrier materials and aggressive permanent PSAs are recommended for these products.

Hard (rigid) and soft (plasticized) PVCs have been suggested as carriers for tapes. Such films are the most used carrier materials for packaging tapes; they do not need release coating and are not noisy. However, generally PVC used as a carrier for tapes is primed. Monoaxially or biaxially oriented transparent or clear, glossy, or embossed films are used with a thickness of 3–90 μm. Embossed films have a lower thickness of 28 μm in comparison with 40 μm for glossy films. In past years, the thickness of the PVC film has been reduced from 40 to 25 μm for packaging tapes; for thicknesses of less than 30 μm, embossed film is suggested. Embossed film is also used as a mechanical release agent. Copolymers of vinyl chloride serve as clear tape for packaging and household applications.

In 1979, about 80% of tapes were manufactured with PVC as a carrier material. In 1995, only about 37% were PVC based. Now, PVC comprises about 25% of the carrier materials for tapes.

Oriented PP (30–50 μm) is recommended for packaging tapes. High mechanical strength and elongation are required for carrier films for common tapes. Therefore, narrow-molecular-weight-distribution HDPE is used also for blown film extrusion for tapes, which is carried out at high blow-up ratios to reduce the imbalance of film's mechanical properties [2]. PE used for tapes is chemically pretreated. Oriented LDPE (80–110 μm) is proposed for insulation tapes. Linear LDPE (LLDPE) and mixtures of LDPE/LLDPE are recommended for tapes as well. Such carrier formulations make use of the deformability of LLDPE and the processibility of LDPE. Extensibility and deformability are general requirements for certain mounting tapes as well. The elasticity of LLDPE is used to avoid blocking during unrolling. Such tapes have a coextruded backing layer of LLDPE. Coextruded ethylene/propylene can be stretched for use as tape carriers as well.

Design and formulation of tapes is discussed in Ref. [22]. The manufacture of tapes was described by Benedek in Ref. [9], and end-use performance characteristics of tapes are discussed by Benedek in Ref. [2]. First, rubber–resin PSAs were used for tapes with a paper carrier. Because of the relatively simple regulation of the adhesive properties in formulations based on natural rubber (NR) or blends of NR with synthetic elastomers (through the use of cross-linkers and active fillers), recipes were developed for almost every application field (e.g., packaging, mounting, and medical tapes) for permanent or removable adhesives. Masticated, calendered compositions have been coated without solvent. The introduction of thermoplastic styrene block copolymers allowed the coating of rubber–resin formulations as HMPSA. Solvent-based and hot-melt rubber–resin formulations are the most common raw materials for inexpensive tape applications. Water-based, carboxylated rubber dispersions replaced acrylics for some packaging applications as a less expensive raw material. However, for certain tapes, rubber–resin formulations cannot be replaced with other raw materials.

1.2.4.2.2 Special Tapes

Tapes can have a special construction and special applications. Special tapes according to their end use are discussed in Chapter 4. In this section, only a short description of the construction of special tapes will be given. Generally, such special tapes are paper or film based. The build-up of special tapes is more complex compared with that of common (e.g., packaging, office) tapes. Although most tapes are monowebs, laminated and double-laminated constructions are also manufactured (see the following). Single-side-coated tapes represent the classic construction of tapes, with a solid-state carrier material coated one side with a PSA. Products that display pressure sensitivity on both sides of the carrier material are double-side coated.

Double-side tapes are PSPs with a soft multilayer (see also Chapter 4); they have a solid-state carrier coated on both sides with PSA. The typical double-side-coated pressure-sensitive tape comprises a flexible carrier that is coated with two different incompatible adhesives, thus enabling the tape to be wound upon itself into a roll. Such tapes may have another coated layer on the back side, for example, a primer. A different release degree is achieved for a double-side-coated release liner using the same release component but single-side PE-coated release paper. The release paper assists in the application of the tape without damaging the coating, which is facing the release paper. Typical double-side adhesive tapes have 30–150 g/cm^2 adhesive per side. These tapes are normally used as the bonding agent when combining two materials. The tape is pulled off its roll and the exposed PSA is placed against the first material. Then the release paper is pulled from the tape and the exposed PSA bonds with the second material. The converting industry uses a large amount of double-side-coated PSA tapes like flying splicing tapes (see Chapter 4). Plastics and woven plastics (e.g., neoprene, PVC, PU, PE) were proposed as carrier materials for double-layered PSA tapes. Release liners used for double-side tapes must fulfill special requirements. Such materials must display adequate unwinding performance, controlled adhesion to the tape, dimensional stability, tear resistance, weather and environmental stability, and confectionability/cuttability. Such tapes possess two active surfaces; therefore, they can bond in plane as common packaging tapes as well as between two surfaces. In their early development stage, double-sided tapes were manufactured with a carrier. Later, carrierless tapes were produced. In principle, such tapes may bond multilaterally, like an overall adhesive profile.

Carrierless tapes (see also Section 1.1.3 and Chapter 4) are PSAs with a temporary carrier, that is, supplied on a carrier as a detachable adhesive layer (e.g., carrier-free splicing tapes; see also Chapter) or reinforced PSAs. Such products were described in detail in Refs. [1–3,9]. Carrierless tapes do not have a solid-state carrier material after their application. These products are so-called transfer tapes, or tapes from the reel. Transfer tapes have a PSA layer inserted between two release liners. They are manufactured as a continuous web, supported by a solid-state carrier. This allows them to be processed and applied as a common tape, but serves as a temporary aid only. The adhesive layer of the tape is detached from the carrier during application. Such tapes without a carrier can be used on multidirectionally deformed substrates with a varied shape up to 150°C because the temperature resistance is provided by the adhesive only. The manufacture of carrierless solvent-free PSA tapes is described by Czech in Ref. [34]. A typical carrier-free transfer adhesive tape has a thickness of 30–200 g/m^2.

Foam-backed adhesive tapes are commonly used to cause an article to adhere to a substrate. To enhance immediate adhesion to rough and uneven surfaces, a resilient foam backing can also be coated on either one side or both sides, with an overall thickness of 0.1 to 2.0 mm. Foam tapes used as structural bonding elements exhibit the following advantages: they do not damage the substrate; no mechanical processing operations (drilling, screwing, riveting, etc.) are needed; they give better and more uniform stress distribution; they are lightweight; they produce a watertight seal; and they protect against corrosion. They might be used, for example, in the design of a thermal break that prevents differences in temperature within a structure, reducing stresses that arise from different rates of expansion. Foam tapes compensate for tolerances in the printing process and help to ensure that dot gain is minimized [35].

Foam tapes are based on a foamed, soft carrier, a foamed adhesive, or both. Such constructions are described in detail in Refs. [1,2,9]. Formulation of foamed tapes can include filling as well (see *Technology of Pressure-Sensitive Adhesives and Products*, Chapter 8). Classic foam tapes are foam-like PSPs with a foamed plastic carrier that is coated with a PSA. Foamed PP films can be used to replace expensive satin and acetate films for tapes. These products are films with a relatively low foaming degree and the geometries of common film carrier material. Common foam carrier materials are thicker, of lower density, softer, and more elastic. Rubber, neoprene, PU, PP, PE, and soft PVC foam are applied for double-side tapes. Double-side-coated foam tapes for industrial gaskets are based on closed-cell PE foams coated with a high tack, medium-shear adhesive. Laminated structures of PET foam are used also. PU and PE foams are the most used foam-like carrier materials for tapes. PE foams are aging resistant and can be applied up to 90°C; PU foams resist temperatures up to 150°C, but are destroyed by UV light. Tensile strength, stiffness, density, flexural modulus, flammability, and impact resistance are the main characteristics of foams used as carrier for tapes.

Carrierless foam-like tapes exist as well. Such products are plastic foams impregnated with PSA or made from a foamed PSA. The foam carrier provides excellent conformability and stress distribution. Traditional foam tapes display the disadvantage that the foam can split from the carrier (adhesive break) at too high peel force. Foam-like transfer tapes have the advantages of softness, thickness, and elasticity of the foam, which allow reliable bonding of uneven and textured surfaces; the ability to dampen sound and vibration; and resistance to temperatures of 90°C in the long term and 150°C in the short term. Glass microbubbles can be incorporated also to enhance immediate adhesion to rough and uneven surfaces. The cohesive (tensile) strength of such a cross-linked adhesive (transferred core) may attain 6000 kPa (depending on the cross-linking conditions). A foam-like pressure-sensitive transfer tape with good shear strength and weather resistance that is useful in bonding uneven surfaces has been prepared by forming foamed sheets from agitated acrylic emulsions, impregnating or laminating with adhesives, and drying. Such a film is coated on a PET release film, such as a 1.2-mm adhesive sheet, with a density of 0.75 g/cm^3. By bonding of acrylic polymer panels it displays an instantaneous shear strength of 4.5 kg/cm and demonstrates twice the resistance of a common PU film after weather-o-meter exposure. Carrierless UV-curable tapes with microspheres that are useful for sealing windows or bonding side mouldings onto automobiles contain microspheres and aerosil. Such tapes are UV cured on PET. For such products hydrophobic silica improves the adhesive properties.

In some applications the PSA is coated first on a flexible PU, acrylic, or other foam and then laminated onto the carrier surface. The opposite surface of the foam may also be provided with a PSA layer. Foam-sealing tapes are made of a foam-sealing tape layer and an interlayer strip, rolled up together in a compressed form. For such tapes there are standard thicknesses, roll lengths, and ranges of width given by the supplier. Special foam-mounting tapes are used for the bonding of soft printing plates for flexoprinting. In this case, foam tapes can compensate for tolerances in the printing process and ensure that dot gain is minimized. Automotive insert tapes used as sealants are based on a foam carrier.

Transfer tapes are carrier-based constructions that allow carrierless application of the adhesive layer. Their construction includes a release liner and the adhesive core. They have a solid-state component only temporarily. The sheet backing is a release liner, and when in use, the exposed adhesive surface of this tape is placed in contact with a desired substrate, the release liner is stripped away, and the newly exposed adhesive surface is bonded to the second surface. Both surfaces of the carrier may have low adhesion coatings, one of which is more effective than the other. The transfer process of the tape from the temporary liner can be regulated by means of the release, but it can also be controlled using different adhesives or adhesives with different degrees of cross-linking. According to the type of adhesive, carrierless tapes may be classified as tapes with an adhesive film or tapes with an adhesive foam core. The adhesive core may be a continuous homogenous adhesive layer or a semicontinuous heterogeneous adhesive layer. The heterogeneity of the adhesive layer in the latter case is due to embedded solid-state, liquid, or gaseous particles (holes), that is, the adhesive layer can be a foam. Double-face tapes were used some years ago only for low-stress applications. Cohesive, high-resistance transfer tapes were developed about a decade ago. They can be used for structural bonding (see Chapter 4).

Transfer tapes with contoured adhesive have been proposed as well. An uneven adhesive surface can be achieved by transferring adhesive during rewinding. This tape has a carrier with recesses on one side that are filled with adhesive. When the tape is unwound, the adhesive transfers from the recesses to the carrier. Like special-label constructions (see Section 1.1), special tapes include products with controlled water resistance as well. For instance, carrier-free splicing tapes must also be water soluble.

Certain tapes ensure a removable adhesive bonding that is required for closure, medical, and masking tapes, among others (see Chapter 4). Like labels, tapes can also be repositionable. For instance, cover tapes for the bathroom are repositionable. Stone impact-resistant automotive decor tapes must also be repositionable. Certain closure tapes, for example, diaper closure tapes, must be removable as well.

Paper is suggested as an extensible carrier for medical tapes (see also Chapter 4). Polymer-impregnated paper and silicone-impregnated paper can also be used as a carrier for tapes. For automotive masking tapes, pleated paper is applied.

A large variety of homogeneous and composite materials, for example HPVC, SPVC, PP, PE, PET, PET, polyimide, PET/nonwoven, PET/glass fiber, paper/PET/glass cloth, PE/EVAc foam, PE foam, polyurethane (PU) foam, aluminium paper, cloth, and nonwovens are suggested as carriers for special tapes depending on the end-use requirements. For instance, electrically insulating tapes are based on PET, polyimide, PET/nonwoven, glass cloth, paper, PET/glass fiber, and other carriers. Acetate films are used as a carrier material for self-adhesive tamper-evident products (see Chapter 4). Soft PVC, which may contain up to 50% plasticizer, is applied for electrical tapes. Ethylene acrylic copolymers with low modulus are recommended as blown or cast film carriers for special elastic tapes.

Fiber-like, fabric (textile) materials were the first carrier materials used for PSPs. Actually, such products are used for applications with high mechanical requirements. Fiber-like materials (water resistant or water soluble) based on cellulose, combined with plastic films (e.g., soft and hard PVC, cellulose derivatives, polyester) or foams (open-pore PU, closed-pore PVC, closed-pore polychloroprene, etc.) are suggested as carriers for special tapes. Cellulose-based nonwoven, PE-coated cloth, polyamide cloth, and siliconized cloth are used for various tapes. Paper and nonwovens can be impregnated with water-soluble or water-insoluble resins; glossy or crepe papers and nonwovens coated or impregnated on one or both sides are used. Special two-layer nonwoven laminate can also be used for tapes. Combinations of materials such as chemically bonded cellulose with cotton and synthetic fibers or thermally bonded PP (50%) with cellulose (50%) have also been proposed for special tapes.

Tyvek, an HDEP-fiber-based carrier material, can also be used as a nonwoven material. It possesses the opacity of paper and the tear resistance of fabric. Generally, there are two different types of nonwovens. Certain materials are manufactured with a special resin as adhesive, whereas others (e.g., Du Pont's Reemay polyester or Monsanto's Cerex nylon) do not have an adhesive matrix. Thermoplastic acrylic films can be laminated with nonwovens as well. Woven Fiberglas scrim has also been recommended as a carrier material for tapes. According to Lin et al. [36], the carriers most commonly used in the manufacture of silicone PSA tapes include polyester, Teflon®, Teflon-coated glass cloth, silicone rubber-coated glass cloth, high-temperature films, such as Kapton®, polyether imides, and woven quartz, among others.

Film/film laminates, film/plastic laminates, film/paper laminates, and metallized film are also applied as carrier materials for tapes. Such materials are manufactured by specialized firms. Metallic foils are suggested for tamper-evident products. Aluminium film is proposed as street-marking tape and insulation tape.

1.2.4.3 Protective Webs

Generally, protective films are web-like, self-adhesive, removable materials that are used in intimate contact with the surface to be protected and require a self-supporting carrier to display the mechanical resistance necessary for the protective function and removability. The web-like attribute refers to the solid-state character of the product, which is manufactured as a solid-state continuum (reel). Such products were introduced in the 1960s for the protection of coated coils. Cold and hot systems are identified. Cold systems use an adhesive-coated film, whereas hot systems apply a thermoplastic film that bonds with the coating layer of the coil (at high temperatures) to produce the desired protection (see Chapters 4 and 7). On the other hand, this bond should allow removability after the coil is processed. Lamination of coils has been used as a coating method for many years. Such (adhesive) lamination of PVD and poly(vinylidene fluoride) films has been carried out at 180–200°C. The manufacture and end use of protective films was described in detail in Refs. [2,9,22]. Coating devices suggested for the manufacture of protective films are listed in *Technology of Pressure-Sensitive Adhesives and Products*, Chapter 10.

On the basis of their their build-up, protective films can be classified as adhesiveless and adhesive-coated films. The adhesiveless products are also called self-adhesive films (SAFs). The SAFs may work physically or chemically, depending on their principle of function, that is, their build-up (see Chapter 7). The SAFs include different chemical compounds as pressure-sensitive components.

Adhesive-coated protective webs are large-surface, tape-like products (generally with a paper or plastic carrier) with a special removable adhesive (see also Chapter 4). Masking tapes are also narrow-web masking (protective) films (see also Chapter 4). Such products can also be manufactured as uncoated monowebs. Protective films with adhesive can be classified according to their adhesive nature, type of carrier film, adhesive properties, bonding/debonding nature, and application domain (see Chapter 4). Tests of special protective films are described in Chapter 8.

1.2.4.3.1 Common Protective Webs

Common protective webs used for static surface production have a paper (protective papers) or plastic carrier (protective films) [2,9]. They are coated with removable rubber–resin or acrylic PSAs. Such adhesives are (with the exception of some water-based acrylics) cross-linked. Their adhesive properties are strongly affected by the coating method (coating device) used (see Chapter 4).

1.2.4.3.2 Special Protective Films

Generally, special protective webs have the same construction as common products. Their processing-technology required properties are provided by the special performance characteristics of the carrier material (e.g., deep drawability, elongation) or the adhesive (see Chapter 4).

1.2.4.4 Forms

Forms are labels that have a multiweb, multilaminate structure, in which continuous and discontinuous carrier materials and PSAs with different adhesive performances are laminated together to allow a time- and stress-dependent controlled delamination (see also Chapter 4). Promotional form labels can have many pages, and their front pages can have a different print quality [37]. Their delamination is carried out in several steps during manufacture and application. Business forms include permanent and temporary laminates as well. For instance, such multilayer self-adhesive labels comprise a carrier label containing silicones, an adhesive layer, a printed message, another carrier layer (e.g., polyester), a release layer (e.g., silicone), an adhesive layer, a third carrier layer (e.g., PE), and a printed message. Such labels may have more than two adhesive–carrier layer–message layer units. The construction and manufacture of forms was described in Refs. [1,8].

References

1. Benedek I., Build up and Classification of Pressure-Sensitive Products, in *Developments in Pressure-Sensitive Products*, Benedek I., Ed., Taylor & Francis, Boca Raton, 2006, Chapt. 2.
2. Benedek I., Manufacture of Pressure-Sensitive Products, in *Developments in Pressure-Sensitive Products*, Benedek I., Ed., Taylor & Francis, Boca Raton, 2006, Chapt. 8.
3. Benedek I., *Pressure-Sensitive Design, Theoretical Aspects*, VSP, Utrecht, 2006, Chapt. 3.
4. Benedek I., Manufacture of Pressure-Sensitive Labels, in *Pressure-Sensitive Adhesives and Applications*, Marcel Dekker, New York, 2004, Chapt. 9.
5. Belgian Patent, 675,420, in Fischer D.K. and Bridell B.J., Ed., Adco Product Inc., Michigan Center, MI, USA, EP 0,426,198A2/1991.
6. *Paper, Film and Foil Converter*, (5) 45 (1969).
7. Benedek I., Adhesive Performance Characteristics, in *Pressure-Sensitive Adhesives and Applications*, Marcel Dekker, New York, 2004, Chapt. 6.
8. Benedek I., Converting Properties of PSAs, in *Pressure-Sensitive Adhesives and Applications*, Marcel Dekker, New York, 2004, Chapt. 7.
9. Benedek I., End Use of Pressure-Sensitive Products, in *Developments in Pressure-Sensitive Products*, Benedek I., Ed., Taylor & Francis, Boca Raton, 2006, Chapt. 11.

10. Czech Z., Synthesis, Properties and Application of Water-Soluble Acrylic Pressure-Sensitive Adhesives, in *Pressure-Sensitive Design, Theoretical Aspects,* Benedek I., Ed., VSP, Utrecht, 2006, Chapt. 6.

11. Feldstein M.M., Platé N.A., and Cleary G.W., Molecular Design of Hydrophylic Pressure-Sensitive Adhesives for Medical Applications, in *Developments in Pressure-Sensitive Products,* Benedek I., Ed., Taylor & Francis, Boca Raton, 2006, Chapt. 9.

12. Feldstein M.M., Cleary G.W., and Singh P., Pressure-Sensitive Adhesives of Controlled Water Absorbing Capacity, in *Pressure-Sensitive Design and Formulation, Application,* Benedek I., Ed., VSP, Utrecht, 2006, Chapt. 3.

13. Kulichikhin V., Antonov S., Makarova V., Semakov A., Tereshin A., and Singh P., Novel Hydrocolloid Formulations Based on Nanocomposite Concept, in *Pressure-Sensitive Design, Theoretical Aspects,* Benedek I., Ed., VSP, Utrecht, 2006, Chapt. 7.

14. Benedek I., Adhesive Properties of Pressure-Sensitive Products, in *Developments in Pressure-Sensitive Products,* Taylor & Francis, Boca Raton, 2006, Chapt. 7.

15. Rikkidain K.K., Japan Pat. 07,278,508/1995.

16. Nakahata H., Japan Pat. 07,316,510/1995, *Adhes. Age.,* (5) 11 (1996).

17. British Pat. 1,102,244, *Adhäsion,* (10) 303 (1972).

18. Benedek I., Chemical Basis of Pressure-Sensitive Products, in *Developments in Pressure-Sensitive Products,* Benedek I., Ed., Taylor & Francis, Boca Raton, 2006, Chapt. 5.

19. Benedek I., Introduction, in *Pressure-Sensitive Design, Theoretical Aspects,* VSP, Utrecht, 2006, Chapt. 1.

20. Larimore F.C. and Sinclair R.A., Minnesota Mining and Manuf. Co., St. Paul, MN, USA, EP 197,622A1/1986.

21. Meinel G., *Papier u.Kunststoff Verarbeiter,* (19) 26 (1985).

22. Benedek I., Pressure-Sensitive Design and Formulation in Practice, in *Pressure-Sensitive Design and Formulation, Application,* Benedek I., Ed., VSP, Utrecht, 2006, Chapt. 6.

23. Czech Z., Removable and Repositionable Pressure-Sensitive Materials, in *Pressure-Sensitive Design and Formulation, Application,* Benedek I., Ed., VSP, Utrecht, 2006, Chapt. 4.

24. Park Y.J., Hot-Melt PSAs Based on Styrenic Polymer, in *Pressure-Sensitive Design and Formulation, Application,* Benedek I., Ed., VSP, Utrecht, 2006, Chapt. 2.

25. Benedek I., Chemical Composition of PSAs, in *Pressure-Sensitive Adhesives and Applications,* Marcel Dekker, New York, 2004, Chapt. 5.

26. Benedek I., Test Methods, in *Pressure-Sensitive Adhesives and Applications,* Marcel Dekker, New York, 2004, Chapt. 10.

27. *Etiketten-Labels,* (5) 16 (1995).

28. *Converting Today,* (11) 9 (1991).

29. *Kunststoffe,* **83** (10) 737 (1993).

30. *Kunststoffe,* **83** (10) 725 (1993).

31. *Finat labelling News,* (3) 29 (1994).

32. Labelexpo Asia, *Labels and Labelling Asia*, Cowise International, Abbey House, U.K., 1998.

33. Tomuschat P., UV-curable s'Silicones for Roll Labels, in *Proc.31st Munich Adhesive and Finishing Symposium 2006, 22–24 Oct. 2006*, Munich, Germany, p. 282.

34. Czech Z., Production of Carrier-Less Solvent-Free PSA Tapes, in *Proc. in Proc. 31st Munich Adhesive and Finishing Symposium 2006, 22–24 Oct. 2006*, Munich, Germany, p. 253.

35. *Label, Labels* (2) 22 (1997).

36. Lin Shaow B., Durfee L.D., Ekeland R.A., McVie J., and Schalau II G.K., Recent Advances in Silicone Pressure-Sensitive Adhesives, *J. Adhes. Sci. Technol.*, **21** (7) 605 (2007).

37. *Labels, Label Int.*, (5/6) 24 (1997).

2

Electrically Conductive Adhesives in Medical Applications

J. Anand Subramony
Dr. Reddy's Laboratories Ltd

The emergence of biomedical devices in the area of diagnostics and drug delivery in the past decade has increased the need to identify newer adhesive materials with unique electrical and mechanical properties. Research in the area of pharmaceutical materials science has been triggered largely due to the advancement in transdermal, implantables, depots, and transmucosal drug delivery systems. In the transdermal realm, the use of medicated patches became a standard of care after the approval of the first patch in 1979 that delivered the potent antiemetic drug scopolamine for motion sickness. Other approaches to deliver therapeutic molecules across the transdermal barrier followed suit in the areas of "enhanced" transdermal, such as iontophoresis, sonophoresis, electromagnetic, and thermally activated systems. The development of various novel drug delivery systems necessitated the need for newer materials that form key components in such devices. Pressure-sensitive adhesive (PSA) products used in medical applications are described in *Application of Pressure-Sensitive Products*, Chapter 4. This chapter will focus on the use of PSAs in transdermal drug delivery systems and particularly on the subset of electroconductive adhesives, a specialty material, used in biomedical applications.

2.1 Introduction: Drug Delivery and Transdermal Patches

Traditional drug delivery methods rely on oral [1] and parenteral routes [2]. Whereas drug delivery via oral administration is time tested and has made several improvements [3], such as controlled release [4] (including immediate release, delayed release, etc.), the oral route has the disadvantages of hepatic first-pass effect, leading to poor bioavailability in some cases and gastrointestinal irritation. Parenteral delivery, on the other hand, is invasive, has potential systemic side effects, and often requires the intervention of a health care practitioner and, hence, affects patients convenience. The delivery of active agents across the transdermal epithelium for systemic and local delivery provides many advantages, including comfort and noninvasiveness, resulting in improved patient compliance. Gastrointestinal irritation and the variable rates of absorption and metabolism, including the first-pass effect encountered in oral delivery, are avoided. Transdermal drug delivery also provides a high degree of control over blood concentrations of any particular active agent and the profile can match that of intravenous (IV) infusion [5], yet is noninvasive.

Transdermal medicated patches release drugs at controlled rates over a period of time ranging from 1 to 7 days. Molecules with certain physicochemical properties that fall below a molecular weight threshold can be delivered transdermally across the stratum corneum [6]. Among adhesive materials that are accepted for medical use, PSAs form an essential component of transdermal patches. The medical applications of PSAs are due to their skin contact properties (given by their viscoelastic behavior; see *Fundamentals of Pressure Sensitivity*, Chapters 1–5) and are largely used in the construction of band aids, wound care dressings, electrocardiogram (ECG) electrodes, and transdermal patches. Medicated PSA patches can be applied on the skin to deliver a therapeutic molecule as a function of time, either locally or for systemic circulation into the blood stream. PSAs, as the name implies, are materials that form a bond upon application of pressure between the substrate and the adhesive. Adhesive contact is achieved by the application of pressure and there is no involvement of solvent or thermal or photo curing in enabling the adhesion. In contrast to structural adhesives where adhesion is mainly due to chemical reaction such as cross-linking between two reactants, most of the PSAs display their bonding properties at room temperature due to the rheological properties of the adhesive material and formation of secondary bonding forces between the substrate and the adhesive.

PSA-based medicated transdermal patches provide enhanced bioavailability of the drug in systemic circulation by avoiding the gastrointestinal (GI) tract and hepatic first-pass metabolism commonly seen with oral administration of drugs. Transdermal patches therefore enhance the efficacy and minimize side effects, in addition to being noninvasive. Since the successful commercialization of scopolamine (Figure 2.1) patches in 1981, several systems followed suit as indicated in Table 2.1. A key requirement of these medicated transdermal patches is the need to formulate the drug in suitable adhesives that are skin friendly and also nonreactive with the drug of interest.

Transdermal patches based on PSAs can be broadly classified into matrix and reservoir systems, as illustrated in Figure 2.2. A multilayer system is also available, which, as the name implies, has another adhesive layer containing the drug. Other components of this patch include a permanent backing layer and a temporary release liner. This type of configuration is depicted in Figure 2.3. In the case of matrix construction (also known as drug-in-adhesive system), the drug layer is embedded in the adhesive layer and is separated via the release liner. Reservoir systems, as the name implies, have a drug layer embedded in the backing layer.

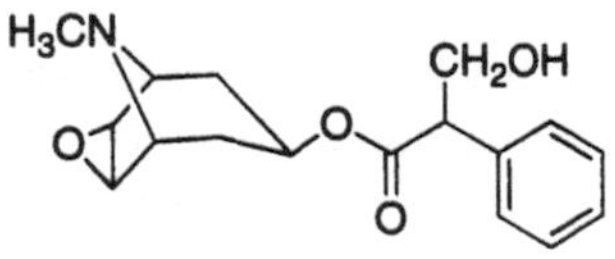

FIGURE 2.1 Chemical structure of scopolamine.

TABLE 2.1 List of Various Commercially Available Passive Transdermal Patches

Transdermal Drug Patch	Indication	Alternate to
Nitroglycerine	Angina	Sublingual nitroglycerine or nitroglycerine oinment
Clonidine	Hypertension	Oral administration
Estradiol	Issues of liver metabolism	Oral administration
Fentanyl	Chronic pain	IV patient controlled analgesia
Nicotine	Smoking cessation	Oral administration
Testosterone	Hormone replacement therapy for men	Injection
Oxybutinin	Anticholinergic-overactive bladder	Oral administration
Methylphenidate	Attention deficit hyperactivity disorder	Oral administration

(Disclaimer: The information provided herein is only for reference purposes. Appropriate medical consultation and precaution should be exercised before using any of the above.)

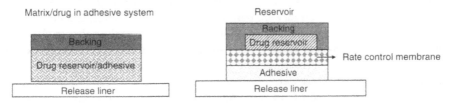

FIGURE 2.2 Schematic representation of various types of adhesives used in transdermal drug delivery.

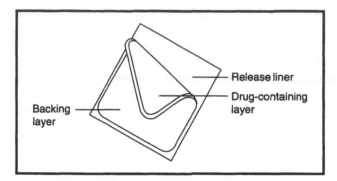

FIGURE 2.3 Schematic representation of a multilayer passive transdermal patch.

2.2 Electrical Conductivity in Polymers

Electrical conductivity is defined by Ohm's law ($V = IR$, where V is the applied voltage, I is the resulting current, and R is the resistance of the material). Conductance, derived from this expression, is the inverse of resistance and depends on the number of charge carriers in the material and their mobility. Electrical conductivity can be divided into two categories, namely electronic and ionic conductivity, depending upon whether the charge carriers are electrons or ions. Although most of the commercially available elastomers/polymers are insulators, there are specialty polymers that have electrical conductive properties due to their chemical structure and other conductive additives imparted during synthesis or processing. Whereas electronically conducting polymers have applications that can potentially replace metals and semiconductors [7], ionically conducting polymers are used in electrolyte applications. Polymer electrolytes are one such family of materials where the ionic conducting properties of the polymers are attributed to the segmental motion of the polymeric chains (that can coordinate cations due to the presence of polar groups with lone pair electrons in their backbone) in the amorphous regions of the polymer matrix [8]. Figure 2.4 illustrates the schematic of cation transport in one such polymer electrolyte, namely poly(ethylene oxide) (PEO) containing polar oxygen groups in the backbone. The cation transport mechanism in polymer electrolytes is believed to be assisted by the motion of polymer chains that enables intra- and interchain hopping of the cation via ion clusters in the amorphous regions (see p. 18 of Ref. [8]). The conductivity in these types of systems can be described mathematically using the Vogel–Tamman–Fulcher (VTF) equation [9].

Electronically conducting polymers are materials in which the electronic conduction is due to the extended delocalized bond network achieved through conjugation [10]. Typical examples in this category are polyacetylene [11], polypyrrole, and polyaniline. The conductivity is dependent upon the molecular structure of the polymer, degree of conjugation, and solid-state arrangements such as crystallinity or ordering. Because conjugated systems have "loosely" bound electrons in the backbone, electron flow can be initiated by adding redox species by a process called doping [12]. Doping typically involves the addition (reduction) or removal of electrons (oxidation). Because the delocalized network in a conjugated system is supported by the continuous overlapping of p orbitals, once doping has occurred, the electrons in the pi bonds become mobile

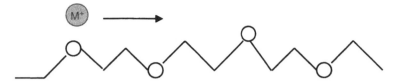

FIGURE 2.4   Structure of ion-conducting polymer, poly(ethylene oxide) (PEO), that can coordinate metal ions (such as Li and Na) using the polar oxygen present in the backbone.

and are free to move along the molecule to enable electrical conduction under the influence of an applied voltage. In this category, the synthesis and characterization of aminonapthoquinone-based electroactive polymers with excellent conductivity has been reported by Pham et al. [13]. The authors obtained polymer-coated electrodes by electropolymerization of naphthalene derivatives containing an amino substituent. Essentially, the poly(5-amino-1-4-napthoquinone) (PANQ) structure is of the polyaniline type, bearing one quinone group per ANQ moiety.

Conductive polymers can also be formed by the process of filing conductive additives such as gold, silver, copper, powdered graphite, and carbon black in polymer matrices. These materials have large-scale industrial applications. Electrically conductive PSAs (electrical resistance less than 100 ohm/square inch) with conductive particles distributed as a monolayer in the adhesive layer and in which the thickness of the particle is less than the thickness of the adhesive layer have been reported [14]. Conductive carbon black, when added to nonconductive PSAs, improves the electrical properties of the PSA in proportion to the degree of carbon loading [15] (see also *Technology of Pressure-Sensitive Adhesives and Products*, Chapter 11). Carbon black (with intrinsic resistivity of 0.01–0.1 ohm-cm) when used in particulate/aggregate form in submicrometer-size diameters (30–100 nm) in polymeric matrices can form conductive networks that are tightly interconnected such that they are not separated during processing of polymers, such as compounding. Wettability and formation of homogenous dispersion are two important aspects during the processing of electroconductive polymers with additives. At smaller particle sizes of the conductive additive, aggregation of the particles will become high, which would require higher shear and mixing energy to obtain improved wettability. Furthermore, to form the conductive network, the primary aggregates distributed within the polymer matrix must be above a certain critical threshold concentration such that the interaggregate distance is short enough for electron transport. The minimum concentration at which the electrical conductive properties begin due to the tunneling phenomenon is called the "percolation threshold," beyond which increased loading will cause a drastic decrease in resistance [16]. The volume percentage of carbon black in the polymer matrix would serve as a good tool to estimate the conductive properties. However, at very high loading of the conductive additive, the adhesive performances of the PSA such as tack, peel, and shear can be adversely affected [17] (for adhesive properties, see also *Fundamentals of Pressure Sensitivity*, Chapters 6–8). One way to circumvent the loss of tack would be to add a secondary tackifying agent or a plasticizer, which can, in turn, affect the manufacturing process of the electrically conductive PSA (see also *Technology of Pressure-Sensitive Adhesives and Products*, Chapter 11).

In addition, the surface area, surface chemistry, and structure of carbon black are considered important for the electroconductive behavior of the material. As a rule of thumb, higher surface area (external and internal, including any porosity) of the conductive additive enables improved wetting and therefore lessened loading to reach the percolation threshold. With regard to surface chemistry, the presence of oxygen-containing groups on the surface of carbon black can act as insulating centers for electroconductivity as they disrupt the electron transfer capabilities. Aggregation of carbon black can also affect its surface structure, which often leads to highly branched and fibroid-like structures. Depending on the pore volume of the structure, the bulk density of the material varies. However, at higher pore volume, more energy is required to wet the material. A highly branched structure has better overlap of particles and is easy to disperse. Ball milling the additive using a solvent vehicle is a well-known method of improving particle dispersion, although there is a risk of disrupting the inherent chain-forming structure of carbon black in such an exercise. Highly structured and long fibroid structures have disadvantages in that they are susceptible to shear breakdown during processing. Akzo Nobel Corporation (www.akzonobel.com) is a manufacturer of a variety of conductive carbon black that can be used as filler materials in the manufacture of conductive plastics. Their products Ketjenblack EC-300J and Ketjenblack EC-600JD are two examples of carbon blacks with high surface areas (800 and 1250 m^2/g, respectively) [18]. A method of making an electrically conductive PSA by incorporating conductive black filler into the PSA at low carbon concentrations to avoid adversely affecting the adhesive properties of the PSA is described by Adhesives Research, Inc. (Glen Rock, PA, USA) [19]. According to this invention, the conductive particles present in the adhesive (PSA is selected from the group consisting of a rubber–resin, acrylic, or silicone-based formulation) are in the form of a carbon structure that optimizes the efficient employment of carbon by providing conductive pathways through the adhesive.

Silver is another commonly used metallic filler to impart electrical conductivity to PSAs. An electrically conductive adhesive tape wherein the adhesive base is impregnated with silver particles has been disclosed [20].

2.3   Key Requirements of Pressure-Sensitive Adhesives in Biomedical Applications

PSAs with electrical conductive properties are a unique subset of materials that have found increased utility in biomedical devices. The components of biomedical devices must pass the stringent requirements set forth by the agency for use of medical devices in diagnostics and drug delivery. The toxic effects of some solvents (because some of them are potentially carcinogenic) have increased the need to identify solventless adhesives in the fabrication of biomedical devices. Photocurable acrylics and epoxies, along with hydrophilic urethane and epoxy-based adhesives, have been commonly used in biomedical devices. Photostability of the device components and any active pharmaceutical ingredients, if present, along with the polymers are important aspects to consider if the curing requires high energies such as ultraviolet (UV) and are done *in situ* in the presence of other components.

Nontoxicity of the adhesive conforming to United States Pharmacopaea (USP) class VI biocompatibility standards is one of the foremost conditions in the choice of adhesives. For solventless adhesives, the ability to cross-link 100% and compatibility with materials used for processing and end use are other features that are considered important. The rheological properties of the adhesives (viscosity, thixotropic index) should be optimized for unique requirements such as conformal coverage in cases in which it is required (see also Chapter 4). The desirable properties of the adhesives are determined by paying attention to the substrates to be bonded (similar vs. dissimilar) and the interface. The safety and potential suitability of a new adhesive in medical applications is evaluated via standard tests such as intracutaneous toxicity, acute systemic toxicity, cytotoxicity, and implantation (in applications that require implantable adhesives).

The selection of a polymer and its physicochemical properties are important parameters to consider in making electroconductive adhesives. The fundamentals of materials science, polymerization technology, and formulation (see also *Technology of Pressure-Sensitive Adhesives and Products*, Chapters 1 and 11) in altering the composition with respect to structure, cross-linking, plasticizer, monomer ratios, etc., to obtain the desirable properties in terms of conductivity, adhesion, and viscosity are essential steps in the design phase. As discussed in detail in *Fundamentals of Pressure Sensitivity*, long-chain polymers demonstrate viscoelastic properties whereby they exhibit both viscous and elastic characteristics when undergoing deformations. Whereas elastic materials reveal an instantaneous stress–strain response, viscous materials resist strain linearly with time when stress is applied. Viscoelastic materials demonstrate properties that are a blend of viscous and elastic classes, in that they exhibit time-dependent strain. A multicomponent PSA composition consisting of both a viscoelastic polymeric adhesive phase and an electrically conductive aqueous phase, containing a hydrophilic polymer, a humectants, and an electrolyte is reported [21]. In addition, processing tools and characterization are also extremely important in designing the final product. The rheological properties of the polymers such as melt viscosity have a profound influence on the electroconductive properties. In metal-filled electroconductive polymers, crystallinity of the semicrystalline polymer is an important factor contributing to the overall electrical properties. The orientation of polymer chains upon curing will also determine the crystallinity of the polymer as well as the proximity between the conductive fillers.

Polyethylene (PE) is a commonly used polymer in metal-filled conductive polymer systems. With variation in its crystalline density (namely low-density PE [LDPE] and high-density PE [HDPE]) and the loading of the metallic filler, several different grades of conductive polymer can be obtained. Other materials commonly used as polymer matrix are styrene–butadiene rubber (SBR) and acrylics.

The morphology of the polymer–additive mix is important in determining the electroconductive properties. For instance, a continuous distribution of the metallic filler is very important to obtain good interconnectivity and electroconductivity, even at reduced loadings.

Compatibility of the conductive filler material with the polymer matrix is also important for processing. Higher compatibility with the polymer improves the homogeneity of the dispersion and will prevent agglomeration of the individual filler particles such as carbon black. Particularly for carbon black, due to its polar nature and higher surface tension, its compatibility is higher with polymers containing polar groups in the backbone such as PEO or any of the polyvinyl-based polymers with polar atoms in the backbone. Compatibility between the polymer and the conductive adhesive will determine the amount of loading of the filler required for electroconductive properties. Normally, the rule of thumb is a linear relationship between loading and compatibility. Particularly for carbon black, poor compatibility means a lower percolation threshold and, hence, a lower loading to activate the conductive pathways.

2.4 Electroconductive Adhesives in Drug Delivery

Many active agents are not suitable for passive transdermal delivery because of their size, ionic charge characteristics, and hydrophilicity [22]. Body surfaces such as skin are a very effective barrier against intrusion by drugs or chemical agents. Generally, hydrophilic agents do not permeate passively through body surfaces such as skin because of the lipid-containing cellular structure of skin. Passive transdermal drug delivery systems also have the drawback of decrease in delivery rate as a function of time due to equilibration and absence of concentration gradient after continued use, primarily due to finite drug loading in the adhesive. Furthermore, passive transdermal drug delivery systems also lack the user-enabled control aspect of drug delivery because the delivery is only proportional to the loading and the surface area of the patch. An alternate way for transdermal delivery of active agents involves the use of electrical current to actively transport the active agent into the body through a body surface (e.g., intact skin) by electrotransport or iontophoresis [23], which differs from passive transdermal delivery in that additional energy is used to drive the movement of the active agent. Electrotransport techniques may include iontophoresis, electroosmosis, and electroporation [24]. Electroconductive adhesives have unique applications in the area of iontophoretic drug delivery.

During iontophoresis, charged particles are transported across tissues by the application of an electrical current [25]. Iontophoretic drug delivery devices typically contain two components, namely the drug/formulation part and the device part. Whereas the drug/formulation unit is typically made of injection-molded plastics that serve as housing units for anodic and cathodic formulations (typically hydrogels), the device part is composed of a battery and a programmable controller and printed circuit board to control the operations. Because electrical connectivity between the drug unit and the device unit is a must for successful functioning of the unit, electroconductive adhesives are typically used at the interface between the gel and the metal tabs of the printed circuit board (PCB) assembly. Figure 2.5 illustrates the schematic of an iontophoretic device and the interfaces where electroconductive adhesives are used, highlighted by arrows. Most of the inventions based on iontophoretic drug delivery across various mucosa [26] require the use of an electroconductive adhesive layer as the key interface between electrode components and electrolyte gels in the donor (drug containing) and counter reservoirs. Examples in this category include the description of an electrotransport device that is useful for transdermal drug delivery and sampling by Kleiner and Scott (ALZA Corporation) [27]. A flexible, electrically powered iontophoretic delivery apparatus using electrically conductive adhesive to deliver water-soluble

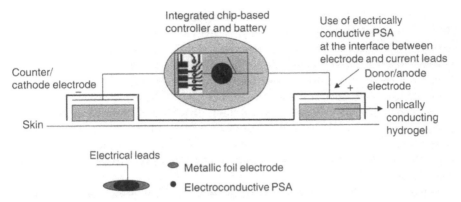

FIGURE 2.5 Schematic representation of an iontophoretic device illustrating the interfaces where an electrically conductive adhesive will be used.

salts of drug, polypeptide, or other molecules has been proposed by Myers and coworkers [28]. The authors claim that the advantage of their apparatus is its simplified structure and ease of manufacturability. The utility of ion-conducting polymer electrolytes as drug-transporting media in iontophoretic devices has been reviewed by Latham and coworkers, which further emphasizes the role of conductive polymers in medical applications [29].

2.5 Biomedical Electrodes—Addressing the Interface

One of the key areas that requires the application of electrically conductive adhesives is the area of biomedical sensing. Biomedical sensors are used to sense either a physiological or a physical response or action. Sensors are used either to quantify a physiological indicator such as glucose sensors or to obtain qualitative information about a physical action such as accelerometers and piezoelectric arrays for detecting physical movements. Biomedical electrodes have been constructed using polyester, polypropylene (PP), or PE with conductive layers (such as Ag ink or metallic Ag) as key components [30].

The ability to deliver a drug in response to an output or readout from a biomedical sensor has been proposed in the construction of closed-loop drug delivery systems [31]. In the area of diagnosis, an iontophoretic device containing an electrically conductive adhesive connecting an electrical component to the reservoir of the active agent for diagnosis of cystic fibrosis was invented by Haak et al. [32]. According to the inventors, the electrically conductive adhesive connects the circuit to the reservoirs and also acts as a reservoir for a drug such as pilocarpine to be iontophoretically delivered. The electrically conductive adhesive is a matrix (mixture of low- and high-molecular-weight polyisobutylene [PIB]) containing conductive carbon filler (fibrous, web, mat, mesh, particles, fibers, or flakes). Many of the interfaces of biomedical devices such as a catheter tip used to map a physiological response require synergistic properties of both electrical conductivity and adhesion. An ocular drug delivery system to deliver medication using iontophoresis with an absorbent foam layer that serves to contain the medication attached to a conducting adhesive has been proposed by Behar and Roy [33].

An electrically conducting interface for biomedical sensors, formed by casting and drying a polymer blend containing poly(methyl vinyl ether–maleic anhydride), glycerin (as plasticizer), polyvinylpyrrolidone (PVP, as viscosity enhancer), and sodium chloride as the conducting electrolyte is described by Woolfson [34]. It has been proposed that the resulting flexible bioadhesive film can be used to adhere to a biological substrate under adverse conditions such as high humidity or when the substrate is immersed in water or a physiological fluid. This kind of material can be used as components of an electrode for the monitoring of fetal heart rate. The use of an electrically conductive adhesive (ECA) to bond a piezoelectric transducer ring to a tungsten carbide tube as two components of the catheter tip has been studied by Tavakoli and coworkers [35].

Polyurethane (PUR) is a hydrophilic polymer that has been widely used in the medical industry (see also Chapter 4). Clinical applications of segmented PUR gels with ionically conducting properties have been reviewed by Shikinami [36]. In another study, Shikinami and coworkers [37] examined the ionic conductivity of blends of segmented PUR with various polyalkalene oxides with added ionic salt from a structure–property perspective. Bioelectrodes for ECG applications were constructed using ionically conducting PUR adhesives by Shikinami and Kaoru [38]. The authors studied the electrical and adhesive properties of the electrode, including contact impedance to human skin at various frequencies. The authors recommended that the ac and dc values of conductivities and voltage change upon initiation were within the limits of the standard for disposable ECG electrodes with good adhesion and high safety to human skin.

Another application of conductive adhesives is their use as biomedical electrodes for diagnosis in the area of cardiovascular applications. Several compositions for conductive biomedical electrodes can be found in the descriptive patent literature. A composition for a conductive adhesive comprising PSA, electrolyte, and an organic humectant has been proposed by Menon and coworkers [39]. The inventors claim that the composition is resistant to the loss of electrical properties following a loss of moisture from adhesive. A biomedical electrode that is a two-phase composite of ionically conductive PSA comprising continuous solid phase of hydrophilic, adhesive conductor, and discontinuous hydrophobic adhesive have been proposed by Dietz and coworkers [40]. According to the inventors, the composite has usefulness as a component in a biomedical electrode because it can act as a means of electrical communication interacting between the conductive medium and the electrical diagnostic, therapeutic, or electrosurgical equipment. In another invention, Engel [41] reported a conductive PSA for a medical electrode prepared from a hydrogen bond donating and accepting monomer in a conductive aqueous solution medium that is useful as a skin-contacting material for biomedical electrodes.

Metal ion contamination from electrodes could be a concern during the use of biomedical electrodes containing metallic fillers for medical applications. This is particularly true in the case of iontophoresis, where use of Ag electrodes, if not controlled properly at the anode interface, can lead to delivery of Ag ions into the tissue, causing argyrosis.

Electrically conductive adhesives based on hydrogels are also known in the literature (see also *Technology of Pressure-Sensitive Adhesives and Products*, Chapter 7). These materials are based on acrylates and quaternary chlorides obtained via UV curing [42].

Some of the commercially available electrically conductive adhesives include the Dow Corning® DA series (DA 6524 and 6533), a silicone-based material that can be thermally cured at 150°C in 60 min. The resultant adhesive has a volume resistivity of 0.0003 ohm-cm and excellent moisture- and water-resistance properties. ElectRelease™ (EIC Laboratories, Norwood, MA, USA) is a family of high-strength electroresponsive adhesives with a special property of release upon the application of a low-power dc current.

2.6 Conclusions and Future Directions

The use of conductive adhesives as a key component in the field of medical devices is constantly on the rise due to the requirement of unique applications that demand a combination of electrical conductivity and adhesive properties. New formulations are being developed to address the need for medical devices in diagnostics and drug delivery. Materials science will play a key role in developing newer compositions based on structure–property relationships. Among transdermal drug delivery applications, the use of adhesives (electroconductive and PSAs) will continue to increase due to the developments in transdermal drug delivery. One such application would be a combination of passive and active (iontophoretic) technologies for applications such as chronic plus breakthrough pain, where the passive component of the patch provides the baseline dose and the iontophoretic part addresses the on-demand dose. This is a unique combination that can be described as an enhanced passive or a turbocharged passive patch [43]. A regular PSA would form the basis of the passive patch, whereas the electroconductive PSA would form a key component of the iontophoretic device side of the combination patch. Conductive adhesives will also continue to be a critical component in hybrid circuits within pacemakers to meet the unique demands of electrical conductivity and adhesion. Future devices might also include a combination of PSAs with piezoelectric and ferroelectric materials. One such invention is that of a flexible tape with bridges of electrically conductive particles with a ferromagnetic core extending through the adhesive layer, wherein the magnetic attraction can serve to form the requisite bridges [44]. Incorporating the selection of the electrically conductive adhesive early in the development plan and understanding its functional and potential toxic properties will be critical points in medical applications for early evaluation and reducing the time to market and commercialization.

Acknowledgments

I acknowledge many of my former colleagues at Alza Corporation (Mountain View, CA, USA) and in particular Dr. Brad Phipps for his input. I also thank Dr. Bill Meathrel of Adhesives Research (Glen Rock, PA, USA) for providing valuable reference information on conductive carbon blacks.

References

1. Sastry, S.V., Nyshadham, J.R., Fix, J.A. 2000. Recent technological advances in oral drug delivery-a review. *Pharm. Sci. Tech. Today* 3(4): 138–145; Langer, R. 1998. Drug delivery and Targeting. *Nature*, 392 (6679 Suppl):5–10.
2. Uchegbu, I.F. 1999. Parenteral drug delivery: 1. *Pharm J.* 263(7060):309–318; Uchegbu, I.F. 1999. Parenteral drug delivery: 2 *Pharm J.* 263(7061): 355–358.
3. Helfand, W.H., Cowen, D.L. 1983. Evolution of pharmaceutical oral dosage forms. *Pharm. Hist.* 25:3–18.
4. Das, N.G., Das, S.K. 2003. Controlled-release of oral dosage forms, Formulation, Fill & Finish 10–16, (from www.pharmatech.com/pharmtech/data/articlesstandard/pharmtech/232003/59302/article.pdf).
5. Gupta, S.K., Southam, M., Sathyan, G., Klausner, M. 1998. Effect of current density on pharmacokinetics following continuous or intermittent input from a fentanyl electrotransport system. *J. Pharm. Sci.* 87:976–981; Gupta, S.K., Sathyan, G., Phipps, J.B., Klausner, M., Southam, M. 1999. Reproducible fentanyl doses delivered intermittently at different time intervals from an electrotransport system. *J. Pharm. Sci.* 88:835–841.
6. Xu, P., Chien, Y.W. 1991. Enhanced skin permeability for transdermal drug delivery: Physiopathological and physicochemical considerations. *Crit. Rev. Ther. Drug. Carr. Syst.* 8:211–236.
7. Novàk, P., Müller, K., Santhanam, K.S.V., Haas, O. 1997. Electrochemically active polymers for rechargeable batteries. *Chem. Rev.* 97:207–281.
8. Armand, M.B., Chabagno, J.M., Duclot, M. 1978. Second International meeting on Solid Electrolytes, (St Andrews, Scotland, 20–22 Sept.), Extended Abstract; Gray, F.M., 1997, Polymer Electrolytes, (Ed. Connor J.A.), RSC Materials Monographs, The Royal Society of Chemistry, Cambridge, U.K.
9. Vogel, H. 1921. *Phys. Z.* 22:645; Tamman, G., Hesse, W. 1926. *Z. Anorg. Allg. Chem.* 165:254. Fulcher, G.S. 1925. *J. Am. Ceram. Soc.* 8:339.
10. Skotheim, T.A. 1986. *Handbook of Conducting Polymers*, Vols. 1 & 2, Marcel Dekker, New York.
11. Shirakawa, H. 1995. Synthesis and characterization of highly conducting polyacetylene. *Synth. Met.* 69:3–8.
12. Baker, G.L. 1986. in *Electronic and Photonic Applications of Polymers*, (Eds. Bowden, M.J., Turner, S.R.). ACS Adv. Chem. Ser. 218.
13. Pham, M.C., Piro, B., Bazzaoui, E.A., Hedayatullah, M., Lacroix, J-C., Novàk, P., Haas, O. 1998. Anodic oxidation of 5-amino-1,4-naphthoquinone (ANQ) and synthesis of a conducting polymer (PANQ). *Synth. Met.* 92:197–205.
14. Stow, R.H. 1969. Electrically conductive adhesive tape, US Patent 3,475,213.
15. Stow, R.H. 1973. Tape (Electric conductive tape-using soft rubber resin containing carbon black as adhesive) US Patent 3,778,306.
16. He, D., Ekere, N.N. 2004. Effect of particle size ratio on the conducting percolation threshold of granular conductive-insulating composites. *J. Phys. D: Appl. Phys.* 37:1848–1852.

17. Folwell, J.H. 1954. Pressure sensitive adhesives containing graphite, US Patent 2,670,306.
18. Akzo Nobel Technical bulletin on polymer additives (Bulletin Black, 96/01).
19. Glackin, R.T. 1992. Method of making an electrically conductive adhesive, US Patent 5,082,595.
20. Coleman, J.J. et al. 1957. Electrically conductive adhesive tape, US Patent 2,808,352.
21. Mruk, N.J., Vaughan, R.C., Eddy, Jr., Arthur, R. 1986. Pressure sensitive adhesive compositions for medical electrodes, US Patent 4,588,762.
22. Cormier, M., Daddona, P.E. 2002. Macroflux technology for transdermal delivery of therapeutic proteins and vaccines (in *Modified-Release Drug Delivery Technology*, Eds. Rathbone, M.J., Hadgraft, J., Roberts, M.S.), Marcel Dekker, Inc., New York, pp. 589–598.
23. Scott, E.R., Phipps, J.B., Gyory, J.R., Padmanabhan, R.V. 2000. Electrotransport systems for transdermal delivery: A practical implementation of Iontophoresis (in *Handbook of Pharmaceutical Controlled Release Technology*, Ed. Wise, D.L.), Marcel Dekker, Inc., New York, pp. 617–659.
24. Phipps, J.B., Gyory, J.R. 1992. Transdermal ion migration. *Adv. Drug Del. Rev.* 9:137–176.
25. Phipps, J.B., Scott, E.R., Gyory, J.R., Padmanabhan, R.V. 2002. Iontophoresis in *Encyclopedia of Pharmaceutical Technology*, Eds. Swarbrick, J., Boylan J.C., 2nd ed. Marcel Dekker Inc., New York, pp. 1573–1587.
26. Higo, N., Mori, K., Katagai, K., Nakamura, K. 1999. Iontophoresis electrode and iontophoresis device using the electrode, US Patent 6.006,130; Talpade, D. 2003. Iontophoretic delivery to heart tissue, US Patent 6.587,718 B2.
27. Kleiner, L., Scott, E.R., Electrochemically reactive cathodes for an electrotransport device EP Patent application EP19990903448, 2004.
28. Myers, R.M., Haak, R.P., Plue, R.W. 1995. Iontophoretic drug delivery apparatus, US Patent No. 5466217.
29. Latham, R.J., Linford, R.G., Schlindwein, W.S. 2003. Pharmaceutical and medical applications of polymer electrolytes. *Ionics* 9(1, 2):41–46.
30. Lyons, J.N. 1998. Positive locking biomedical electrode and connector systems. US Patent 5,465,715; Ferrari, R.K. 1998. Multifunction electrode, US Patent 5,824,033.
31. Subramony, J.A., Sharma, A., Phipps, J.B. 2006. Microprocessor controlled drug delivery. *Int. J. Pharm.* 317(1):1–6.
32. Haak, R.P., Gyory, J.R., Meyers, R.M., Landrau F.A., Sanders, H.F., Kleiner, L.W., Hearney, L.M. 2001. Iontophoretic drug delivery apparatus, US Patent 6317629.
33. Behar, F., Roy, P. 2004. Device for delivering medicines by transpalpebral electrophoresis, US Patent application number 20040267188.
34. Woolfson, D.A. 1996. Moisture-activated, electrically conducting bioadhesive interfaces for biomedical sensor applications. *Analyst* 121(6):711–714.
35. Tavakoli, S.M., Nix, E.L., Pacey, A.R. 1995. Joining components of a cardiac catheter tip with electrically conductive adhesives, Annual Technical Conference—Society of Plastics Engineers 53rd (Vol. 3):3362–3366.

36. Shikinami, Y. 1993. Adhesive polymer gel for medical use. *Kagaku to Kogyo* 67(4):141–150.
37. Shikinami, Y., Hata, K., Morita, S., Kawarada, H. 1991. Synthesis and ionic conductivity of adhesives composed of complexes from segmented polyether-urethane and lithium perchlorate. *Nippon Setchaku Gakkaishi* 27(7):266–274.
38. Shikinami, Y., Kaoru, T. 1991. Bioelectrodes consisting of ion-conducting polyurethane adhesives. *Iyo Denshi to Seitai Kogaku* 29(3):178–185.
39. Menon, V.P., Kumar, K, Nelson, C.T., Rizzardi, D.A. 2007. Conductive adhesives and biomedical articles including same, US Patent application 20070032719.
40. Dietz, T.M., Asmus, R.A., uY, R. 1994. Two-phase composites of ionically-conductive pressure-sensitive adhesive, biomedical electrodes using the composites, and methods of preparing the composite and the biomedical electrodes, US Patent 5,338,490.
41. Engel, M.R. 1989. Electrically-conductive, pressure-sensitive adhesive and biomedical electrodes, US Patent 4,848,353.
42. Perrault, J.J. 1998. Electrically conductive adhesive hydrogels, US Patent 5,800,685.
43. Haak, R.P., Theeuwes, F., Gyory, J.R., Lattin, G.A. 1995. Transdermal delivery device, US Patent 5464387.
44. Hartman, R.B. 1985. Flexible tape having bridges of electrically conductive particles extending across its pressure-sensitive adhesive layer, US Patent 4,548,862.

3

Pressure-Sensitive Adhesives for Electro-Optical Applications

Daniel L. Holguin
E.-P. Chang
Avery Research Center

3.1 Introduction

Light management materials began with the use of glass windows to allow light into dwellings and then moved into glass lenses for microscopes and telescopes. Although adhesives predated the development of glass, both electro-optical devices (from calculators to televisions) and pressure-sensitive adhesives (PSAs) were developed and matured in the twentieth century. The advent of electro-optical applications created a diverse need for PSAs, including optically clear PSAs for displays. This chapter largely focuses on compositional and performance aspects of functional optically clear PSAs.

PSAs are a distinct category of adhesives that are aggressively and permanently tacky (see also *Fundamentals of Pressure Sensitivity*, Chapter 10, and Chapter 8 of this book) and adhere to a variety of substrates without more than finger or hand pressure contact (with the exception of certain removable products; see also Chapter 4). PSAs do not require activation by heat or solvents. The primary mode of bonding for a PSA is not chemical or mechanical but, rather, a polar attraction to the substrate. Such a definition may not include some of the adhesives (such as the hydrogels) described in the following pages (see also *Technology of Pressure-Sensitive Adhesives and Products*, Chapter 7). From a functional point of view, the described adhesives may be considered specialized PSAs.

From a functional perspective, the first optically clear PSA presented here can simplify the process for manufacturing a component of a display (i.e., liquid crystal display [LCD] color filter), but the adhesive requirements to transition from a removable PSA to a permanent PSA and finally to a near structural adhesive are complex. The second type of optically clear PSA can improve light distribution by using a low refractive index (RI) PSA to reduce the light distortion when using the adhesive to adhere a light management film to a display (i.e., flat-panel display). The low-RI PSA required the development of unique polymers in a common solvent. Finally, the third optically clear adhesive is a hydrogel with low RI that can serve as a gap-filling material to adhere to and support lenses (i.e., Fresnel lens to lenticular lens), with minimal distortion of the traversing light. These hydrogel formulations are easy to process and cost effective.

3.2   Curable Optically Clear Pressure-Sensitive Adhesives

A functional PSA simplified the process of applying a protective epoxy coating to LCD color filters. Using different chemistries and curing processes, this optically clear adhesive was developed for the required complex transition from a removable PSA to a PSA and finally to a near structural adhesive for this application [1]. The first stage utilized ionic cross-linking of the acrylic polymer during the coating and drying phase. The second stage used ultraviolet (UV) curing for cross-linking the urethane oligomer forming the interpenetrating network, and the third stage used thermal curing to completely cross-link the acrylic polymer. Such progressive transformation to different types of adhesives is monitored and confirmed by the shift in the viscoelastic window proposed by Chang [2] (see also Chapter 8).

Full-color LCDs have become very common in recent years as flat-panel displays. They generally have a structure comprising a color filter with a plurality of colors, for example, red (R), green (G), and blue (B), on a glass plate, a counter electrode substrate facing the color filter, and a liquid crystal layer in a gap between the color filter and the counterelectrode substrate, as illustrated in Figure 3.1 [3]. The color filter requires a protective epoxy layer to protect the colored layer, as illustrated in Figure 3.2.

Currently, the protective epoxy layer is formed using a photo-curable resin coated directly onto the color filter that permits portions to be cured so they can be easily limited through a mask. This usually involves an organic solvent used in development after

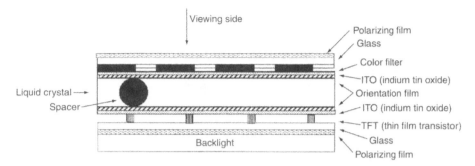

FIGURE 3.1 Schematics of multilayer construction and components in an LCD. (From Chang, E.P. and Holguin, D.L., *Curable Optically Clear Pressure-Sensitive Adhesives*, Taylor & Francis, London, 2005. With permission.)

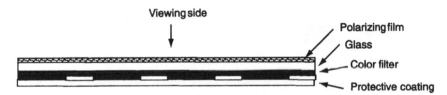

FIGURE 3.2 Protective coating for color filter. (From Chang, E.P. and Holguin, D.L., *Curable Optically Clear Pressure-Sensitive Adhesives*, Taylor & Francis, London, 2005. With permission.)

exposure of the radiation-curable protective coating material to UV light. The handleability and wastewater treatment are troublesome and the method lacks profitability and stability. The preferred alternative is a curable label construction comprising a curable PSA layer (1–2 μm thick) and curable epoxy layer (0.5–1.5 μm thick) coated onto a transparent carrier film.

3.2.1 Requirements of the Color Filter Adhesives

The optically clear curable PSA layer should have adequate adhesion to the epoxy layer throughout the manufacturing process. The coated PSA should be repositionable or easily removable from the transparent substrate on which the color filter is formed. The UV-cured PSA should adhere to the color filter and to the epoxy layer so that the UV-cured epoxy layer may be removed from the carrier film (Figure 3.3). Finally, the thermally cured PSA adhesive should anchor the thermally cured epoxy layer to the color filter and transmit light.

3.2.2 Material Selection

The main component of the curable PSA layer is a PSA, prepared in solvent, that contains acid and epoxy functionality. This adhesive was combined with an acrylated urethane oligomer, a methacrylated silane thermoset cross-linker, a photoinitiator, and a metal

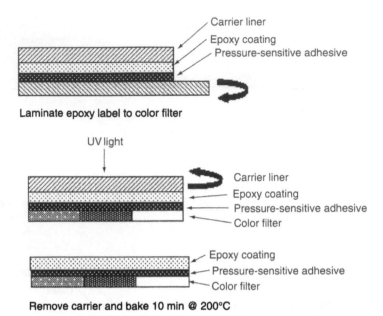

FIGURE 3.3 Stages of preparing a protective epoxy coating label. (From Chang, E.P. and Holguin, D.L. *Curable Optically Clear Pressure-Sensitive Adhesives*, Taylor & Francis, London, 2005. With permission.)

chelate ionic cross-linker, forming the adhesive blend for the curable PSA layer. A typical curable PSA blend has the following composition:

70.0 parts acrylic PSA (dry weight basis)
30.0 parts aliphatic urethane diacrylate
1.4 parts benzophenone (photoinitiator)
1.4 parts methyldiethanolamine (photoinitiator)
0.6 parts aluminum acetoacetonate metal chelate ionic cross-linker
0.6 parts methacrylated silane cross-linker

Stage 1. The adhesive blend was coated and dried, yielding a curable, optically clear, removable (or repositionable) PSA consisting of ionic cross-linked acrylic polymer with epoxy functionality (Figure 3.4) that was plasticized with the acrylated urethane oligomer (along with the methacrylated silane cross-linker and the photoinitiator).

Stage 2. The curable, optically clear, removable PSA was then UV cured (with Fusion lamp bulb D at 5400 mJ/cm² total dosage), yielding a curable, optically clear, permanent adhesive consisting of ionic cross-linked acrylic polymer with epoxy functionality that formed an interpenetrating network with the tough cross-linked acrylated urethane oligomer (copolymerized with the methacrylated silane cross-linker), as illustrated in Figure 3.5 (for fundamentals and practice of UV curing, see also *Technology of Pressure-Sensitive Adhesives and Products*, Chapter 8).

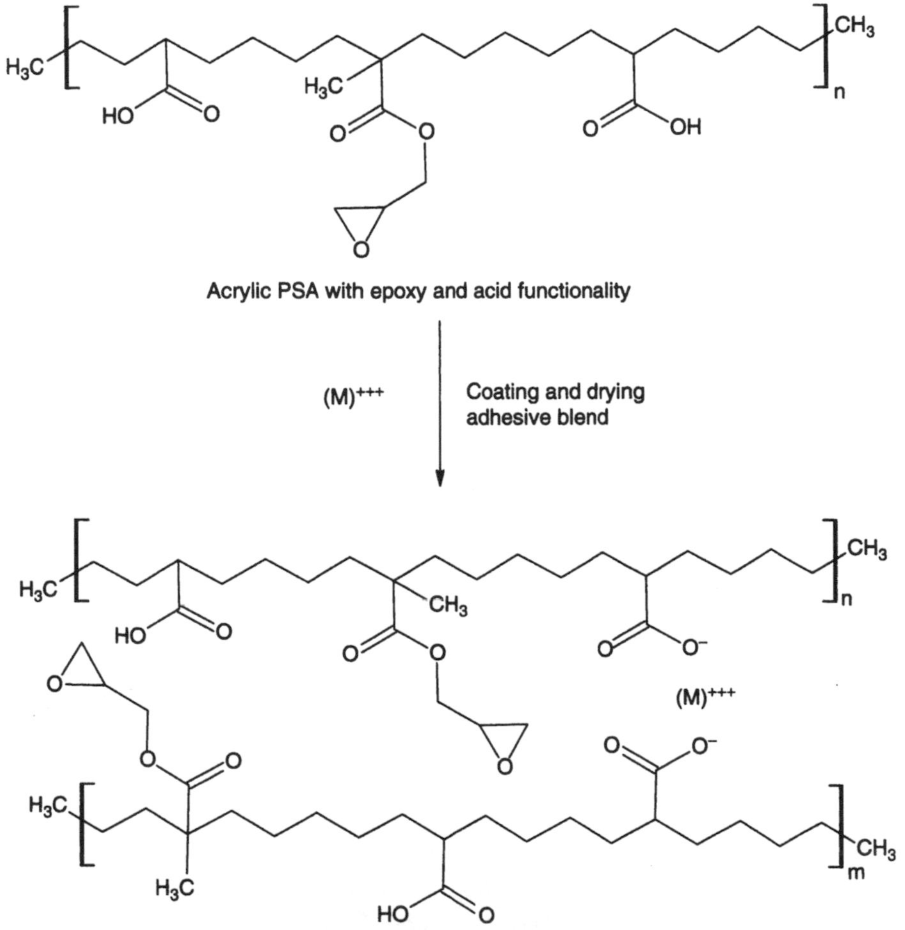

Acrylic PSA with epoxy and acid functionality

(M)$^{+++}$ Coating and drying adhesive blend

FIGURE 3.4   Cross-linking of an acrylic PSA. (From Chang, E.P. and Holguin, D.L. *Curable Optically Clear Pressure-Sensitive Adhesives*, Taylor & Francis, London, 2005. With permission.)

Stage 3. The curable, optically clear, permanent adhesive was then thermally cured, yielding an optically clear near-structural adhesive consisting of an interpenetrating network of ionic, epoxy, and silane cross-linked acrylic polymer with a cross-linked urethane polymer, as illustrated in Figure 3.6.

3.2.3   Viscoelastic Windows of Investigated Adhesives at Different Curing Stages

Figure 3.7 illustrates the viscoelastic window [2] of the acrylic PSA, whereas Figure 3.8 presents the viscoelastic window of its blend with urethane oligomer. The rheologic footprints of the two adhesives at the successive stages of transformation from an uncured

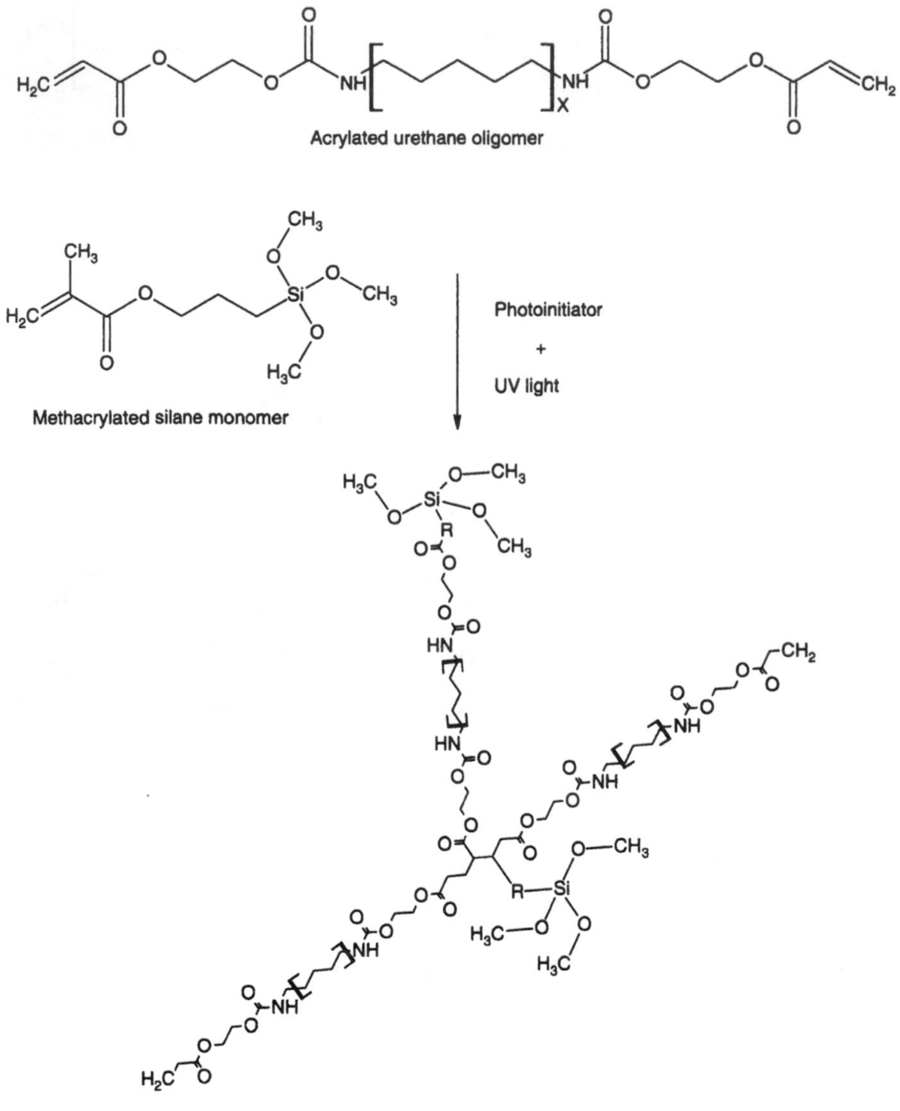

FIGURE 3.5 Cross-linking of acrylated urethane oligomer. (From Chang, E.P. and Holguin, D.L. *Curable Optically Clear Pressure-Sensitive Adhesives,* Taylor & Francis, London, 2005. With permission.)

removable adhesive to a UV-cured permanent adhesive to an ultra-high-adhesion adhesive are well illustrated by the movement of the viscoelastic window. The rheologic footprint of the adhesive blend correlates well with lap-shear testing. Lap-shear testing using an Instron 5542 Tension/Force Testing System at 10 mm/min, with 25 μm of adhesive blend bonding the aluminum test panel to a glass test panel with a 1 × 1 in. overlap demonstrated a value of 11.0×10^4 Pa for samples pre-UV curing, 26.3×10^4 Pa after UV curing, and 34.7×10^4 Pa after thermal cure (see also Chapter 8).

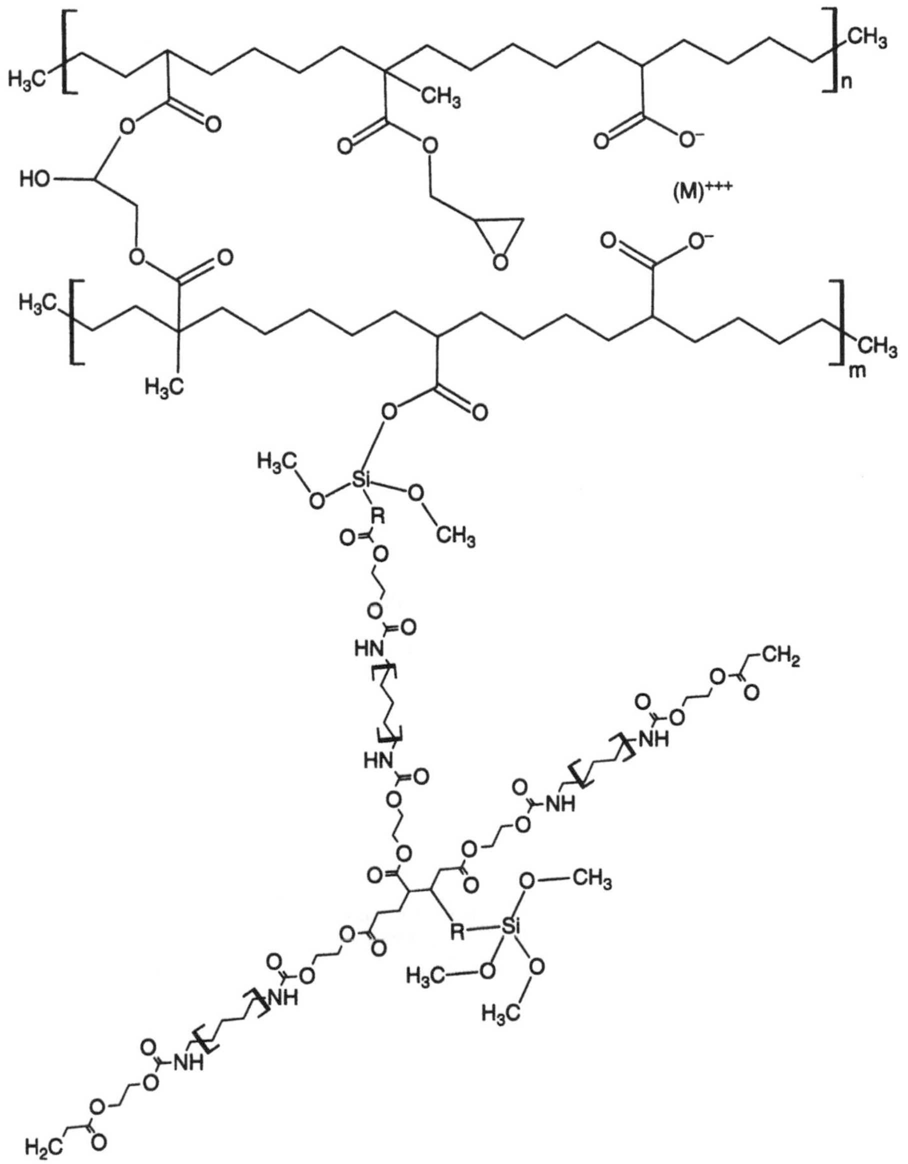

FIGURE 3.6 Thermal epoxy cross-linking of an acrylic PSA and silane cross-linking of urethane oligomers, yielding an interpenetrating network. (From Chang, E.P. and Holguin, D.L. *Curable Optically Clear Pressure-Sensitive Adhesives*, Taylor & Francis, London, 2005. With permission.)

Table 3.1 presents the viscoelastic properties of both the pre-UV-cured PSA and the adhesive blend. Compared with the acrylic PSA, the adhesive blend containing the urethane oligomer generally demonstrates lower G' and G'' values at the bonding and debonding frequencies. This can be attributed to the plasticization effect of the urethane oligomer. Table 3.2 correspondingly illustrates the viscoelastic properties of the PSA

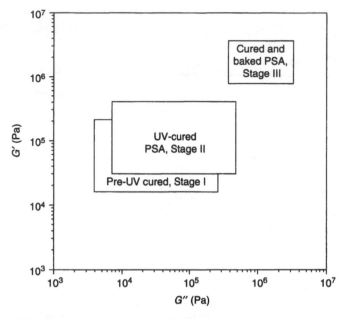

FIGURE 3.7 Viscoelastic windows of an acrylic PSA at different stages of cure. (From Chang, E.P. and Holguin, D.L. *Curable Optically Clear Pressure-Sensitive Adhesives*, Taylor & Francis, London, 2005. With permission.)

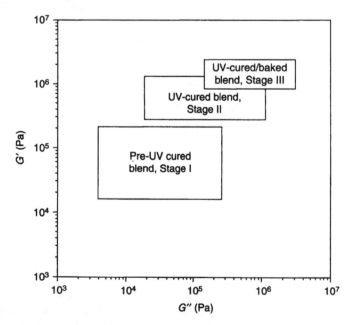

FIGURE 3.8 Viscoelastic windows of an acrylic PSA/urethane blend at different stages of cure. (From Chang, E.P. and Holguin, D.L. *Curable Optically Clear Pressure-Sensitive Adhesives*, Taylor & Francis, London, 2005. With permission.)

TABLE 3.1 Viscoelastic Properties of Pre-UV-Cured Acrylic PSA and Its Blend with Acrylated Urethane Oligomer

Property	Acrylic PSA	Acrylic/Urethane Blend
T_g (°C)	0	−5
Tan δ at T_g	1.41	1.19
23°C G′ at 0.01 rad/s (Pa)	2.8×10^4	1.7×10^4
23°C G″ at 0.01 rad/s (Pa)	3.3×10^3	3.4×10^3
23°C G′ at 100 rad/s (Pa)	3.7×10^5	2.1×10^5
23°C G″ at 100 rad/s (Pa)	3.5×10^5	2.3×10^5

Source: From Chang, E.P. and Holguin, D.L. *Curable Optically Clear Pressure-Sensitive Adhesives*, Taylor & Francis, London, 2005. With permission.

TABLE 3.2 Comparison of Viscoelastic Properties of UV-Cured Acrylic PSA and Its Blend with Acrylated Urethane Oligomer

Property	Acrylic PSA	Acrylic/Urethane Blend
T_g (°C)	1	8
Tan δ at T_g	1.39	0.82
23°C G′ at 0.01 rad/s (Pa)	5.3×10^4	2.6×10^5
23°C G″ at 0.01 rad/s (Pa)	8.5×10^3	1.9×10^4
23°C G′ at 100 rad/s (Pa)	6.3×10^5	1.4×10^6
23°C G″ at 100 rad/s (Pa)	6.8×10^5	1.2×10^6

Source: From Chang, E.P. and Holguin, D.L. *Curable Optically Clear Pressure-Sensitive Adhesives*, Taylor & Francis, London, 2005. With permission.

TABLE 3.3 Comparison of Viscoelastic Properties of Thermally and UV-Cured Acrylic PSA and Its Blend with Acrylated Urethane Oligomer

Property	Acrylic PSA	Acrylic/Urethane Blend
T_g (°C)	2	10.5
Tan δ at T_g	1.37	0.45
23°C G′ at 0.01 rad/s (Pa)	1.9×10^6	9.4×10^5
23°C G″ at 0.01 rad/s (Pa)	6.7×10^5	1.2×10^5
23°C G′ at 100 rad/s (Pa)	7.8×10^6	4.7×10^6
23°C G″ at 100 rad/s (Pa)	2.6×10^6	2.3×10^6

Source: From Chang, E.P. and Holguin, D.L. *Curable Optically Clear Pressure-Sensitive Adhesives*, Taylor & Francis, London, 2005. With permission.

and the adhesive blend after UV curing. The originally plasticized lower G′ and G″ adhesive formulation now demonstrates significantly higher G′ and G″ values than the acrylic PSA. Such an increase can be rationalized from the intra- and intercross-linking of both the adhesive components, most probably forming an interpenetrating network. Table 3.3 presents the subsequent viscoelastic properties of the two cured adhesives after thermal cure. With thermal curing, the G′ and G″ values further increased, approaching the values of a structural adhesive [2].

TABLE 3.4   RI of Various Materials

Material	RI
Air (vacuum)	1.00
Water	1.33
Poly(tetrafluoroethylene)	1.35–1.38
Poly(dimethylsiloxane)	1.43
Poly(oxyethylene)	1.45
Poly(acrylates)	1.46–1.47
Poly(ethylene)	1.51
Poly(butadiene-co-styrene)	1.53
Glass	1.53
Polycarbonate	1.59
Poly(ethylene-co-terphthalate)	1.64

Source: From Chang, E.P. and Holguin, D.L. *Electro-Optical Light Management Material: Low-Refractive-Index Hydrogel,* Taylor & Francis, London, 2007. With permission.

3.3   Low-Refractive-Index Adhesives

A functional PSA with improved optical properties was developed by reducing the light refraction of the adhesive [4]. Light refraction of a material expressed as the RI is a key feature in the application of optical polymers. This PSA is a fluoro-substituted mono-acrylate adhesive. Fluoropolymers have a low RI, but typically the fluoropolymers are not sticky and are opaque due to crystallinity. The design of this novel fluoro-substituted monoacrylate adhesive polymer was based on the rheologic marriage of fluoropolymers and PSAs with the desired physical properties of RI, glass transition temperature (T_g), light transmittance, and adhesion.

An example of an application for which a low-RI adhesive is useful is the adherence of an optical film to a full-color LCD (a flat-panel display) to obtain the desired light distribution (i.e., antiglare, anti-iridescence, low reflectance, or uniform brightness). Typically, PSAs used to adhere optical films to LCDs are acrylic polymers with an RI of 1.46–1.47. A significant reduction in light distortion would require a PSA with an RI less than 1.36. Materials with an RI less than 1.4 are limited to air, water, and fluoropolymers (Table 3.4) [5].

3.3.1   Refractive Index of Fluoropolymers

Groh and Zimmerman [6] plotted the ratio of molar refraction, R_L, to molar volume, V_L, for different atoms present in organic polymers using previously published data [7–9]. Despite the broad range of values for each atom due to different binding structures and varying chemical environments, it is apparent that fluorine and, to a lesser extent, oxygen lower the RI of a compound. Figure 3.9 [10] illustrates the correlation of the RI with the fluorine content (percentage) in the polymer. The higher the fluorine content, the lower the RI. To achieve a polymer with an RI <1.36, the percentage fluorine content must be >57%.

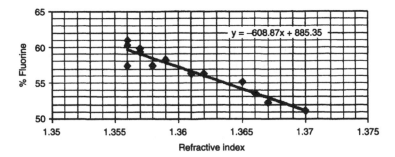

FIGURE 3.9 Correlation of experimental RI with fluorine content. (From Chang, E.P. and Holguin, D.L., *Electro-Optical Light-Management Material: Low-Refractive-Index Adhesives*, Taylor & Francis, London, 2005. With permission.)

The relationship between the RI with molar refraction and the molar volume is demonstrated in the Lorentz–Lorenz equation [11,12]:

$$\frac{(N_D^2 - 1) \times M_G}{(N_D^2 + 2) \times \rho} = R_L \tag{3.1}$$

where N_D is the RI, M_G is the repeating unit molecular weight, ρ is the density, and R_L is the molar refraction.

Equation 3.1 demonstrates that a low RI can be achieved by either lowering the molar refraction or increasing the molar volume (M_G/ρ).

3.3.2 Material Selection

Based on the above concept, as well as the material property requirements of PSAs, the preferred molecular structure has fluorination on the bulky polymer side chain (i.e., to increase the molar volume and decrease the molar refraction by fluorination), rather than the more typical fluorinated polymer main chain (e.g., polytetrafluoroethylene, which is hard and crystalline). Also, a fluoropolymer with fluorinated side chain is required for the polymer to be soluble in a nonhalogenated solvent for processing.

The side chain fluoroacrylate monomers used in our work were 1H,1H-pentadeca-fluoro-octyl acrylate (PDFA) plus 1H,1H-heptafluorobutyl acrylate (HFBA) and a non-fluorinated comonomer acrylic acid (AA) was used for organic solvent solubility, ionic cross-linking, and a specific adhesion promoter (see Schemes 3.1 through 3.3).

$$\tag{3.1}$$

1H,1H-Pentadecafluoro-octyl acrylate (PDFA)

(3.2)

1H,1H-Heptafluorobutyl acrylate (HFBA)

Acrylic acid (AA)

(3.3)

A typical polymerization reaction is as follows [4]: Ethyl acetate solvent (20 g) is added into a 100-mL reactor equipped with a nitrogen purge, an agitator, and a reflux condenser. The solvent is heated to reflux with a jacket at 85°C. A monomer mixture of 44.0 g of PDFA, 24.0 g of HFBA, and 1.0 g of AA with 0.084 g of 2,2'-Azobis(2-Methylbutyronitrile) (AMBN) polymerization initiator was added to the reactor over a period of 2 h, followed by the addition of 0.15 g of AMBN in 5.1 g ethyl acetate over 5 h. The polymer solution at the end of the reaction contains 73.9% solids. The polymer in the solvent appears slightly hazy, with no gel or precipitation present.

3.3.3 Correlation of Rheologic and Surface Properties With Adhesive Performance

Table 3.5 presents a series of experiments using PDFA and HFBA at various ratios, along with AA. The data in Table 3.5 demonstrate that for the target RI (RI <1.36), the upper limit of HFBA was 30% along with 2% AA (Experiment 5). Figure 3.10 also illustrates the effects of PDFA/HFBA/AA composition on T_g as well as RI. As the RI decreases, the T_g increases correspondingly.

Figures 3.11 and 3.12 compare, respectively, the temperature dependence of the dynamic shear storage modulus G' and viscoelastic index and tan δ for three PDFA/HFBA/AA polymers with varying ratios of 93/5/2 (i.e., Experiment 1), 49/49/2 (i.e., Experiment 6), and 23/75/2 (i.e., Experiment 7). The Experiment 5 polymer demonstrates a T_g of 30°C based on the tan δ peak (Figure 3.8), and the room temperature (20°C) G' is 3.3 × 10⁶ Pa, which does not satisfy (i.e., is higher than) Dahlquist's criterion of 3 × 10⁵ Pa. On the other hand, although the T_g and room temperature G' of Experiment 7 appear to be the most favorable for a PSA, the RI is slightly too high (1.366). In view of this, the Experiment 6 polymer was chosen for further improvement of the RI and adhesion performance.

TABLE 3.5 Physical Properties of PDFA/HFBA/AA with Varying Composition

Experiment	1	2	3	4	5	6	7	8
PDFA/HFBA/AA								
Weight ratio	93/5/2	88/10/2	83/15/2	78/20/2	68/30/2	49/49/2	23/75/2	49/49/0.2
Fluorine (%)	60.9	60.4	59.8	59.4	58.3	56.3	53.6	57.4
Polymer RI	1.356	1.356	1.357	1.357	1.359	1.362	1.366	1.358
T_g (°C) by tan δ max	30	25	22	13	13	6.5	1	1
G' at 20°C (Pa)	3.3×10^6	8.0×10^5	2.0×10^5	1.5×10^5	1.3×10^5	2.0×10^5	1.6×10^5	4.0×10^4
Surface energy (dyne/cm)	9.6	7.6	5.5	3.9	4.2	4.1	5.1	
Transmission (%)	94.2	92.7	94.2	93.9	93.1	93.7	94.0	93.5
Initial peel on glass (N/m)	0	70	88	105	123	228	210	228
50°C peel on glass (N/m)	18	53	70	88	298	385	263	280
50°C peel on HDPA (N/m)	0	18	18	18	18	35	35	53
50°C peel on steel (N/m)	18	88	88	298	280	333	193	245
50°C peel on teflon (N/m)	0	18	18	18	18	18	18	70

Source: From Chang, E.P. and Holguin, D.L. *Electro-Optical Light-Management Material: Low-Refractive-Index Adhesives*, Taylor & Francis, London, 2005. With permission.

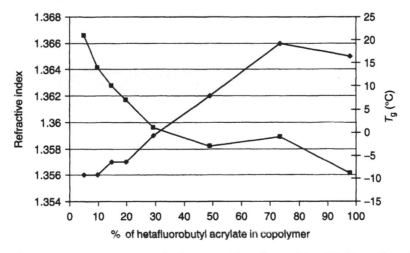

FIGURE 3.10 PDFA/HFBA/AA copolymer composition effect on RI and T_g. (From Chang, E.P. and Holguin, D.L. *Electro-Optical Light-Management Material: Low-Refractive-Index Adhesives*, Taylor & Francis, London, 2005. With permission.)

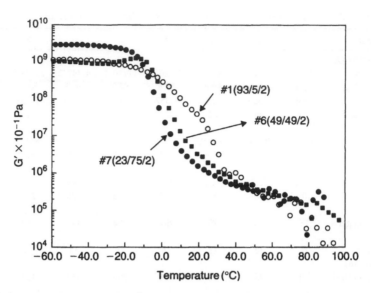

FIGURE 3.11 Temperature dependence of G' for three different PDFA/HFBA/AA composition polymers. (From Chang, E.P. and Holguin, D.L. *Electro-Optical Light-Management Material: Low-Refractive-Index Adhesives*, Taylor & Francis, London, 2005. With permission.)

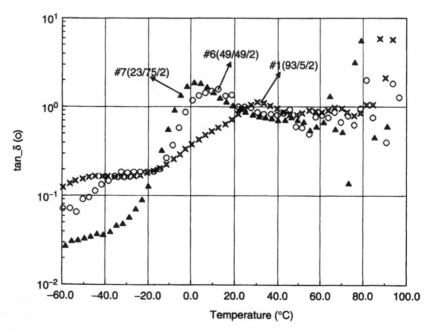

FIGURE 3.12 Temperature dependence of tan δ for three different PDFA/HFBA/AA composition polymers. (From Chang, E.P. and Holguin, D.L. *Electro-Optical Light-Management Material: Low-Refractive-Index Adhesives*, Taylor & Francis, London, 2005. With permission.)

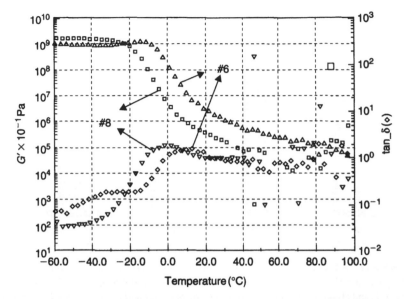

FIGURE 3.13 Temperature dependence of G' and tan δ for 49/49/2 versus 49.9/49.4/0.2 PDFA/ HFBA/AA polymers. (From Chang, E.P. and Holguin, D.L. *Electro-Optical Light-Management Material: Low-Refractive-Index Adhesives*, Taylor & Francis, London, 2005. With permission.)

To further reduce the RI, the concentration of AA was decreased from 2 to 0.2% (because AA is nonfluorinated and has a higher RI than fluorinated acrylics), with a 0.1% AA cross-linker to provide good cohesive strength. AA was required for improved polymer solubility, cross-link sites, and specific adhesion to polar surfaces like glass and stainless steel. The rheologic properties of these two copolymers (i.e., Experiments 6 and 8) are compared in Figure 3.13. In Experiment 8, with lower AA, not only was the RI reduced (1.362–1.358) but also both the T_g (from 6.5 to 1°C based on tan δ peak) and the room temperature G' (from 3.3 × 10^6 to 4 × 10^4 Pa).

From the bulk properties viewpoint, the lower T_g and lower room temperature G' would predict a better PSA. This was reflected in the increased peel resistance against low surface energy substrates like high-density polyethylene (from 0.1 to 0.2 × 10^4 Pa) and Teflon (from 0.1 to 0.3 × 10^4 Pa). However, lower peel adhesion against high-surface-energy substrates like glass (from 1.5 to 1.1 × 10^4 Pa) and stainless steel (from 1.3 to 0.9 × 10^4 Pa) was also observed (Figure 3.14). This reversed trend is consistent with the 10-fold decrease in AA, a high-surface-energy substrate promoter.

3.4 Low-Refractive-Index Hydrogels

The RI is a key feature in the application of optical polymers. A functional gap-filling hydrogel was developed [13] with the desired functional physical properties of RI, light transmittance, and mechanical strength to help support and adhere light-transmitting devices. Currently, for applications in which there is a significant distance between the

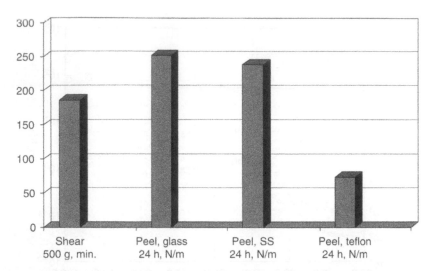

FIGURE 3.14 Adhesion performance of a low-RI fluoroacrylate PSA. Testing used a 50-µm polyethylene terephthalate film as carrier and adhesive coating weight of 25–30 g/m². (From Chang, E.P. and Holguin, D.L. *Electro-Optical Light-Management Material: Low-Refractive-Index Adhesives*, Taylor & Francis, London, 2005. With permission.)

components of the light-transmitting device, lens (or lenses), and other components, the medium used is air and the components are mechanically supported. Air is a poor supporting medium. Replacing the air gap with supportive material is achieved by employing hydrogels with a low RI (<1.35) to obtain the desired light distribution.

An example of an application in which a low-RI material is useful for filling an air gap is in supporting and adhering the lenticular lens to the Fresnel lens of a rear-projection screen for a television back-lit projection screen (Figure 3.15). The Fresnel collimating lens along with the lenticular lens for use with a projection screen should maintain the ability to effectively collimate light to provide more uniform brightness across the screen. It is generally advantageous to increase the difference between the refractive indices of the lenses and medium or support material to help control light distribution. The lens material is glass or poly(methyl methacrylate) with an RI of 1.53 and 1.49, respectively. An RI less than 1.4 for the supportive material would be a significant difference. As mentioned earlier, materials with an RI less than 1.4 are limited to air, water, and fluoropolymers (Table 3.4).

3.4.1 Background

There is one light-receiving device that uses water as a medium—the eye. The eye actually uses a natural hydrogel as the medium and has an RI of 1.336. We also investigated and developed the use of hydrogels as a low-RI medium. Figure 3.16 illustrates the use of polymeric hydrogels: a 2-hydroxyethyl methacrylate/methacrylic acid copolymer [14], a high-molecular-weight poly(ethylene oxide) [13], and a polyacrylamide [13]. Water is

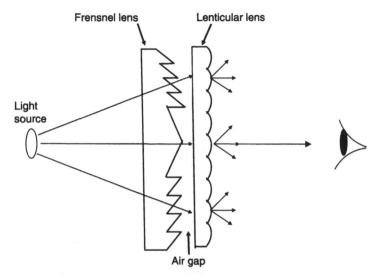

FIGURE 3.15 Typical component setup of a rear-projection screen. (From Chang, E.P. and Holguin, D.L. *Electro-Optical Light Management Material: Low-Refractive-Index Hydrogel*, Taylor & Francis, London, 2007. With permission.)

very effective in making a low-RI medium, but to obtain an RI below 1.35 the hydrogel would have to contain more than 85% water (Figure 3.16). Hyrogels are a cost-effective product, but these polymers are not likely candidates because their mechanical strength is poor and it is difficult to make bubble-free hydrogels.

3.4.2 Experimental

Based on the need for a cost-effective low-RI material, a hydrogel was developed that was easily processed and had good mechanical strength. The formulation included UV-curable poly(ethylene glycol) (PEG) acrylate oligomers, a UV photoinitiator, and a surfactant. Twelve grams of formulation were weighed into an aluminum dish (57-mm diameter × 14-mm height) and formulations were UV irradiated with two passes at 0.254 m/s under a Fusion Systems D-bulb at 850 mJ/cm². The UV curing raised the formulation temperature by 5°C (from 23 to 28°C), but there was negligible water loss to the hydrogel.

3.4.3 Materials Selection

The selected materials are described in Schemes 3.4 through 3.7. Figure 3.17 illustrates the use of UV-curable, water-soluble oligomers for making hydrogels at three different ratios of oligomer 1 to oligomer 2, as described in Table 3.6. The RI of these oligomer blends is hardly affected by the blend ratio, confirming that water is the

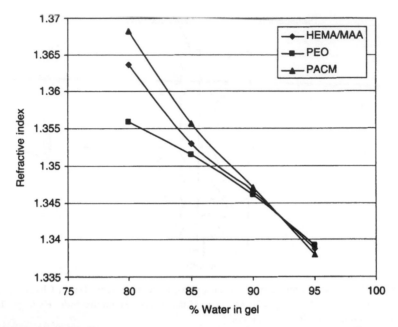

FIGURE 3.16 RI of polymeric hydrogels as a function of water content. (From Chang, E.P. and Holguin, D.L. *Electro-Optical Light Management Material: Low-Refractive-Index Hydrogel*, Taylor & Francis, London, 2007. With permission.)

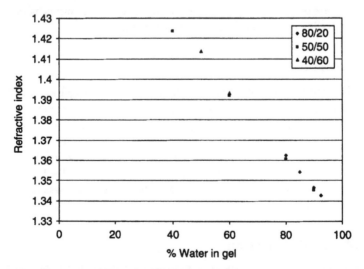

FIGURE 3.17 RI of UV-curable PEG hydrogels as a function of water content. (From Chang, E.P. and Holguin, D.L. *Electro-Optical Light Management Material: Low-Refractive-Index Hydrogel*, Taylor & Francis, London, 2007. With permission.)

TABLE 3.6 Formulations and Dependence of RI with Percentage of Water in Hydrogels

Oligomer Ratio	80/20	80/20	80/20	80/20	40/60	40/60	100/0	50/50	50/50	50/50	50/50
Oligomer 1 (g)	6	8	12	16	16	20	5.0	5	10.0	20.0	30.0
Oligomer 2 (g)	1.5	2	3	4	24	30	0	5	10.0	20.0	30.0
UV initiator (g)	0.2	0.3	0.50	0.6	1.5	1.5	0.15	0.3	0.6	1.2	1.8
Surfactant (g)	0.01	0.02	0.020	0.03	0.04	0.03	0.015	0.06	0.12	0.24	0.36
Water (g)	92.5	90.0	85.0	80.0	60.0	50.0	95.0	90.0	80.0	60.0	40.0
Total (g)	100.2	100.3	100.5	100.6	101.5	101.5	100.2	100.4	100.7	101.4	102.2
Water in gel (%)	92	90	85	79	59	49	95	90	79	59	39
RI	1.3427	1.3466	1.3543	1.3625	1.3933	1.4136	1.3399	1.3452	1.3607	1.3919	1.4237

Source: From Chang, E.P. and Holguin, D.L. *Electro-Optical Light Management Material: Low Refractive-Index Hydrogel*, Taylor & Francis, London, 2007. With permission.

dominating RI contributor. A gel with at least 85% water is required to achieve an RI of about 1.35.

(3.4)

(3.5)

(3.6)

(3.7)

3.4.4 Rheological Properties

Figures 3.18 and 3.19 illustrate, respectively, the frequency dependence of G' (dynamic storage modulus) and tan δ (viscoelastic index) as a function of the water content of gels. For the samples containing 40 and 60% water, there is a significant frequency dependence of increasing G' with increasing frequency, indicative of its viscoelastic nature. This is borne out by and consistent with the rather high tan δ (>1.0). As the water content increases to 80%, the G' value is only slightly affected but remains fairly constant with frequency, suggestive of a rubbery plateau consistent with the sharp drop in tan δ values (<0.2). Upon further increase of the water content to 90%, there is an almost two-decade drop in G' values that is again frequency independent, consistent with the further drop in tan δ values to <0.1. At 95% water content, the G' values further drop one decade, but still maintain reasonable structural integrity and cohesive strength, with G' values close to 10^2 Pa. The G' and tan δ profiles are relatively flat and low. The sharp drop in G' values at over 80% water content most probably can be attributed to the lower cross-link density of the gel, whose cohesive strength is dominated by the water matrix. To prepare a gel with a reasonable mechanical strength, the water percentage in the gel should be around or less than 90% because this would provide some reasonable structural integrity for the attachment of the secondary component to the Fresnel lens, as depicted in Figure 3.15.

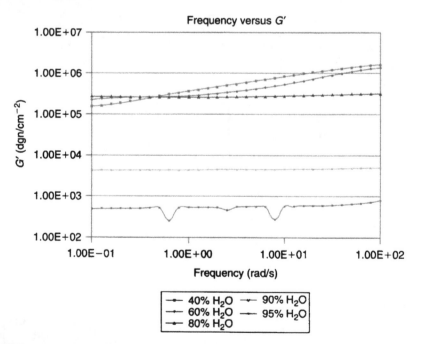

FIGURE 3.18 Frequency dependence of G' as a function of the water content of the gel. (From Chang, E.P. and Holguin, D.L. *Electro-Optical Light Management Material: Low-Refractive-Index Hydrogel*, Taylor & Francis, London, 2007. With permission.)

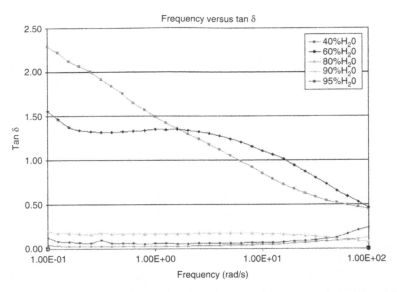

FIGURE 3.19 Frequency dependence of tan δ as a function of the water content of the gel. (From Chang, E.P. and Holguin, D.L. *Electro-Optical Light Management Material: Low-Refractive-Index Hydrogel*, Taylor & Francis, London, 2007. With permission.)

TABLE 3.7 Rheological and Optical Properties of a 50/50 Oligomer-Blend Gel at Different Water Content (% bw)

Oligomer Ratio	Water Content (wt %)	G′ at 1 rad/s (Pa)	tan δ at 1 rad/s	RI	Light Transmission (%)
50/50	40	3.30E + 04	1.6	1.4237	93.8
50/50	60	2.70E + 04	1.4	1.3919	97.6
50/50	80	2.70E + 04	0.19	1.3607	100
50/50	90	4.50E +.04	0.028	1.3452	100

Source: From Chang, E.P. and Holguin, D.L. *Electro-Optical Light Management Material: Low-Refractive-Index Hydrogel*, Taylor & Francis, London, 2007. With permission.

3.4.5 Optical Properties

Table 3.7 presents the percentage light transmission of the 50/50 UV-curable oligomer. There is a gradual increase in the percentage light transmission with an increase in the percentage of water in the gel. At 80% water and higher, the gel is virtually transparent. Table 3.7 also summarizes the rheologic and optical properties of a 50/50 oligomer ratio hydrogel containing different weight percentages of water.

3.5 Conclusions

This chapter describes three types of PSAs used in electro-optical applications. The first, a transfer label construction that can be used to adhere a protective epoxy coating onto a color filter, was developed using an optically clear, curable PSA layer that utilized different chemistries and curing processes. The curable epoxy and the curable PSA label are applied to the color filter and then exposed to UV radiation. The radiation transforms the PSA from a removable (or repositionable) adhesive to a permanent adhesive. The color filter with the UV-cured adhesive is then thermally cured to fully cure the adhesive layer, which further transforms the adhesive into a high-adhesion laminating adhesive. Such progressive processing to the optically clear adhesive can transform the adhesive from removable to permanent and finally to an ultra-high-adhesion laminating adhesive, as monitored and confirmed by the shift in viscoelastic windows noted by Chang [2] (see also Chapter 8).

The second type, a functional PSA with improved optical properties achieved by reducing the light refraction of the adhesive (RI <1.36), along with optical clarity and PSA, was achieved through the interplay of polymer chemistry (employing side-chain fluoroacrylates) and physics (material properties of T_g, rheology, and surface energy). Polymer crystallization (main or side chain) was overcome to reduce the predicted RI as well as improve the contact efficiency of PSAs. These fluoroacrylated polymers can be made in nonhalogenated solvents with a viscosity range that enables them to be coated with industrial coaters.

The third type demonstrates the feasibility of making a low-RI gap-filling material. The UV-curable hydrogel containing more than 90% water (plus water-soluble oligomers, photoinitiator, and surfactant) cures instantly with 425 mJ/cm^2 UV radiation dosage at a thickness of up to 10 mm, yielding a firm hydrogel with a modulus of 2.3×10^2 Pa. Along with low RI (<1.35), the hydrogel has excellent light transmission and reasonable cohesive strength for supporting and attaching a light management lens.

References

1. Holguin, D. and Chang, E.P., U.S. Patent **6,372,074 B1**, *Method of Forming a Protective Coating for Color Filters*, April 16, 2002.
2. Chang E.P., *J. Adhesion*, 34,189–200, 1991.
3. Chang, E.P. and Holguin, D.L., *Curable Optically Clear Pressure-Sensitive Adhesives*, Taylor & Francis, London, 2005.
4. Holguin, D. and Chang, E.P., U.S. Patent 6,703,463 B2, *Optical Adhesive Coating Having Low Refractive Index*, March 9, 2004.
5. Chang, E.P. and Holguin, D.L. *Electro-Optical Light Management Material: Low-Refractive-Index Hydrogel*, Taylor & Francis, London, 2007.
6. Groh, W. and Zimmermann, A., *Macromolecules*, 24, 6660–6663, 1991.
7. *American Institute of Physics Handbook*, 3rd ed., McGraw Hill, New York, pp. 5–124, 1972.
8. Van Krevelen, D.W. and Hoftyzer, P.J., *J. Appl. Polym. Sci.*, 13, 871, 1969.
9. Van Krevelen, D.W., *Properties of Polymers*, 3rd ed., Elsevier Scientific Publishing Co., Amsterdam, pp. 293–294, 1997.
10. Chang, E.P. and Holguin, D.L., *Electro-Optical Light-Management Material: Low-Refractive-Index Adhesives*, Taylor & Francis, London, 2005.

11. Lorentz, H.A., *Wied. Ann. Phys.,* 9, 641–665, 1880.
12. Lorenz, L.V., *Wied. Ann. Phys.,* 11, 70–103, 1880.
13. Barker, P.H., Chang, E.P., and Holguin, D., U.S. Patent 6,668,431 B2, *Optical Coating Having Low Refractive Index,* February 3, 2004.
14. Holguin, D., U.S. Patent 6,706,836 B1, *Hydrophilic Polymers, Pressure Sensitive Adhesives and Coatings,* March 16, 2004.

4

End-Use Domains and Application Technology of Pressure-Sensitive Adhesives and Products

István Benedek
Pressure-Sensitive Consulting

End-use properties, together with adhesive properties, are the most important performance characteristics of pressure-sensitive products (PSPs) [1–3]. The end-use properties of PSPs must fulfill the different requirements of a number of application fields. Some are general requirements, related to the application technology of a product class (e.g., labels, tapes, protective webs), to the substrate, or to the processing of the labeled, taped, and protected products. Some products are manufactured with different technologies that are proposed for the same application (e.g., certain adhesive-coated and adhesiveless protective films). On the other hand, some classic products have the same build-up, but different application, or vice versa (see also Chapter 7). The adhesive properties of pressure-sensitive adhesives (PSAs) can be considered end-use performance characteristics as well. The influence of the adhesive properties on the end-use properties was discussed by Benedek in Ref. [4]. Convertability of the adhesive is a function of adhesive properties [5]. Design and formulation of PSAs for various applications in the main adhesive classes (labels, tapes, and protective films) was described by Benedek in Ref. [6]. End-use-related formulation is discussed in *Technology of Pressure-Sensitive Adhesives and Products*, Chapter 8, in addition to formulation for environmental resistance (e.g., for water resistance/ solubility, temperature resistance, and biodegradability), special end-use characteristics, and application domain-related formulation. The fundamentals of pressure sensitivity that allow the science-based synthesis, design, and formulation of PSAs and the construction of sophisticated PSPs are described in detail in *Fundamentals of Pressure Sensitivity*.

Chapter 1 discussed the principles of build-up of the main PSPs, which allow their classification according to their construction; as discussed in *Technology of Pressure-Sensitive Adhesives and Products*, Chapter 10, the construction of PSPs determines their manufacturing technology and application domain (label, tape, protection film, etc.). Product construction-related formulation (i.e., carrier-related, release liner-related, and PSP build-up-related formulation) is discussed in *Technology of Pressure-Sensitive Adhesives and Products*, Chapter 8.

PSPs can be divided into classes according to their end-use. Some classes contain well-known products with a broad range of representatives (e.g., labels, tapes); others are special or one-of-a-kind products. Because of the continuous expansion of their application field, the number of special products is increasing.

PSAs have been used since the late nineteenth century, mainly for medical tapes and dressings. Industrial tapes were introduced to the market in the 1920s and 1930s and self-adhesive labels were introduced in 1935 (see also *Technology of Pressure-Sensitive Adhesives and Products*, Chapters 3 and 11).

A comparison of the requirements for various product classes (see Chapter 1) demonstrates that labels and tapes have similarities with respect to the nature of the adhesive. Protective films and tapes exhibit common features concerning the nature of the carrier. Although the end use of tapes and protection films is quite different, their conversion properties demonstrate similarities due to their monoweb character. On the other hand, conversion properties and end-use properties are correlated (see Table 4.1). End-use properties determine the application technology of PSPs. Table 4.2 summarizes the application conditions for major PSPs. This chapter serves as a short guide that allows the correlation of various special products discussed in Chapters 2, 3, 5, 6, and 7 with the common application domains of PSPs and the test methods described in Chapter 7.

TABLE 4.1 The Main Conversion and End-Use Properties of Principal PSPs and Their Mutual Influence

PSP	Converting Performance		End-Use Performance
		Influences	
Label	Slitting ability		
	Cutting ability		Labeling behavior
	Die-cutting ability		
			Delabeling performance
	Printability		Aesthetics
			Postprintability/writability
			Adhesion Labeling behavior
Tape	Slitting ability		Delabeling performance
	Cutting ability		Application
			Tearing ability
	Printability		
			Special electrical characteristics
			Special thermal characteristics
			Special dosage characteristics
Protective film	Slitting ability		Laminating ability
	Printability		
			Mechanical and thermal processability
			Adhesion Delaminating ability

4.1 Grades, End-Use Domains, and Application of Labels

Stanton Avery developed Kum-Kleen labels in the 1930s; these were the first pressure-sensitive labels. Other types of labels have been used for almost 100 years. The growth in the use of PSA labels is the result of substitution (see also Chapter 7). Label manufacture was discussed by Benedek in detail in Refs. [1,7,8] (see also *Technology of Pressure-Sensitive Adhesives and Products*, Chapter 10). Build-up and classification of labels was described in Ref. [9].

Labels are discontinuous items that serve as a carrier of information and can be applied on different substrates. Various end uses are known for labels, but a label is always an information medium. Some of the end uses are general, classic application fields, in which PSA labels replaced wet-adhesive labels (see also Section 4.1.5.1). Other products are designed to place information or aesthetic, decorative elements on a product surface. Some products are used in new domains in which only a PSP can meet practical requirements. Their geometry, material, build-up, processing, and application technology can vary widely. Carrier materials weighing up to 350 g/m² are

TABLE 4.2 Application Conditions for Major PSPs

	PSP			
	Label			
	Reel	Sheet	Tape	Protection Film
Lamination Conditions and Delamination Conditions				
Impact	•	•	•	•
Nonimpact	•	—	—	—
Pressure				
High	—	—	—	•
Low	•	•	•	—
Application speed				
High	•	—	—	—
Medium	—	—	—	•
Low	—	•	•	—
Application temperature				
Low	•	—	—	—
RT	•	•	•	•
High	•	—	•	•
Stress				
Low	•	•	•	—
Medium	—	—	•	•
High	—	—	•	•
Application method				
Manual	•	•	•	•
Automatic	•	—	•	•
Substrate surface				
Dry	•	•	•	•
Wet	•	•	•	—
Readherability	•			
Repositionability	•	•	—	—
Converting				
During application	•	•	—	—
Postapplication	•	•	—	—
Postapplication processing	—	—	—	•
Delamination method				
Manual	•	•	•	•
Automatic	—	—	—	•
Chemical	•	•	—	—
Thermal	•	•	—	—
Dry	•	•	•	•
Wet	•	•	•	—
Delamination speed				
Low	•	•	•	•
High	—	—	—	•
Variable	—	—	—	•
Delamination stress				
High	—	—	•	•
Low	•	•	—	—
Delaminated surface				
Global	•	•	•	•
Partial	•	•	—	—

used as labels. Fan-fold material can also be processed as labels. Single labels, strip labels, and label strips can be printed. Labels between 30.2 and 164 mm wide cover almost all areas of application. The diversity of labels with respect to construction and end use is increasing. Label grades according to their construction are described in Chapter 1—this chapter discusses their variances according to their end use and application technology.

Labels can be considered a special case of soft laminates (see also Chapter 1) that allow reciprocal mobility (shrinkage and elongation) of the solid-state components. The nature, geometry, and build-up of the carrier material; the nature and geometry of the adhesive; the nature and geometry of the release liner; and the build-up of the finished product may vary greatly. Labels can be round, square, rectangular, elliptical, or irregular in shape. Their purpose may be to impart essential information (i.e., company address, product description, or weight statement). In addition, depending on the product, they may give instructions and warning statements. Base labels are almost without exception printed in one color and coated with permanent or removable adhesive (see *Technology of Pressure-Sensitive Adhesives and Products*, Chapter 11). Decorative labels are printed in many colors and on various materials. The larger ones are coated with a so-called semipermanent adhesive with low initial tack.

Label markets and applications include food, toiletries, cosmetics, pharmaceuticals, and computer and industrial goods, among others. Routing labels (for mail and goods transport), inventory control labels, document labels (for libraries and documentation centers), individual part and product labels (e.g., for the automotive, aerospace, and electrical and electronic industries), supermarket shelf labels, health sector labels (blood bags, drugs, medical equipment, and surgical utensils) comprise the most important sector of nonimpact printing. Labels for variable data and short-run jobs like price labels, short-run product labels, labels and tags for bulk packaging, tickets, baggage tags, and other transport documents have been introduced.

New labeling methods like in-mold labeling (IML) and sleeve labeling (see also Sections 4.1.4.3 and 4.1.5.5) have been developed. Extended text labels were introduced (see also Section 4.1.2.5). Magnetically encoded mailing labels are used by the German postal service for the automatic identification and tracking of packages. They contain a bar code and magnetic stripe to allow optical and magnetic processing.

Various principles are used to categorize labels according to their build-up (see also Chapter 1), dimensions, labeling methods, end uses, and so on. Labels can be classified according to their main performance characteristics as related to application conditions (e.g., temperature-resistant labels, water-resistant/water-soluble labels) or special application field (e.g., price-weight labeling, bottle labeling). In their application field they may have different end uses (e.g., automotive labels or temperature labels in the category of technological labels and tamper-evident labels or nondestructible labels in the category of antitheft labels, etc.). Some products (e.g., temperature-resistant, temperature-display, freezer labels) must support special climatic or temperature conditions. It is also possible to classify labels according to carrier material and application domain. For instance, according to Waeyenbergh [10], self-adhesive labels with an oriented polypropylene (OPP) carrier are used for metal cans (e.g., preserves, paints, and drinks), glass bottles (e.g., champagne, wine, and cosmetics), cardboard containers (e.g., foodstuffs), flexible packaging (e.g., food and nonfood items), technical domains (e.g., labels, stickers, and rating plates), and special domains (e.g., pocket labels).

Taking into account the increasing number of requirements concerning environmental considerations (see also *Technology of Pressure-Sensitive Adhesives and Products*, Chapter 8), labels could also be classified as repulpable and nonrepulpable. Recyclability is desired for many applications (e.g., address and franking labels, magazine supplements).

The existing range of labels—printed circuit board labels, library labels, warehouse labels, floor labels, retroreflective labels, magnetic labels, asset labels, tamper-evident labels, decorative labels, multifunctional labels, bar code labels, imprinted labels, scented labels, rub-off labels, labels with a no-label look, aluminum labels, harsh environment labels, ceramic labels, titanium labels, UV-protective labels, textile labels, laboratory labels—will broaden as the result of new label applications and new engineering concepts. Taking into account the growing importance of hydrophilic, biocompatible hydrogel-based pressure-sensitive systems [11–15], this product class is discussed in detail in *Technology of Pressure-Sensitive Adhesives and Products*, Chapter 7, by Feldstein et al.

Because of the broad range of bonding forces required in label application, the differences in types of bonding, de-bonding speed, and bond–break nature, and various possible types of coating, a number of different adhesives are used for labels. The main PSAs applied in this field are acrylics and rubber–resin formulations (see also *Technology of Pressure-Sensitive Adhesives and Products*, Chapter 8). Although ethylene–vinylacetate copolymer (EVAc)-based adhesives are produced as hot-melt, solvent-based, or water-based formulations, because of their unbalanced adhesive performance they have been used more in applications other than web coating. Water-based EVAc have been tested for labels as a less expensive alternative to acrylics. Formulation of vinyl acetate copolymers (VAc) was discussed by Benedek in Ref. [16].However, the adhesion–cohesion characteristics of such copolymers are not balanced, and they cannot be balanced by tackification [17] (see also *Technology of Pressure-Sensitive Adhesives and Products*, Chapter 8). Two decades ago, Benedek [18] compared tackified EVAc PSAs with acrylates and concluded that, with the exception of shear strength, the adhesive properties of EVAc and styrene-butadiene-rubber (SBR)-based PSA dispersions are inferior to those of acrylic dispersions. The adhesives for labels are described in detail in Refs. [3,6,16].

Generally, the adhesive for labels must have a good adhesion on critical substrate surfaces; chemical, temperature, and UV resistance; plasticizer resistance; dimensional stability; noncorrosivity; adequate instantaneous/final adhesion; and uniform adhesion toward the release liner and it must allow easy labeling. For special labels, it must possess removability, repositionability, adhesion on wet surfaces, transparency/opacity, no emission of gaseous components, and Food and Drug Administration (FDA) approval (see also *Technology of Pressure-Sensitive Adhesives and Products*, Chapter 8). By evaluating these properties according to their priority, one can note that, in general, well-defined adhesive performance characteristics and the stability of these characteristics in time and environment are needed. As discussed in *Technology of Pressure-Sensitive Adhesives and Products*, Chapter 8, solvent-based, water-based, hot-melt, and radiation-cured adhesives can be used for labels. Some years ago, solvent-based adhesives were preferred in label manufacture. Now water-based formulations are the most common. The use of hot-melt PSAs (HMPSAs) for labels is described by Hu and Paul in *Technology of Pressure-Sensitive Adhesives and Products*, Chapter 3.

Because of the different substrates used and the various requirements concerning the aesthetic, mechanical, and chemical characteristics of the label material, various materials have been developed as face stock (carrier) for labels. Material combinations are also used. For instance, cast vinyl marking films are used for producing labels, emblems, stripes, and other decorative markings for trucks, automobiles, and other equipment. They are durable, conformable, and dimensionally stable. Special polyolefin-based films were developed as well (see also Chapter 1).

Principally, labels can be grouped as common labels and special labels. Within each class there is a wide variety of products. The adequacy of a label is judged for a particular use by the customer. It is not the aim of this book to list or describe all types of labels or their end use. However, certain products will be discussed to clarify some technical aspects related to their manufacture.

4.1.1 Common Labels

Common labels are products based on common, inexpensive carrier materials and PSAs used as competitors for common, paper-based, wet-adhesive labels. In such applications the adhesive properties of the label constitute its most important end-use performance characteristics. Price-weight labels, closure labels, and bottle labels are the main representatives of this product class.

4.1.1.1 Price-Weight Labels

Price-weight labeling is one of the most dynamic application fields. Many applications for labels, such as bottles, flexible packaging, and textiles, constitute special cases of price-weight labeling. The technical functions of labels have changed with the development of labeled products. As an example, price labels, in addition to communicating such information as product name, weight, composition, expiration date, and manufacturer's name, must be able to capture the potential buyer's attention. Thus, price-marking and price-weight labels include components of computer labels as well (see Chapter 3). The computer-label business is one of the fastest growing segments of the label market. Electronic data processing (simple and high-quality) labels and variable data bar-coded labels have been introduced. The application requirements for products in this range vary. Simple and complex labels were developed for different classes of products. Even the simplest products must meet various criteria. In principle, the "same" price-weight labels should be usable on unpacked items (i.e., banana label), items packed in film, or items with a plastic plate label substrate (hang tags). Some price labels require perforation for transport.

Papers without a top coating are used for price-weight labels of perishables and quick sale items, tags, and tickets.

Removable price labels were introduced almost exclusively with a paper carrier coated with rubber–resin adhesive based on natural and synthetic rubber and polyisobutene, together with soft resins and plasticizers as a tackifier. For these solvent-based adhesives, gasoline, toluene, acetone, or ethyl acetate was used as solvent. The rubber has been calendered and masticated, so legging and migration (see also Chapter 7), due to the small molecular weight, are characteristics of the formulations. The adhesion build-up on soft polyvinyl chloride (PVC) and varnished substrates may become too high after long storage times. No clear adhesions failure was observed through debonding. Later, water-based acrylic formulations with cross-linkable comonomers (e.g., acrylo- or methacrylonitrile, acrylic or methacrylic acid, *N*-methylol acrylamide) and plasticizers (10–30% w/w) were proposed. Common commercial products (e.g., alkylesters of phthalic, adipic, sebacic, acelainic, or citric acid and propylene glycol methyl vinyl ether) have been suggested as plasticizers.

4.1.1.2 Other Common Labels

Other common labels include bag closure labels and bottle labels. Both were developed as special products. In the range of closure labels, bag closure labels are the most used products.

4.1.1.2.1 *Bag Closure Labels*

Bag closure products are applied to plastic bags that contain stacks and buckets that are stored outdoors. Labels are applied to the packages at the end of the production process. They must be able to withstand outdoor temperatures and moisture. Acrylic adhesives were proposed for such labels. Bag re-closure labels (or tapes) are used as well, mainly for tobacco, coffee, textiles, and dry and wet tissues.

4.1.1.2.2 *Bottle Labels*

Organization and logistic problems for food and beverages are solved using bottle labeling. Direct printing on bottles is limited by the quality of the image; in this case labeling is recommended. Although glass possesses a polar surface, labeling of glass bottles with pressure-sensitive labels has been developed only in recent decades. Economic and technical problems made this development difficult. As mentioned previously, a chemically treated glass surface, condensed water on the bottle, water resistance of labels, and water-soluble adhesives call for specially tailored adhesive properties. Bottle labels need high scratch resistance, chemical resistance, mechanical resistance during filling, and resistance against pasteurization and washing.

Pressure-sensitive labels are also used for wine and champagne bottles. This is a special application in which different water-based adhesives are used on a large range of (mainly) paper face stock materials. Such labels have been developed since 1985 [19]. Their design must meet certain special requirements. These products include permanent labels with cold-water resistance (ice water) and permanent labels that are removable with hot water. Items in all of these categories must possess good adhesion on moist, condensation-covered surfaces [20]. The number of paper qualities used for wine-bottle labeling is high; for instance, a single supplier of labels uses 25 different paper qualities coated with various water-based PSAs. Such labels use glossy, matte, antique, and aluminum laminated papers with permanent and washable adhesives. According to Ref. [21], such labels need high scratch resistance, chemical resistance, mechanical resistance during filling, and resistance against pasteurization and washing.

In 1977, polyethylene terephthalate (PET) was introduced for carbonated soft drinks, and glass bottle labeling was replaced, at least partially, with plastic bottle labeling.

Polypropylene (biaxially oriented), as a white, matte, pretreated material with a thickness of 30–80 μm, is used as a carrier for labels for plastic bottles and containers. A white opaque polypropylene film overlaminated with PET is used for labeling soft drinks. The film has a low coefficient of friction to metal, which allows trouble-free high-speed application.

IML is a special case of plastic bottle labeling in which first the label is manufactured and then the plastic bottle (see also Section 4.1.4.3).

A combination label-folded information booklet is used for bottle labeling of pharmaceutical products; thus, no supplementary instructions are needed (see also Section 4.1.5).

Water-soluble compositions (see also *Technology of Pressure-Sensitive Adhesives and Products*, Chapter 8, and Chapter 7 of this book) are also required for labeling special bottles. The adhesive must display good adhesion to wet as well as dry (polar or nonpolar) surfaces and removability with hot or cold water or commercial detergents and alkalies.

4.1.2 Special Labels

Special labels can be categorized into one of the previously discussed general label classes (e.g., price weight, bottle), but they must satisfy certain special end-use criteria (e.g., water resistance or water solubility, mechanical resistance, or loss of mechanical strength). Different application fields require labels with water resistance or water solubility [3,6] (see also Sections, 4.1.2.3, 4.1.2.5, and 4.1.2.10). Solubility or dispersibility in alkaline or neutral aqueous solutions at hot or room temperature and resistance to humid atmosphere or immersion in water are desired (see also Chapter 7). Applicability on wet, condensation-covered, or frozen surfaces is related to water solubility of PSAs (see also Section 4.1.2.2). Although they are more expensive than classic wet glues in such applications (see also Section 4.1.4.1), the industry appreciates the fact that PSA labels can be applied at high speed from roll stock and they do not need cut-label inventories and glue applicators. Formulation, properties, and manufacture of water-resistant/water-soluble PSAs were discussed in Refs. [6,22,23] and in *Technology of Pressure-Sensitive Adhesives and Products*, Chapter 8. Test methods for water resistance/solubility are discussed in Chapter 7.

Special requirements for various PSPs are discussed in *Technology of Pressure-Sensitive Adhesives and Products*, Chapter 8, and in Chapter 7 of this book. Special labels can also be classified according to the type of adhesive, such as permanent, repositionable, semipermanent, or removable (see also Chapter 1. The formulation, properties, and manufacture of removable PSPs were discussed in Refs. [6,22,24,25] (see also Chapter 8). Special labels can also be classified according to their processibility (e.g., postprocessibility), application field, function, and so on. The main class of postprocessible labels includes computer labels and copyable labels.

4.1.2.1   Computer Labels

The computer-label business is one of the fastest growing segments of the label market. Variable data bar-coded labels (routing labels for mailing, inventory control labels, document labels, individual part and product labels, supermarket shelf labels, health sector labels) are the most important sector of computerized nonimpact printing. Generally, writability and printability are common features of labels. Postprintability using a computer, that is, digital printing, is a performance characteristic of a separate label class called computer labels. Other special labels, like table and copy labels, may belong to this product group as well. Printability and the ability to lay flat are required for table labels. Such labels are computer imprintable. Computer-imprintable labels may be transferable or nontransferable and based on paper, Tyvek, PVC, cardboard, aluminium, or other carrier materials. Different digital postprinting methods exist (see also *Technology of Pressure-Sensitive Adhesives and Products*, Chapter 10); therefore, the construction of computer labels can vary as well. Laser-printable labels are a special class of computer-printable labels. Computer labels can be classified as label sheets, endless labels, and folded labels (with pinholes for transport). They are manufactured as printed or blank (nonprinted) labels. Their printing is carried out using dot matrix, ink jet, laser or thermotransfer printers, or a copying machine. In 1995 in Europe, about 40% of labels were computer labels; now they comprise more than 50% of the market. Various end uses are addressing, marking, organizing, logistics, and other fields.

Requirements for computer labels include machining and printing quality, application-related properties, and environmental performance. Computer labels are designed for a service temperature of −20 to +90°C. Electronic data systems primarily use a paper engineered for strength, with high absorbency to capture ink. Latex-impregnated paper is used for data system applications requiring moisture resistance and improved flexibility. Such papers provide smudge-resistant ink absorption. The most important development concerns printing quality; modern labels must accept 600 dots per inch. Drop-on-demand printing devices for product coding are able to print logos, bar codes, autodating, numbering, and graphic programs and have a printing speed of up to 5 m/s. Suppliers for labeling technology offer software for label printing, label printers, and printing and labeling systems and materials.

Release properties for such products have also changed. Some years ago, slow-running machines allowed slight release forces of 0.1 N/25 mm. At high speed, a better mutual anchorage of the laminate components is required, so the release force must be increased to more than 0.2 N/25 mm.

Copy labels must allow light to penetrate the face stock to achieve a copy of the image on the liner.

4.1.2.2   Freezer Labels

Some products are labeled at room temperature and subjected to lower temperatures later in their life cycle. Others are labeled at low temperatures and work at low temperatures. For instance, refrigerated meat requires a PSA that can be applied at low temperatures. Such products are wrapped in plastic film (polyethylene [PE], polypropylene [PP], PVC, polyamide [PA], ionomer/PA, or coated cellophane) or top-coated papers.

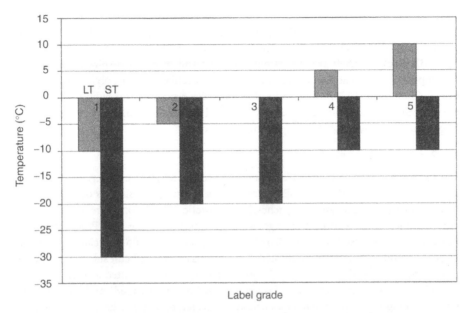

FIGURE 4.1 Labeling temperature (LT) and service temperature (ST) of various pressure-sensitive labels.

The service temperature of the film is −50 to +150°F. Blood bag labels must be water resistant and able to withstand very low temperatures, so they are overlaminated. Generally, there is an interdependence between the labeling temperature (LT) and the service temperature (ST) of labels (see Figure 4.1).

Low-temperature labeling requires labels with a special PSA composition that ensures adhesive flow for bonding at low temperatures. The surface temperature of the adherend influences the adhesive properties. Most general-purpose adhesives are formulated for tack at room temperature. If the adherend temperature is lower, a higher degree of adhesive cold flow is required to provide adequate wet-out. At low temperatures, tack and peel resistance may be too low. This behavior is illustrated by the data presented in Table 5.1 of *Fundamentals of Pressure Sensitivity*, Chapter 5, where special block copolymer formulations are examined by Chang at room temperature and 3°C. The T_g of the styrene block copolymer (SBC) is lower than that of the acrylic polymers, but through tackification T_g shifts toward higher temperatures. As a result, HMPSAs tend to show more aggressive room temperature adhesion, but have relatively poor adhesion at extremely low temperatures compared with solution and emulsion acrylic PSAs (see also *Technology of Pressure-Sensitive Adhesives and Products*, Chapter 3). In practical tests at 0°C, the peel of an untackified acrylic is almost 25% less than that at 23°C. For tackified formulations, peel reduction at 0°C may attain 300%. Owing to the increase in the modulus of the adhesive at low temperatures, a loss of wetting and jerky debonding may be observed for certain adhesive formulations. Freezer labels should display the same viscoelastic properties between −40 and +20°C. For common labels at 20°C a storage modulus value of $10^{4.5}$ Pa is suggested and for freezer labels a value of 10^4 is required.

Thus, for the soft formulation, the dwell time, that is, the time necessary to build a full-surface contact, is shorter. Formulation of such adhesives requires special knowledge; 180° peel resistance at 0°C should be tested by formulation screening (see also Chapter 7). Tackified adhesives for labels are softer (more viscous) and therefore display better wetting of nonpolar surfaces at low temperatures. Low-molecular-weight polyisoprenes with a low T_g (-65 to -72°C) have been used as viscosity regulators for rubber–resin-based hot-melt formulations for freezer labels. They replace common plasticizers and improve migration resistance and low-temperature adhesion.

4.1.2.3   Medical Labels

The main application fields for PSPs include medical and hygienic uses (e.g., plasters, tapes, and dosage systems). Medical PSPs cover various applications, such as operation tapes, dosage tapes/labels, transdermal patches, iontophoretic systems, skin care products, tooth care products, and wound dressings. Medical tapes were the first PSPs introduced to the market (see also Section 4.2.2.6). These products must fulfill special requirements imposed by their application under sterile conditions and on difficult substrates. End-use requirements for medical PSPs were discussed in Ref. [3] and are listed in Table 4.3.

A short presentation of the use of water-soluble PSAs in the medical domain is given by Czech in Ref. [19], where OP tapes (the aim of development, market situation, quality assurance of OP tapes, and their biodegradabilty) are discussed. Bioelectrodes are presented as well. Types of bioelectrodes, development in the area of bioelectrodes, and the composition of PSAs for bioelectrodes are described. Taking into account the similar main requirements for various medical PSPs, the most important representatives will be described in Section 4.2.2.6.

4.1.2.4   No-Label-Look Labels

Special film labels provide a no-label look. Such labels are needed for cosmetics and office use, that is, in applications that require that the label itself virtually disappears when applied to the background. Such products require special carriers, special release liners, and PSAs. Polystyrene films were the first "no-label-look" materials. Later, BOPP and nonoriented, transparent PP with a thickness of 60 μm were recommended for no-label-look labels. Only plastic-based liners provide a really glossy, no-label look for the adhesive layer. A paper carrier gives textured surfaces due to the paper fibers.

No-label-look transparent labels require a plasticizer-resistant PSA (see also Chapter 8). Such PSAs must display improved holding power with a plasticizer content of 15%. A common label-grade PSA demonstrates a holding power of less than 1 h (with a 2 kg/in.² load) and a plasticizer-resistant grade demonstrates a holding power of more than 3 h.

As indicated in Ref. [15], in a similar manner to no-label-look labels, the optical appearence of hydrocolloid dressings used in medical applications, that is, their translucency, is one of the main requirements.

4.1.2.5   Pharmaceutical Labels

The pharmaceutical industry requires the use of so-called wraparound labels (see also Section 4.1.5.5), with a flag (i.e., an overlapping part). Such multilayer labels may display

TABLE 4.3   End-Use Requirements for Medical PSPs

Requirement	PSA	Carrier	Whole Product	Medical Product
		PSP Component		
Medicine dependent				
Physiologically compatible	—	—	•	Label, tape, TDD, electrode
Body fluid nondegradable	—	—	•	Label, tape, TDD, electrode
No skin irritation	—	—	•	Label, tape, TDD, electrode
Sterilization resistant	—	—	•	Label, tape, TDD, electrode
Ethylene oxide resistance	—	—		
γ-Radiation resistance	—	—		
Low cytotoxicity	—	—	•	
Substrate dependent				
Low skin adhesion and adhesion build-up	•	—	—	Label, tape, TDD, electrode
Detachable/painless removal	•	—	•	Label, tape, TDD, electrode
Low adhesive transfer to skin (cohesion)	•	—	•	Label, tape, TDD
Long-term wear	•	—	—	Label, tape, TDD, electrode
Conformability	—	—	•	Label, tape, TDD
Breathability (50 s/100 cm³/in.²)	—	—	•	Label, tape, TDD
Moisture–vapor transmittance (MVTR; 500–21,100 g/m², 24 h, 38°C)	—	—	•	Label, tape, TDD
Liquid transmittance (6–36 h)	—	—	•	Label, tape, TDD
Application dependent				
Moisture insensitivity	—	—	•	Label, tape, TDD, electrode
Cold-water-insoluble, hot-water-soluble (65°C) adhesive	•	—	—	Label, tape
Light resistance	—	—	•	Label, tape, TDD
Heat resistance (350°F)	—	—	•	Label, tape, TDD, electrode
Electrical conductivity	•	—	—	Electrode
Adhesion to permeable backing	•	—	—	Label, tape, TDD, electrode

supplemental information, with a constant message on one side and a variable message on the other side. On the other hand, the flag can be used for the bill. Pharmaceutical labels can have a sandwich structure (i.e., a secondary, removable label attached to the main label). The secondary label is printable by thermal printing. The laminated structure possesses perforations as well to allow multiple detachment (i.e., multiplication) of the labels and instructions, that is, to allow multiple information transfer. Parts of the label remain on the drug or on the patient; such a label works like a business form (see also Section 4.1.3.4). The detached part of the label should have a medical adhesive so it can be applied to the skin (see also Section 4.1.2.3). In another case, multiply labels double as a purchase record for insurance purposes [26].

Pharmaceutics and toiletry demand 100% retraceability to the individual production steps and the raw material. Security labels for the pharmaceutical industry are used to protect product integrity. Changes in the product could involve indicators about time and temperature, light sensibility, moisture, impact sensibility, contamination, loss of vacuum, etc. Various constructions exist for tamper-evident pharmaceutical labels (see also Section 4.1.2.7), such as time–temperature indicators, peel-away labels, reverse-peel labels, and peel-apart closures. Time–temperature indicators change color to reveal an underlying graphic when exposure to a specific temperature exceeds a specified time limit. A special ink can be printed flexographically on the label surface. Using a simple testing kit, it can be validated within 2–3 min [27]. Security face stock materials can work using color change in a peel-away, reverse-peel, or peel-apart closure version. The peel-away label is standard, placed across the opening to be closed. Removal activates color loss and exposes a hidden message that indicates tampering. Unlike many fragile security materials, it leaves no deposit on the adherend. Reverse-peel labels are used on transparent surfaces as a uniformly colored overlayer. When removed, the label splits apart, losing its color to reveal the hidden message. With peel-apart closures, the hidden message is incorporated in a laminated pouch or envelope with a transparent overlay. Opening activates the color loss and exposes a hidden message. A pharmaceutical label that cannot be removed from a container with steam or water once the label is applied is built up as a combination of an image-producing self-contained carbonless label overlaminated with an opaque carrier material. According to FDA recommendations [28], tamperproof packaging can be achieved using a packaged product wrapped in a transparent, boldly labeled printed foil, packaging sealed with foil or adhesive tape, or shrink film. According to Post [29], acrylic PSAs provide nonflagging characteristics to labels on pharmaceutical containers during autoclave sterilization.

4.1.2.6 Radiofrequency Identification Labels

Labels with appropriate security features make up part of the various proprietary Electronic Article Surveillance (EAS) systems. The most popular are still standard labels and tags with radiofrequency (RF) or electromagnetic circuits and strips. RF tags do not harm sensitive magnetic media. This is not true for magnetic deactivation. So-called RF EAS/identification (ID) tags will provide a portable database for system solutions, including manufacturing logistics, supplier and retailer distribution management, self-service checkout, warranty and return fruad, etc. RF-based product identification systems can use pressure-sensitive labels as well. Advances in RFID require the use of special labels including the RFID chip. Their printing requires a raster of 1000 L/cm. Unlike common bar codes, such labels are rewritable [30]. RFID labels are used mostly for product identification in the pharmaceutical industry [31,32], library systems [33], and the automotive industry [33], where the main applications include warranty management, asset management, and return management. According to Ref. [34], the RFIDs that are embedded in a substrate material have been given a self-adhesive coating for temporary and sometimes even permanent identification. Lamination of the RFID into the compound layers of a ticket to make it copyproof and tamperproof can only be achieved using reactive adhesives that undergo cross-linking after application. In the United States the pharmaceutical industry, together with FDA, forced the introduction of RFID labels since 2004. Unfortunately, the identification level is too low (70% instead of 95%), and the temperature of the product may increase during code-reading; thus, temperature-sensitive biological products may be damaged [35]. RFID labels for medical use must resist sterilization and, in special cases, $5000 \times g$ gravitational acceleration.

For PSP designers and manufacturers the manufacture of RFID labels impose special printing skills due to the different requirements of common technology used for labels (in printing, priming, lacquering, etc.; see also *Technology of Pressure-Sensitive Adhesives and Products*, Chapter 10) and for printing the electronic (functioning) layer on the label. Common printing ink layers possess a fine structure of more than 20 µm, a passer of about ±5 µm, and a layer thickness of about 1 µm. The homogeneity of the printed layer is not important in this case. Printed electronic layers for RFIDs must have a fine structure $\ll 20$ µm, a passer of less than 5 µm, and a very homogenous coated layer of 30–300 nm. On the other hand, flexoprinting, commonly used for printing of labels, is not adequate for RFID because of the variable layer thickness (center/margins).

4.1.2.7 Security Labels

Advances in document security and identification affect the label market (identification systems, document security, anticounterfeiting, corporate fingerprinting, copyable labels, etc.) and label construction. Pressure-sensitive tamper-evident labels are widely used for tamperproof packaging and sealing. Tamper-evident security labels are special permanent labels with a sophisticated construction that does not allow de-bonding. Their design demonstrates similarities to special closure tapes (see also Section 4.2.1.2). This market includes labels for traceability and identification, shock damage or temperature change, and security. Such labels are used for ware hanging, advertising or product information text, bar codes, and electronic security elements. In the health care industry the goal is a zero-defect product, that is, zero-defect packaging and documentation (see also Section 4.1.5). Product integrity, safeguarding against theft or improper use, childproofing, and user friendliness are necessary as well, and safeguarding against mishandling and abuse must be ensured. As a technical solution, nonremovable and nontransferable constructions were manufactured as antitheft labels.

Generally, permanent labels are tamperproof; they tear when any attempt is made to remove them. However, for special requirements tamper-evident seals, security tags, tactile labels, special backgrounds, special (e.g., magnetic) inks, controlled front or reverse delamination, security threads and numbering, materials sensitive to visible light (e.g., laser beams), and holograms (see also *Technology of Pressure-Sensitive Adhesives and Products*, Chapter 10), have been used. The choice of an adequate technical solution for tamper evidence is a complex problem. Principally, security marks or signs or security constructions can be used.

4.1.2.7.1 Security Signs

Special foils that can be applied using the hot-foil stamping technique (see also *Technology of Pressure-Sensitive Adhesives and Products*, Chapter 10) can incorporate optically variable features such as holograms and kinegrams on a single change in color, depending on the angle of view. A special method was developed to permanently bond holograms onto a substrate, and such products can also be used for identification. Hologram labels that need low light were developed [36]. Advances in hologram technology allowed the use of a diffractive variable image device for the design of computerized hot-stamp holograms that are difficult to copy. Papers with watermarks, papers that react to ultraviolet (UV) light or heat, and iridescent papers are used. Another possibility is the use of iridescent multilayer (some hundreds) films that change color according the light direction. Films with a so-called "floating image" (the printed image moves according to the light direction) were also developed. Lenticular films allow 3-dimensional images and motion images, in which vertical or horizontal vibrations of the image are possible [37].

Conventional gold and silver foils provide excellent antcopy features. Pressure-sensitive labels with holograms play a decisive role in publicity and promotion. They are die-cutted on the roll, have various carrier materials, and are used for direct mailing. If there is sandwich build-up, using a transparent cover film, the label can be separated into two parts, one carrying the technical (previously hidden) message [38].

4.1.2.7.2 Security Constructions

Nonremovable and nontransferable constructions were manufactured as antitheft labels. Design of such security labels is based mainly on the choice of special constructions, which use special (e.g., fragile) carrier materials manufactured and converted (e.g., security cuts) through special methods.

Pressure-sensitive labels displaying security peel-off must have either exceptional strength to prevent removal or be so fragile that they cannot be removed in one piece. In this case, excellent conformability, good printability, and destructibility are required for the carrier. Fragile pressure-sensitive materials either are very thin or have low internal strength. To achieve tamper-evident properties, special face stock materials and mono- and multilayer construction are used for security labels and tapes. In such construction the tamper-evident cast films ensure that the printed label will fracture if removal is attempted. For this purpose, low-strength carrier materials are combined with high-strength adhesives and optically working printed elements. Destructible, computer-imprintable vinyl film and a very aggressive PSA have also been used for this purpose. The adhesive must bond in less than 24 h [30]. The vinyl film fractures when removal is attempted. For instance, a 50-μm cellulose acetate film was modified to be a brittle film with a low tear strength, but it must be die-cut [39]. Tamper-evident polyester label stocks provide an alternative to PVC and acetate films. Such a product exhibits tamper-evident performance on chipboard cartons, glass, painted metal, PE, and PP. They can be used for security foiling.

Security cuts enhance performance. Light-weight machine-coated paper face stock combined with strong adhesives can be used, together with security edge cuts and perforations applied on the press. Aluminium base layers are also used for special laminates for security film technology. In some cases, the construction includes a matte top-coated transparent film. This film is coated on its back side with a release layer that has moderate adhesion to the film. This release should be printable. A primer is used to achieve good anchorage of the printing ink (graphics in mirror image) and to ensure effective splitting during removal. The adhesive film should be slightly narrower than the web width to prevent possible adhesive contamination of the laminating roll. During manufacture, nip pressure should be light and web tension should be adjusted to avoid label curling. Adequate die-cutting is provided by a uniformly densified paper liner. Making slits or perforations in such face stock materials improves tamper-evident properties [40].

Generally, these tamper-evident products are made using a common direct- or transfer-coating procedure of a face stock material such as film or paper. Unfortunately, the manufacture, coating, and conversion of special carrier materials involves technical and economical problems because of the low tear resistance of these materials. Certain security labels have a complex construction. Their conversion includes a range of operations. Tamper-evident constructions must stay put as the waste matrix is removed, yet break up or tear when removed from the substrate. Such products require special liners; the mechanical properties of the liner complement the low mechanical properties of the face stock material during manufacture and application. Special labeling machines (see also Section 4.1.4.1.3) allow the labeling of pharmaceutical products with labels and a closure for originality.

A different construction can be manufactured as well. A tamper-proof label is based on chrome polyester [41]. Upon removal the word "Void" appears on both the marked item and the label, preventing their reuse. So-called "security foiling" (i.e., overlamination of the label with a destructible upper layer) is a common technology to manufacture security labels. A purple/pink film alters according to the angle of light and changes from purple to clear upon removal or attempted peeling [42].

Another tamper-evident construction includes a release liner and a composite layer coated on the release liner. The upper part of the composite layer is nontacky. It is a film-like layer with low tear resistance. The bottom layer anchored on the face stock layer is a common PSA. The film-forming polymer should be manufactured (cast) by the same coating procedure used for the PSA layer, but first the liner should be coated with an aqueous dispersion. This dispersion-made face stock layer of the tamper-evident PSA layer is manufactured using a low-flexibility, low-tack, medium-cohesion polymer on a basis of acrylate, styrene–acrylate, or EVA, with or without filler. To prepare labels that can be dispensed with a labeling gun, 6–250 g/m^2 coating weights are used.

4.1.2.8 Removable Labels

The principles of removability were discussed in detail in Ref. [24]. The criteria for removability, the ways and means to achieve removability, and the influence of the viscosity/modulus on removability were examined. The formulation, properties, and manufacture of removable PSPs were discussed in Refs. [6,22,24,25]. A short discussion of removable pressure-sensitive formulation is given in *Technology of Pressure-Sensitive Adhesives and Products*, Chapter 8. Removable PSPs cover a wide range of end uses. Such products can be used as special labels, tapes, etc. Therefore, they are presented as special products, for example, medical, mounting, and insulating tapes and labels. Their main segment includes protective webs (see Section 4.3 and Chapter 7).

4.1.2.9 Temperature Labels

The principles of formulation for temperature resistance are discussed in *Technology of Pressure-Sensitive Adhesives and Products*, Chapter 8. As noted in Ref. [2], the standard temperature resistance of a PSA formulation is given by the rheological characteristics of the components. The stability in time of such performance characteristics is affected by the chemical resistance of the raw materials. For instance, butyl systems should perform well at low temperatures, whereas acrylics are superior in a room-temperature environment. Low-temperature peel adhesion for silicone increases at −75°C; for common organic PSAs the peel value is zero below −25°C. Hence, silicone PSAs are recommended for low-temperature conditions.

Generally, there is a difference between LT and ST (see also Section 4.1.2.2). Suggested LTs are situated between −10 and +10°C. However, in almost every label application field, there are products that must be applied on substrate surfaces at extreme temperatures. Fuel and heat make special demands on plastics and labels used in automobiles. Five-hundred-hour tests at temperatures typically occurring in the engine compartment (up to 120°C) are carried out [43]. Computer labels are designed for an ST of −20 to +90°C. Solventless rubber-based adhesive for filled high-density PE (HDPE) milk and fruit juice containers must operate between −25 and +70°C, with a minimum surface application temperature of −5°C. Bag closure labels must be able to withstand outdoor temperatures, and labels for butt splicing have a high temperature stability up to 117°C. For high-temperature splicing tapes, shear resistance at 250°C has been tested [44]. Blood bag labels must be water resistant and able to withstand very low temperatures, so they are overlaminated [45]. Flame-resistant advertising labels must resist temperatures up to 170°C [46].

The data regarding the temperature domain of application for such labels are varied. According to Dobmann and Planje [47], office labels must work at +23°C, labels for refrigerated meals at +40°C, and labels that are applied to frozen foods at −18°C. According to Dunckley [48], temperature-resistant labels are used generally between +40 and +260°C. According to Ref. [49], temperature-resistant labels are used between +40 and +126°C. Temperature-display labels comprise a special class of such temperature-resistant labels. In such temperature-resistant labels the optical display of temperature sensitivity is applied. Common end-use areas for such labels are airline tags, caution labels, automotive labels, police forces, and gaming machines. Temperature-display labels are used as thermometers. They possess a conformable, heat- or cold-resistant, water- and oil-resistant carrier material and a special temperature-indicating component. Such reel labels are used between 40 and 260°C and display the measured (substrate) temperature values (37.8–260°C) with an exactness of ±1%. Thermochromic inks change color with variations in temperature. Such inks are currently printed on shrink labels [42].

4.1.2.10 Water-Resistant/Water-Soluble Labels

Water resistance was discussed in detail in Refs. [2,50]. The principles of water-resistant/water-soluble adhesive formulation were discussed in Refs. [6,22] (see also *Technology of Pressure-Sensitive Adhesives and Products*, Chapter 8, and Chapter 7 of this book). Generally, the water resistance of a PSA is the sum of the synthesis-given water resistance and of formulation-given water resistance. The water resistance of PSPs depends on the water resistance of the solid-state components of the laminate. A wide range of products require water resistance. For instance, textile, automotive, and bottle labels (see also Section 4.1.1.2.2) require water resistance. Labels used for marking textiles or clothing should resist 50 laundry cycles and 10 dry-cleaning cycles. Water resistance for more than 72 h is required for ice-proof gums. Bag closure labels (see also Section 4.1.1.2.1) must be able to withstand outdoor temperatures and moisture. Labels used on PA substrates (that absorb up to 30% water) require moisture-independent peel resistance. Blood bag labels must be water resistant and able to withstand very low temperatures. Logging tags must withstand long exposures to salt water; therefore, they are printed on a special synthetic face stock material. For surgical tapes used for fastening cover materials during surgery, the adhesive must be insensitive to moisture and insoluble in cold water and display adequate skin adhesion and skin tolerance (see also Section 4.2.1.6). The water solubility of the product is reached at 65°C and at a pH value of more than 9 [51]. Taking into account the multiple uses of water resistance/solubility, water-resistant labels and tapes are discussed in the sections covering special products for various applications (see also Sections 4.1.1.3, 4.1.2.6, and 4.1.2.10).

4.1.2.11 Other Special Labels

A wide range of special labels exists. There are various grades of instruction labels. Some are removable. They have a PET or PVC carrier and are used in clean rooms and on delicate electronic parts. Such labels for technological use are used in various fields, such as the automotive and textile industries. Iron-on labels used in the textile industry must resist temperatures up to 180°C (10–20 s) during application and be chemical and water resistant. A German label manufacturer has succeded in placing original perfume oils in a special two-layer label structure (scent labels). Baggage tags for airlines, promotion labels, tickets, and address labels are other label applications that have been developed. Special antiskid, chemical, and moisture proof, polycarbonate (PC) film floor labels are used for the identification of pallet or bulk location in warehouses, distribution centers, or factories [52].

4.1.3 Label-Like Products

Various PSPs have a label-like build-up, but applications fields that are different from labeling. For instance, pressure-sensitive booklets, coupons, and piggybacks can also be considered labels. A thermal tag is produced as a reinforced laminate construction for subsequent conversion by label producers. Taking airline luggage as an example, a central plastic layer provides strength, the self-adhesive layer allows the label to wrap around luggage handles, and the upper layer is a top-coated thermal paper onto which the data are printed as a bar code. The most important label-like products include decals, nameplates, decoration films, and forms.

4.1.3.1 Decals

Decals are large surface sheet-like labels used for various applications, mostly in advertising. Decal films have been developed for marking functional machines, building parts, etc. For instance, two-way decal films are used for push/pull glass doors. Security decals and association membership decals have been produced. Collections are manufactured also. Decals used for clear overlays must possess an adhesive that is perfectly transparent before and after drying. For advertising decals, weather resistance is required. These products are applied on buses, trucks, and trains and must possess chemical resistance to washing agents (see also Section 4.2.2.1).

Plotter films (see also Section 4.4), which are cut decals with various forms, are used as chablones that are cut from the film material. The negative letter signs are applied on the substrate. Thus, spray lacquering of the chablones is eliminated. The decorative and informational elements are placed and temporarily fixed with application tapes. Such tapes can be considered temporary mounting tapes (see Section 4.2.2.1).

4.1.3.2 Nameplates

Equipment, apparatus, electric appliances, and other durable goods have always been given serial numbers. Traditionally, this has been done using nameplates (rating plates, type plates, inventory labels) to identify the manufacturer and state conditions (voltage, amperage, year of fabrication, etc.). Nameplates must provide information over a period of years on serial numbers, date of manufacture, article number, and performance data. Such plates are used in varied industrial and commercial domains. They must be light, chemically resistant, and weatherproof and be able to withstand abrasion. Generally, they must be printed on the spot. These data once had to be stamped by hand, but stamping was expensive and time-consuming. Up until the mid-1960s, these plates were mounted with screws or rivets. The application of adhesives allowed (required) the use of thinner metal plates and faster mounting. The data could be processed with high-impact type writers. At first, only paper and cellulose acetate could be computer printed, and they were not suitable for nameplates. The development of synthetic films, especially computer-imprintable PET, provided a solution to this problem.

Special performance characteristics are required for carrier materials for nameplates. Such films must have a matte, metallic appearance, similar to that of anodized aluminium. They must display the handling characteristics of hard aluminium. No curl should occur after removal. A heat resistance of 150°C is required. The matte surface must absorb ink, maintaining sharp graphic contrast, good color intensity, and excellent legibility. Films used for nameplates should be printable via flexo-, letterpress-, and computer-printing methods (see also *Technology of Pressure-Sensitive Adhesives and Products*, Chapter 10). Abrasion, chemical, and environmental resistance are needed for such materials. Dimensional stability and abrasion resistance are important in maintaining bar code scanner and visual readability.

Polyester and vinyl films and special heat-resistant materials have been suggested for this purpose. Metallized 90-µm PET has been used. To be printed with flexographic-, letterpress-, thermal-, laser-, or computer-printing methods, a film must have a matte, metallic appearance, similar to that of anodized aluminum. Computer-printable top-coated PET films are used for stiff labels such as nameplates. Top coating or treatment of PET makes it printable by different methods, including impact-, thermal transfer-, and flexoprinting. Laser-engraved acrylic films for nameplates are tamperproof (see also Section 4.1.2.7). Such labels are ideal for applications where either contact-free data recording is needed or when direct laser inscription of work-pieces is not possible. Formulation of "laser–resistant" adhesives, that is, products that exhibit enhanced temperature resistance, requires special tackifier resins (see also *Technology of Pressure-Sensitive Adhesives and Products,* Chapter 8). The adhesive for nameplates must bond to difficult surfaces, high- or low-surface-energy plastics, textured or curved surfaces, and cold surfaces. The liner must exhibit trouble-free release, dimensional stability, and good die-cutting, perforating, pre-punching, and fanfolding.

For printing such products, letterpress- or screen-printed presetting is used as well. The printer should have a comprehensive library of types and standard bar codes, electrical symbols, etc. If thermal transfer printing is used, the resolution is 7.6–11.4 dots/mm and printing speed reaches 175 mm/s. Rating plates can also be printed by sandwich printing, where the reverse side of a transparent material is printed with mirror lettering. Easy-to-print destructible roll-label materials can be used as tamper-evident products for name plates (see also Section 4.1.2.7). The internal strength of the label stock totally disappears once applied, and if any attempt is made to remove the film the label breaks into tiny fragments. This is a faster and more cost-effective system compared with costly-to-produce laser-etched labels and nameplates [53].

4.1.3.3 Decoration Films

Decoration films are used in household and technical domains [54]. They are manufactured as adhesive-coated or self-adhesive films. The first generation of decoration films consisted of soft PVC films, coated with PSA and printed on the back side with a special design. Soft PVC may contain up to 50% plasticizer; thus, they required special formulation to avoid interaction (migration) between the components of the carrier material and the adhesive (18–28 g/m^2 HMPSA), and it was difficult to print on them. For such decor films, water-based gravure printing is suggested. They can have a printed release liner as well. Abrasion-resistant automotive decoration films also have a separate release liner. Such products are made on a polyurethane (PU) or EVAc/PVC basis. The carrier is coated with a PU primer.

4.1.3.4 Forms

Forms (which, according to their construction can be considered special labels) are very complex, multilayer constructions that have exactly controlled adhesion forces in the laminate layers and are discontinuous/finite elements, or inserted in a web-like product, which allows detachment of one or many parts of the forms, making simultaneous, multipurpose information transfer possible. They are manufactured like multilayered labels by coating and laminating of "prefabricated" webs [3]. In the first manufacture step, a paper carrier is coated with a release agent or adhesive. The adhesive-coated web can be laminated with a film, and the adhesive-coated release liner can be laminated with another release liner to give a temporary laminate. The sandwich structures manufactured in this way can be combined to yield a PSP. In the coating practice adhesives with different bonding strength and release liners with different abhesive characteristics are combined to achieve a partially detachable product. Their formulation includes a range of permanent or removable adhesives with different, exactly controlled adhesion. Some have very low adhesion (allowing the detachment of die-cut label parts; e.g., cards).

They are generally formulated from very hard, high-T_g acrylates, high-T_g VAc–acrylate copolymers, or low-T_g cross-linked acrylates.

Business forms are designed to carry information to be distributed to different end users at different times in different forms. Form-label combinations, cut-sheet and laser-imprintable forms, security forms, etc., were developed. A business form with removable label, with tape pieces adhesively mounted on the paper substrate, is described in [55]. As usual, the business form has die-cut label areas that can be removed. This laminate is made of paper, printed continuously in a press. Pieces of adhesive tape are applied at spaced points of the printed paper. The formed binary areas with a greater thickness than the carrier include the tape in relief to the substrate. One of the layers is die-cut.

Piggyback mailing labels are a typical example of forms; they are sandwich labels. Such products can contain two or more peel-off layers over a single backing paper. The complete label, which is die-cut and personalized in one pass, is affixed to the consignment, and two small numbered slips are peeled off and attached to control forms during the routing process [45]. Forms are applied by detaching, separating the parts of the form. This is carried out manually. Detachment and refixing of parts of forms (e.g., cards) is carried out by hand, but certain parts of a form (e.g., main web) can be processed automatically.

4.1.4   Label Application Technology

Generally, end-use properties of PSPs must be evaluated as "instantaneous" application characteristics and storage- or aging-dependent characteristics (see also Chapter 7). The conditions under which PSPs are applied differ according to the PSP classes and their special uses. In comparison with other PSPs, such as tapes or protective films, which are manufactured and applied as a continuous web, labels constitute the temporary component of a laminated web, applied as discrete, discontinuous items. Therefore, their construction must allow them to be separated from the rest of the laminate and applied on the adherend surface. To allow them to be separated, labels must be die-cut (see *Technology of Pressure-Sensitive Adhesives and Products*, Chapter 10) and must have an abhesive layer on the back side. They must possess sufficient mechanical stability and adhesivity to be applied. As information carriers, labels must fulfill aesthetic, coating, and mechanical demands. The main requirements of labels are flexibility, aesthetic value, low cost, quality, ease of application, high speed of application, strength, and moisture resistance. Of the above-listed technical criteria, ease and speed of application have been the most difficult to fulfill. Application and deapplication characteristics of the main PSPs were examined comparatively by Benedek in Ref. [8]. As noted, labels allow impact or nonimpact labeling, high-speed application, room temperature or low-temperature application, repeated application (i.e., readherability), very low application pressure, manual or automatic application, and conversion during and after application.

The application conditions for labels depend on the product to be labeled, the labeling method, and the type of label. The product to be labeled influences the instantaneous and final bond strength by its surface temperature, surface structure, and geometry. The labeling process is influenced by the label and by the product to be labeled. The main label-dependent parameters of labeling are the adhesive type, coating weight, release degree, labeling system, labeling speed, label geometry, labeling tolerance, label material, label printing, and converting. Application conditions include the application temperature, environment, and speed of application; they may also include readherability.

Substrate surfaces used in label-laminating (labeling) vary widely (see also Sections 4.1.1.2 and 4.1.1.9). Their diversity is due to the raw materials used for their manufacture and the processing technology used for them. Special application fields need special application (adherend) surfaces. Therefore, the label designer must have information about the adherend surface to tailor the product to serve its intended purpose. The influence of the substrate on the adhesion properties of PSAs was discussed in detail by Benedek in Ref. [4].

The product form and flexibility/rigidity influence the application and labeling conditions. Application on flexible surfaces is difficult. On the other hand, the flexibility of packaging materials is increasing due to the development of new materials and downgauging. For instance, new vinyl alloys allow blow-molders to create cosmetic bottles and tubes with a frosty squeeze effect, which gives the product a satin look, a soft touch, and the ability to recover its shape after compression. New packaging films may be manufactured with a reduced thickness. As an example, biaxially oriented ethylene–vinyl alcohol copolymer (EVOH) film with a thickess less than 10 µm can be manufactured with sufficient strength for many packaging applications.

Generally, the label application technology is determined by the label application process (labeling), by the labeling device used for delamination (from release liner) of the label and its relamination on the item to be labeled, and by the labeling machine.

4.1.4.1 Labeling

Label application is carried out by labeling, which is a cold-laminating procedure. Labels are applied manually or by labeling machines. Labeling technology includes the labeling methods, labeling devices, and labeling machines. It involves the use of adhesive-backed carrier on a release liner. Equipment for this method varies from hand-fed units to sophisticated equipment, with automatic feeders and cutoffs. Requirements concerning the mechanical properties for dispensing are probably higher than that for coating and converting, and dimensional stability-related requirements for printing are more severe than that arising during conversion (see also *Technology of Pressure-Sensitive Adhesives and Products*, Chapter 10). Release from the liner should be adequate to prevent predispensing of labels, but not so tight as to cause the web to break or the liner to tear.

Tear-sensitive labels with low mechanical resistance are dispensed by machine as well. Pressure-sensitive labels for tamper-evident applications must be applied automatically. Generally, labeling quality depends mainly on the machine, the label components, and the substrate and is influenced by the labeling system, substrate surface, label material, adhesive, and transport and storage conditions.

4.1.4.1.1 Labeling Methods

Labels are manufactured as web-like products, but can be applied in other forms. One main classification of labels concerning their application technology divides them into roll and sheet labels (see also Chapter 1). Roll labels are those manufactured in roll form and applied from the roll, where their continuous web-like backing allows them to be handled as a continuous material after their confectioning (die-cutting) as a separate item. Sheet labels are manufactured in roll form and cut into a discontinuous sheet-like finite product (such as a finite laminate) that cannot be applied with a common labeling gun. Requirements are generally less critical for roll applications. Sheet application requires PSAs with sufficient anchorage to prevent pulling out, but sufficiently low tack to prevent gumming of the guillotine (see also *Technology of Pressure-Sensitive Adhesives and Products*, Chapter 10).

Labeling methods differ according to the sheet or reel form of labels, their dimmensions, and the characteristics of the item to be labeled. The raw materials (carrier and adhesive) for sheet and reel labels differ; they were discussed in Ref. [2]. Different application (labeling) technologies are used for small labels and (large) decals, for machine (gun) labelling, and for hand-applied labeling. In most cases the adhesive tack and release force (peel force from the release liner) must be closely monitored. Reel labels require an agressive tack, whereas sheet labels need more cohesion and less tack. Therefore, the base elastomer for reel/sheet labels will be selected in a different manner. Cohesive, low-tack, high-molecular-weight polymers of short (side) chain acrylics and EVAc are adequate for such sheet applications. Soft, very tacky polymers of ethylhexyl acrylate or butyl acrylate (BA; i.e., long-side-chain acrylics) are preferred for reel labels.

Many pressure-sensitive labels are still affixed by hand, but a growing percentage are applied by automated equipment. Labeling machines for roll and sheet labels have been developed. Labeling speeds as high as 240 labels/min are achieved.

Adhesives used for products with automatic labeling should have a level of quick stick that will provide for secure adhesion with minimal application pressure. Very good PSAs do not need a measurable application pressure. Application of small roll labels needs almost no mechanical lamination. As PSA testing practice demonstrates, a loop tack test simulates the real application conditions of labels (see also Chapter 7). Pneumatic (blow-touch) label transport and application has been available for many years. In blow-touch labeling operations labels are blown onto a surface using air pressure. They fly through the air and land without pressure on the substrate. This is possible because of their high instantaneous adhesivity.

The choice of the labeling method depends on the labeled item. For instance, hang tabs can be laminated manually or by an automatic dispenser or automatic machine application.

4.1.4.1.2 Labeling Devices

Various labeling devices exist according to the principle of labeling used (e.g., mechanical, air-driven label transfer). Conventional dispensers for paper-based pressure-sensitive materials do not require a special applicator system because of the inherent stickiness of the paper. Special systems are needed for conformable and squeezable labels. Tight label placement tolerances of 0.015 in. are typical for application systems.

A hand-labeling device contains a storage roll, a pressure roll, and a cutting device. Generally, a handheld labeler used to print and apply PSA labels releasably laminated on a carrier web comprises a housing with a handle, label-roll printing device, delaminating device, applying device, and some means of advancing the carrier web. A hand-operated printing and labeling machine may possess a carrier band on which the labels are temporarily positioned. The machine has a conveyor system that draws the label carrier head around a label-detaching edge. Hand held labelers are computerized and can be used for price marking and remarking, coding, etc. [56].

Labeling devices depend on the labeled item. For instance, a handheld labeling device is used for marking wires and cables [57]. Sometimes the labeling device is part of bag-making equipment, corrugated and folding carton-converting machinery, envelope machinery, or engraving equipment.

4.1.4.1.3 *Labeling Machines*

There are various labeling machines, classified mostly according to the item to be labeled. Label application machines (labelers) of quite different constructions and capabilities have been developed. Suppliers of such machines offer hardware and consumables.

Labeling speed depends on the dimensions of the label. The maximum labeling capacity is a function of the length of the label, but is strongly influenced by the construction of the labeling machine. Common labeling systems allow the application of 50 labels/min with label dimensions of 120 × 1450 mm. The same equipment ensures a labeling speed of 30 labels/min for labels of 13 × 350 mm.

Labeling speed is a function of label stiffness as well. For high-speed labeling, films with a minimum thickness of 50 μm are suggested [58]. Microprocessor-controlled PSA label application offers variable speed settings to match product requirements and may have a special dispensing edge and a collapsible backing paper rewind. Where a high degree of conformability is desired, flexible label stock such as saturated paper should be used rather than a high-gloss paper. The layout should be in the long-grain direction to provide maximum conformability. Labeling is a function of the release as well. According to Ref. [59] with new with UV-cured silicone (0.8 g/m^2) on 30-μm BOPP, high-speed dispensing is possible up to 40,000 labels/h (i.e., 11 labels/s).

The productivity of labeling machines depends on the item to be labeled and the tape of label. Developments in machine construction allow a continuous increase in labeling speed. For instance, a decade ago no-label-look labeling of glass bottles was carried out with a productivity of 450 pieces/min [60]; now productivity is higher than 500 pieces/min. Control systems check and separate mislabeled bottles.

Labeling machines with tandem labeling provide better productivity. These machines can simultaneously apply two labels at a running speed higher than 250 containers/min. Round-around labeling allows the labeling of the front and the back sides and the use of multilayer labels (e.g., info-roll labels); such machines with three-roll systems allow labeling with two labels, on the front and back sides, or the application of overlapping labels, dispensing up to 400 labels/min [61]. Special machines allow labeling of up to 140 tubes/min [62].

4.1.4.2 Delabeling

Delabeling is the debonding of labels from a substrate. As noted in Ref. [8], labels allow mechanical, chemical, or thermal deapplication. Partial deapplication is also possible. Such operations are carried out using dry or wet technology. Dry delabeling is ensured by the removable character of the PSA; wet delabeling is based on the water solubility of the PSA or PSP (see also *Technology of Pressure-Sensitive Adhesives and Products*, Chapter 8).

4.1.4.3 Postlabeling Technology

As discussed previously (see also *Technology of Pressure-Sensitive Adhesives and Products*, Chapter 10), certain conversion operations are transferred from web manufacture in the end-use phase of labels. Postdesign, postprinting (writing), and postcutting are carried out by the end user. Therefore, end-use properties include label versatility for these operations. Writability, printability, computer printability, laser printability, and cuttability are special characteristics of labels in development. For instance, promotional form labels can have many pages, and their front pages can have a different print quality [63]. Laser beams can be used for cutting, punching, sealing, and printing labels (see also *Technology of Pressure-Sensitive Adhesives and Products*, Chapter 10). The label material (adhesive and carrier) must fulfill the related requirements.

4.1.5 Competitors of Pressure-Sensitive Labels

In Europe, traditional labeling techniques include roll-feed labeling (wraparound labeling), cut-and-stack labeling, and self-adhesive labeling. The types of labels produced include wet-glue applied, pressure-sensitive, gummed, heat-seal, shrink-sleeve, and in-mold labels.

4.1.5.1 Wet-Adhesive Labels

Labels and tapes are the main PSPs. Both were developed by replacing classic wet adhesives with PSPs.

According to CEN/TC 193/WG3N12 European Norm, the adhesives for coating can be classified into five categories:

Solvent reactivatable: nontacky adhesive coatings to be reactivated by a solvent
Remoistenable: nontacky adhesive coatings to be reactivated by moisture
Heat sealable: nontacky adhesive coatings to be reactivated by heat
Cold sealable: nontacky adhesive coatings to be reactivated by pressure
PSAs: permanently tacky adhesive coatings

Remoistenable adhesives include wet-adhesive labels. Wet-adhesive labels are suggested for application fields with very high-speed but low-quality requirements (metal cans, glass and plastic bottles, etc.). Generally, the main features of wet labels with good labeling performances are the ability to lay flat (see also Chapter 7), narrow die-cut tolerances (see also *Technology of Pressure-Sensitive Adhesives and Products*, Chapter 10), good mechanical properties (tear resistance in wet and dry states), and low stiffness. Although some of these requirements are valid for PSA labels, other requirements are different. Because of the differences in labeling systems used for wet and PSA labels, different carrier characteristics are required for each class.

Labeling speed for common pressure-sensitive labels is lower than that for wet labels. Pressure-sensitive labels have a higher manufacturing cost than wet-adhesive labels, but their application is easier. Certain label functions such as removability can be provided only by pressure-sensitive labels.

4.1.5.2 Heat-Seal Labels

Heat-adhesion (seal) labels and papers have been used in the packaging industry. Some heat-adhesion labels adhere immediately, whereas others have a delayed adhesive affect. The advantages of the heat-activatable or heat-applied labels are that they can be applied at high speed, they require no preliminary operations before application, they are rapidly activated, their use requires no solvent (no pollution) and no humidity, there is no chemical yellowing, they give good permanent adhesion, they require short sealing times, and they exhibit no edge splitting. The same principle is used for hot stamps, that is, labels without a solid-state carrier (see also *Technology of Pressure-Sensitive Adhesives and Products,* Chapter 10). Some special labels are applied as PSPs, but use other bonding mechanisms (e.g., iron-on labels) or are applied as alternatives to pressure-sensitive labels (e.g., in-mold labels). For iron-on labels, used mostly for marking textile products, the final bonding is achieved via molten plastomer (PE, PQ, etc.) embedded in the PSA. The PSA layer can contain 5–60% uniformly distributed, powdered thermoplastics. Such construction can be used for a label or strip bearing a legend to be bonded onto textiles such as garments, which will later be ironed to form a permanent bond, resistant to washing and dry cleaning.

4.1.5.3 In-Mold Labels

IML inserts the label in the mold prior to molding. IML is a special case of injection molding with in-mold parts. In-mold fixing techniques include (1) insert molding, (2) outsert molding, (3) back melting of textiles, (4) removable tool technology, (5) in-mold decoration, and (6) IML. In these cases part of the finished product is fixed in or out of the mold by the molten polymer. In-mold decoration and IML are used to fix a nonfunctional part of an item. The principle of the method is simple. The label is placed in the mold, and its back side is fixed with the molten polymer. IML allows a plastic item to be simultaneously manufactured by injection molding and labeled with a special label placed in the mold. No supplemental postlabeling or postprinting of the finished item is necessary. The label is made from the same polymer as the labeled item, and both are recycled together. However, the material structure may change due to the cooling or crystallizing conditions. This special category of labels is applied on injection-molded, thermoformed, or blow-molded containers or soft spreads. These labels are used for toiletries and chemicals. PP and metallized films and papers are applied as carrier material paper and synthetic paper. Film-insert molding can add a soft-touch, soft-feel surface, switching fom PC to PU film [64]. One of the strengths of the elastic thermoplastic polyurethane (TPU) is that in many cases they must not be formed prior to injection molding. Such aliphatic TPU films can be brightly colored.

Label insertion and molding techniques require special knowledge. IML allows the manufacture of the plastic item by injection molding (or other thermoforming procedures; form fill and seal cups can also be labeled using this procedure) and its labeling in the mold, with a special label placed in the mold before the plastic is added. The handling of this label is difficult. A special apparatus is used to place the labels in the mold. Stacker systems have been developed to feed the label into an in-mold production machine.

In-mold labels are not pressure sensitive. They are coated with a heat-activatable lacquer (210–230°C for injection molding and 100–110°C for deep drawing). New in-mold labels that work like self-adhesive films have also been developed.

However, the PSP manufacturer or user must be informed about this technology because it is competitive with pressure-sensitive technology and the manufacture of in-mold adhesives is a hot-melt formulation. For instance, for such labels paper of 100–120 g/m² is printed, overlacquered, and adhesive coated (15–20 g/m²) with a hot melt. Film-insert molding can add a soft touch, soft-feel surface using a flexible TPU film.

4.1.5.4   Shrink Labels

Puncture-, impact-, and tear-resistant shrink and stretch films (based on linear low-density PE (LLDPE), very-low-density PD, and LLDPE/low-density PE (LDPE) blends and coextrudates) have replaced craft paper wrapping. Shrink labels use a shrinkable 40- to 70-μm film applied using special machines. According to FDA recommendations, tamperproof packaging can be achieved using packagings sealed with shrink film. Bilayered heat-shrinkable insulating tapes for anticorrosion protection of petroleum and gas pipelines have been manufactured as well (see the following).

4.1.5.5   Wraparound Labels

Wraparound (sleeve) labeling is recommended if the items to be labeled are not rigid enough, if a horizontal transport of the product is required, and if high-quality positioning of the labels is needed. Label applicators in the pharmaceutical industry affix wraparound labels on ampoules at a speed of 30,000 items/h [65].

In wraparound labeling, labels made of plastic or paper are bonded to a cylindrical container. The overlap bonding is especially stressed because at high temperatures the plastic bottles can expand and the bond is subject to shear stresses. Labeling with hot melts and subsequent shrinkage (roll-on–shrink-on) can be a low-cost alternative to sleeve labeling for containers requiring a profile shrinkage of 10–15% [66]. After an initial application of the plastic label to the container, the package passes through a heated (hot air up to 340°C) tunnel. The subsequent shrinkage of the label around the container leads to high stress on the bonded joint, and at the same time this heat prevents the hot melt from cooling, which delays the formation of cohesion. To solve this problem, reactive hot melts were adopted.

4.2   Grades, End-Use Domains, and Application of Tapes

PSA tapes have been used for more than half a century for a variety of marking, holding, protecting, sealing, and masking purposes. Industrial tapes were introduced in the 1920s and 1930s, followed by self-adhesive labels in 1935. Single-sidecoated PSA tapes were used in the automotive industry in the 1930s. Self-adhesive packaging tapes replaced gummed tapes on a paper basis.

Tapes constitute the largest group of PSPs. The volume of PSAs used for coating of tapes is higher than that of labels (due to the higher coating weight). The main tape market segments are packaging tapes (mostly HMPSA coated), electrical tapes (solvent based), industrial tapes (HMPSA and solvent based), health care tapes (HMPSA and water based), masking tapes (water and solvent based), and consumer tapes (HMPSA, solvent based, and water based).

Tapes are continuous web-like PSPs, applied in continuous form. Generally, their role is to ensure the bonding, fastening, or assembling of adherend components. They provide structural integrity, dimensional stability, and shape retention or display another particular property (thermal, sound, chemical, or electrical insulation, etc.). In Ref. [67], PSA tape is defined as a PSA-coated substrate in roll form, wound on a core, at least 0.305 m (12 in.) in length. The PSTC developed a guide for pressure-sensitive tapes that covers their significance, standard width, labeling, test, units of measurement, and tolerances. Tapes as an adhesive system offer the advantages of being positionable, having a controlled coating weight, allowing automatic use through die-cut parts, distributing stress, and exhibiting a low level of cold flow. In comparison with labels, where the role of the nonadhesive, coated-carrier surface is more important for long-term end use and the adhesive must primarily allow only such instantaneous adhesion that ensures the contact between the label and the labeled item, the mechanical resistance of the carrier is the most important parameter for tapes and the adhesive acts as an assembly tool between the carrier and the taped product. Tapes function as an adhesive in film form, that is, a prefabricated glue. As a glue in web form on a solid-state carrier, tapes exhibit an exactly regulated thickness and allow fast processing. Taping requires only low-energy use, requires no clean-up, and produces no pollution.

Although common tapes are used mainly as packaging materials, tapes have special properties as well. Certain products require water resistance, water transmission, oil resistance, or tear and dart drop resistance. Some standards for tapes specify requirements with respect to unwinding resistance and unwinding behavior, water permeation, and water vapor permeation. Electrical conductivity or nonconductivity may be required for insulating, packaging, and medical tapes.

Tapes have a very broad domain of application. Their performances and build-up are varied. Table 4.4 presents the peel resistance values for various tapes.

As Table 4.4 illustrates, various special tapes display different peel resistance values, but peel values differ for the same grade (e.g., packaging tapes) as well. Their raw material basis and coating technology strongly differ according to their general or special character. Rubber–resin and acrylic adhesives are mainly used for common tapes. Both can be applied as hot melts, solutions, or water-based dispersions. Their coating technology includes the corresponding equipment. Design and formulation of tapes was discussed by Benedek in detail in Ref. [6]; the build-up, manufacture, and end use of tapes were discussed by Benedek in Refs. [3,8,9] (see also *Technology of Pressure-Sensitive Adhesives and Products*, Chapter 11, as well as Chapter 1 of this book). There is a broad raw material basis for various special tapes. Rubber–resin formulations and acrylate-based formulations compete with special raw materials (e.g., silicones, polyvinyl ethers (PVEs), polyesters, radiation-curable recipes). Monomer-, oligomer-, and polymer-based compositions are used. Tackifiers, plasticizers, and cross-linking agents are the usual components of such formulations.

TABLE 4.4 Peel Resistance Values for Selected Tapes

	PSP	Base Polymer	Adhesive				Peel Resistance (g/25 mm)
			WB	SB	HMPSA	Cross-Linked	
Tape	Packaging	SIS	—	—	•	—	200
		SEBS	—	—	•	—	800
		Rubber–resin	—	—	•	—	1060
		Rubber–resin	—	—	•	—	420
		Acrylic	—	•	—	•	925
		Acrylic	—	•	—	•	1200
		Acrylic	—	•	—	•	1000
		Acrylic	•	—	—	—	1450
		Acrylic	—	—	•	—	1425
	Insulation	SBC	—	—	•	—	370
	Sealing	Acrylic	—	•	—	• UV	4100
	Removable	Acrylic	—	•	—	•	200
		Acrylic	—	•	—	• EB	400
		Rubber–resin	—	—	—	•	100
	Masking	Acrylic	•	•	—	—	300
		Acrylic	—	•	—	•	150
	Double-side coated	Acrylic	•	—	—	—	1626
	Temperature resistant	Silicone	—	•	—	•	3500
			—	•	—	•	1150

Water-based acrylics are preferred for common packaging tapes; they are also formulated as removable PSAs (e.g., application tapes). For such tapes, a coating weight of 5–35 g/m² is generally applied. Water-based chloroprene dispersions exhibit improved shear performances for tapes. Copolymers of maleic acid with acrylic and vinyl monomers, styrene, vinyl pyrrolidone, and acrylamide are used as release coating for tapes. The use of HMPSAs for tapes is discussed by Hu and Paul in *Technology of Pressure-Sensitive Adhesives and Products*, Chapter 3).

When tape was first manufactured, paper was used as a carrier. Different types of paper, such as dimensionally stable or deformable, tear-resistant and fragile materials, glossy, and crepe, have been used. The tear resistance of the tape must be as high as that of the paper to be bonded. For special tapes, kraft paper is reinforced with a mid-PP film. Polymer-based films constitute the main carrier material for tapes.

The carrier for tapes must support and absorb stress during application to allow bonding of the adhesive to the surface. The carrier is chosen according to the nature, magnitude, and direction of the forces. It is important to know the yield and tear resistance of the material and elongation in the machine direction and in the cross direction. It is imperative to know whether the carrier splits parallel to the machine direction. Dart drop resistance and foldability are required. Elmendorf tear resistance also plays an important role. The ability of a film to retain a crease is an important property in packaging and is also required for some special tapes. The mechanical properties of the carrier materials used for various PSPs were described in Ref. [68]. Tensile strength, tear strength (surface tear strength and bulk tear strength), impact properties, stiffness, elongation, and modulus are described. The mechanical properties (e.g., force at break, elongation, maximum force, and elongation at maximum force) of carrier materials for common (e.g., packaging) tape and special tapes (e.g., insulating, masking, application, medical) are examined comparatively. The characteristic values are very disperse, ranging from thicknesses of 25–90 µm, tensile strength of 3.5–35 kg/10 mm, and elongation of 25–250% (MD).

In the mid 1960s, cellulose hydrate, cellulose acetate, PVC, crepe paper, and fabric were the main carrier materials used for tapes coated with solvent-based adhesives. In the 1980s PVC, PET, and OPP were the most used carrier materials for tapes. Cellulose acetate and hydrate have been introduced as carriers for clear tapes in household and office applications. Nonoriented and oriented mono- or multilayer films and film/film or film/fabric reinforced laminates are recommended for tapes (see also Chapter 1). Fiber-reinforced films (applied for wet-adhesive tapes) are used for PSA tapes also. Chemically pretreated PE (octane-based LLDPE) is recommended for tapes. Biaxially oriented PP with a thickness of 30–35 µm is suggested also; it costs 20–30% less than PET. Because of the wide range of materials, tape manufacturing methods vary widely.

4.2.1 Common Tapes

Common tapes include office, marking, and packaging tapes. A large portion of tapes for common use are formulated with hot melts. Such formulations can be coated by slot-die coating (see also *Technology of Pressure-Sensitive Adhesives and Products,* Chapter 10). The classic type of tape used for fixing and fastening is known as packaging tape.

4.2.1.1 Packaging Tapes

Packaging tapes are still the main grade produced (about 80% of tapes), and productivity developments in the past decade relate mostly to these products. They account for most European tape consumption. Although their volume is increasing, their proportion of the global PSP production is strongly decreasing. Classic tape construction, a mechanically resistant, nondeformable, but flexible carrier material coated with a high coating weight of an aggressive but inexpensive adhesive, is valid for packaging tapes without special requirements. The high processing speed for these tapes requires a decreased noise level. Low-noise, color-printable, PP-based packaging tapes have been developed.

Generally, packaging tapes are based on PVC or PP, and they may also have a reinforcing layer. Such products are generally some shade of brown. Copolymers of vinyl chloride serve as clear tape (with a thickness of 30 and 50 µm) for packaging and household applications; they possess the advantages of high temperature resistance, good environmental resistance, conformability to the substrate shape, high transparency (for homogeneous tapes), low and constant thickness tolerances, and variable adhesive thicknesses. Kraft paper, cloth, and oriented plastic carrier materials are also used. Biaxially oriented cast films are recommended as carriers for packaging tapes as well. Such prereleased BOPP is coated with hot melts (21–22 g/cm^2). Oriented PP with a thickness of 30–50 µm is recommended. In the United States, biaxially oriented PP tapes are the most important packaging tapes.

Extensible, reinforced packaging tapes have been developed. Such tapes use a fiber-like reinforcing material, laminated on or embedded in the tape carrier material. Like high-strength wet tapes, pressure-sensitive packaging tapes may also have a reinforcing textile carrier.

HMPSAs, water-based acrylics, and rubber–resin formulations were used for such tapes. Common HMPSA systems for tapes are based on SIS, a resin that is partially compatible with the mid-block of the SBC (see also *Technology of Pressure-Sensitive Adhesives and Products*, Chapter 8) and a plasticizer (oil). Generally for such tapes there is no need for a very light color, so a color number of 6 is adequate for the resin. For a 1100/100/10/2 resin/SIS/oil/antioxidant composition, rolling ball tack values of 11 cm, polyken tack of 1400 g, loop tack of 16 N/25 mm, and peel resistance value of 18 N/25 mm (on stainless steel) have been obtained. High-solid content (50%), solvent-based rubber–resin adhesives (with a viscosity of 500 Pa s) were coated on PVC carrier for tapes. Packaging tapes are manufactured with water-based formulations as well. For such products with OPP, a carrier of medium coating weight (20–30 g/m^2) is suggested. Acrylic adhesives for packaging tapes offer the following advantages: water-borne, superior instantaneous cardboard adhesion, good shear resistance, good tack, clarity, UV stability, temperature resistance, and ready-to-coat status.

Packaging tapes are manufactured by coating. Their coating technology is similar to that of common labels (see also *Technology of Pressure-Sensitive Adhesives and Products*, Chapter 10). The data concerning current production speed for machines manufacturing packaging tapes are contradictory. A decade ago, coating machines for such tapes were running at 60–220 m/min (for solvent-based PSA) and 100–250 m/min (for water-based PSA); in fact, they are run at 250–500 m/min depending on the adhesive nature. The maximum production speed of such tapes increased from about 250 m/min (1985) to about 600 m/min (1995) and the coating machine width increased from ca. 1400 mm (1985) to about 2000 mm (1995). The average coating speed increased from about 100 m/min to about 300 m/min. The average coating machine width increased from 1300 to 1500 mm. On a new coating machine for tapes with a width of 1200 to 2400 mm, a 25-μm BOPP film is coated with HMPSA with a speed of 400 to 600 m/min.

For such tapes, special end-use and adhesive properties are required. For instance, shrinkage and elongation of the tape during unwinding and deapplication and low unwinding resistance (i.e., low peel resistance from the back side, good roll-down properties) are important. High tack toward corrugated paper board, high peel resistance (fresh and aged), and high shear resistance at various temperatures is also required. Weathering at low temperature is necessary. For instance, for packaging tapes a peel resistance value of 18 N/25 mm (on steel), loop tack values of 20–30 N/25 mm, and rolling ball tack values below 5 cm are preferred. Packaging tapes require shear resistance values higher than 680 min.

Packaging tapes for corrugated cardboard use PP as a carrier material and have a silicone release coating. Such coating is necessary because hot melts do not release easily from the nonadhesive side of the carrier material. For such packaging tapes on an HMPSA basis, adhesion (excellent tack and shear resistance on cardboard) and open-face aging are of special importance (see also Chapter 7). For adhesion on cardboard a flap test is carried out; this is a combination of shear, peel resistance, and tack and the preferred adhesion value is 100 min. The application temperature is situated between −40 and +100°F. Although at normal temperatures rubber–resin adhesives have better shear resistance than water-borne acrylics, acrylic adhesives display superior performance at 120°F.

There is a trend on the market to provide customers with printed packaging tapes. The text or graphics is imprinted on the nonadhesive side of the carrier material before the tape is made.

4.2.1.2 Closure Tapes

Various common and special tapes are used for closure. For bag closure tapes, sealability is a main requirement. Sealing tapes based on PVC carrier are printable. Sandwich printing (flexoprinting on the top side and gravure printing for the back side, between the film and the adhesive) is also possible. Such tapes are applied by hand, or by using a sealing machine. These products are used for closing, manufacturer identification, anti-theft protection, copyright protection, quality assurance, advertising, information, classification, and administration.

Bag lip tape (or bag closure tape) is applied to the lip of the bag. Bag closure tape must be cut with the same hot knife (hot wire) that is used to form the bag. It should be easily unwound for automatic applications. Tapes for closing PE bags, such as that used for nonwoven envelopes, must be reclosable. A double-side-coated PE tape was developed for such applications. A closure tape based on cross-linked acrylate terpolymer uses a reinforced web and a reinforcing filler. Such adhesive is useful for bonding the edges of a heat-recoverable sheet to one another to secure closure.

Tapes on cellulose hydrate or PVC as a carrier material were designed for box closures as competitors to wet tapes (see also Sections 4.2.2.9 and 4.2.4).

4.2.2 Special Tapes

The range of special tapes is varied. They can be classified according to their construction or their use (see also Chapter 1). Carrierless tapes and paper-, film-, or foam-based tapes and uncoated, one-side-coated, or double-side-coated products are manufactured. Insulation tapes, splicing tapes, mounting tapes, and tapes of many other types are produced. Such tapes could also be classified according to their carrier material. For the same carrier class, writable and printable tapes are considered special products. Textile-based tapes are applied on various substrates; insulation, medical applications, and packaging are their main end-use domains. The end use of tapes requires special adhesion properties according to the surface to be bonded. For instance, tapes for rubber profiles in automobiles must adhere to difficult surfaces such as ethylene–propylene–diene multipolymer (EPDM), CR, and SBR, according to VW Specification TL 52018. EPDM foams are difficult to tape; they provide only 2.4–3.6 N/25 mm peel resistance compared with foam failure with PE or PU foams. Tapes used in automobiles must fulfill special norms concerning volatiles and fogs (e.g., the PV3341 VW/Audi norm and the PBVWL709 Daimler norm) [69].

Acrylic adhesives are also used for special tapes. They are applied as solvent-based, water-based, or hot-melt formulations for tapes of various constructions and end uses. Solvent-based acrylics are used for special applications (e.g., medical [25], insulation, mounting, and removable [23]). They allow easy regulation of conformability, porosity, and removability. Adhesives that are resistant to perspiration and body liquids are suggested for personal-care products. Hot-melt acrylics are expensive products and are used mostly for special medical, sanitary, and mounting tapes. Radiation-cured 100% acrylics are recommended for transfer tapes and medical applications. The development of such cross-linkable and highly filled acrylics allowed the design of carrierless tapes with a foam-like character (see also Chapter 1). Some of the long-established applications for silicone PSAs are in industrial operations (masking, splicing, roller wrap) as well as the electrical and electronics, medical care and health care, and automotive markets [70]. Silicone PSA-based tapes are used in applications that take advantage of the unique features of both the substrate and the silicone PSA. Reference [70] lists the segments, applications, and key silicone adhesive requirements that have been the mainstay in this industry for the past 30 years. Such applications include packaging, electrical, and splicing tapes, among others.

4.2.2.1　Application Tapes

Application tapes are special paper- or plastic-based mounting tapes, used together with plotter films (see also Chapter 3) as removable mounting aids. Such mounting tapes help in placing signs, letters, decorative elements, etc., that are die-cut from plotter films. They are applied to ensure the transfer and temporary fixing of another permanent pressure-sensitive element (letters or written text) Such products are special carrier-based removable tapes. Paper and film carriers are used for the manufacture of application tapes. Special adhesive formulation and special coating methods are applied to ensure removability.

Paper- and film-based application tapes are manufactured. Paper premasking/application tapes are made from a latex-saturated paper, which is coated with a PSA.

In most cases, because low-adhesion adhesives are utilized, the decal premask/application tape does not require a release coating. Adhesion levels of 130 to 300 g/15 mm (PSTC-1) are typical. Because the adhesionto decal is usually a direct function of the area of the decal, larger decals commonly require lower adhesion products, whereas smaller decals, such as individual letters, require higher adhesion decal premask/application tapes. Premask/application tape has several functions. When applied to the decal, the tape protects the finished sign surface during succeeding stages, it provides a means of maintaining absolute registration of the prespaced lettering, whether die-cut or cut on computerized letter cutting equipment, and the application tape is essential for successful application of the decal. The application tape provides the "body" with which the thin, stretchy, expensive decal can be handled and protects the decal during the adhesion process when the decal is squeegeed to increase its adhesion to the final surface and eliminate trapped air.

As mentioned previously, application tapes are produced as paper- or film-based PSPs. Although they are used for the same application, the build-up of paper-based and film-based products is quite different. Products with a film carrier exhibit better conformability than products with a paper carrier. For excellent ability to lay flat, the carrier contains HDPE and is manufactured by coextrusion. To avoid curl if unwound by application, adhesion on the back side should be minimal. Because of the different surface quality of plotter films (based on calendered or cast PVC, cast PE, etc.), which function as an adherent surface for application tapes, a broad range of application films has been developed with different coating weights.

Solvent-based acrylics were suggested for application tapes. Water-based acrylics are also formulated as removable PSAs for application tapes. The choice of hard acrylic dispersions (see also *Technology of Pressure-Sensitive Adhesives and Products*, Chapter 1), as a base formulation for application films allows the design of an uncross-linked formulation for film-based products. Paper-based products are formulated with natural rubber (NR)–latex as the main component. Both grades of tape can be produced with or without a supplemental release layer.

The plastic-based tapes have PVC, PE, or PP carrier material that can be embossed to lower unwinding resistance. Plastic-based application tapes can also have paper release liners: these tapes are used for signs with small dimensions. Special application papers can also be used as stencil paper (see also Section 4.4). Film-based application tapes must be conformable but also removable. Therefore, gravure printing is favored as the manufacture (coating) method for such products. The pattern-like, striped surface structure of the adhesive provides less contact surface and better removability. The decreased contact surface influences the build-up of statical electricity. Such a coating can be unwound with less build-up of static charges.

4.2.2.2 Diaper Closure Tapes

Diaper closure tapes are refastenable closure systems for disposable diapers, incontinence garments, etc. Such systems must allow reliable closure and refastenability. Diapers include standing gather, frontal tape, and foamed waist elastics. Postexpansion adhesive sheets are used to cover odd-shape spaces between parts. These products include two- or three-tape systems. The two-tape systems include a release tape and a fastening tape. The fastening tape comprises a carrier material such as paper, PE, or PP. The preferred material is PP (50–150 µm) with a finely embossed pattern on each side. Such tapes allow reliable closure and refastenability from an embossed, corona-treated PE surface used for the diaper cover sheet.

Adhesives for diaper tapes must display high shear resistance. Cross-linking monomers (e.g., *N*-methylol acrylamide) are added in their formulation to increase shear resistance. Unsaturated dicarboxylic acids can be included also as cross-linking agents; polyfunctional unsaturated allyl monomers and di-, tri-, and tetrafunctional vinyl cross-linking agents have been suggested. The polyolefinically unsaturated cross-linking monomer enhancing shear resistance should have a T_g of -10°C or lower. Shear resistance is improved in water-based polymerization using a special stabilizer system comprising hydroxypropyl methyl cellulose and ethoxylated acetylenic glycol.

For such adhesives, tack values of 16–24 N/25 mm are acceptable. These tapes must exhibit a maximal peel force at a peel rate between 10 and 400 cm/min and a log peel rate between 1.0 and 2.6 cm/min. Tapes exhibiting these values are strongly preferred by consumers.

4.2.2.3 Electrical Tapes

Industrial tapes were introduced on the market in the 1920s and 1930s [71]. A large segment of these products is electrical tapes. Electrical tapes are used as conductive, insulation (wire wound), mounting, or packaging tapes for electrical products and electronic technologies.

The principles of formulation of electrically conductive PSPs are discussed in *Technology of Pressure-Sensitive Adhesives and Products*, Chapter 8. Generally, electrical conductivity is achieved using special fillers. Adhesives containing so-called extrinsic conductive polymers may exhibit anisotropic conductivity. Tapes with adequate electrical conductivity are manufactured using a metal carrier (nickel foil) and an adhesive with nickel particles as filler (3–20%) in an acrylic emulsion [72]. Such electrically conductive adhesive tapes were manufactured by coating an acrylic emulsion containing 3–20% nickel particles onto a rough flexible material (e.g., Ni foil). They can also contain silicium carbide as a filler [73]. Silver-coated glass powder (20%) or glass microbubbles (33%) used as filler lead to a specific surface conductivity of 1.1×10^{-11}–7.6×10^{-5} G [58]. The adhesive may also include a metallic network to ensure electrical conductivity [74]. Electrolyte-based conductive formulations are used in the medical domain (see also Section 4.2.2.6).

PVC with silicium carbide as a filler is recommended as a carrier for electrically conductive tapes. Electrically conductive polyolefins with a high loading of carbon black (30% w/w) have also been proposed. Transparent electroconductive films must be capable of the smooth writing demanded by pen input screens and touch panels. They have good transparency and surface conductivity.

Electrical tapes based on silicone PSAs are used for coil winding, high-temperature duct sealing, and lead and ground insulation due to their heat resistance up to 180°C, high shear resistance and holding power, and dielectric properties [70]. Electric insulation tapes will be described in Section 4.2.2.5.

Applications for PSAs in the electronics industry were the subject of a paper by Ding and Zhang [75], which describes the use of silicone PSAs and tapes in the printed circuit board (PCB) production and assembly process. Silicone PSAs are primarily used because of their unique ability to withstand high temperatures, conform to the topography of PCB boards, maintain adhesion to wet low-energy surfaces, and exhibit excellent dielectric properties. The use of silicone PSA tapes in various steps of PCB production was described. Excellent chemical resistance to plating solutions was another critical attribute of silicone PSAs for use in the gold-plating process (commonly known as "gold finger plating").

Dicing tape is adopted to tightly hold silicon wafers during dicing in the fabrication of semiconductor chips. Upon mounting the silicon wafer on the dicing tape, it must hold the wafer tightly otherwise the diced chip scatters in the dicing step. In the picking-up process, the adhesion strength of the dicing tape decreases so that the diced chip can be picked up easily [76].

4.2.2.4 Freezer Tapes

The application temperature and atmospheric humidity during lamination and end use of the tapes vary widely. Special application conditions are given for freezer, medical, insulating, sealing, and splicing tapes. Generally, tapes should be applied between 20 and 30°C, but butt-splicing tapes, for instance, must resist up to 117°C. Products designed for short-term outdoor application should not be used for more than 3 days under such conditions. If the adhesive is to be used in a bathroom (e.g., for adhering soap dishes and utility racks to ceramic tile), the adhesive must adhere well to tile, have high cohesive strength, withstand temperatures in the 110–120°F range, and endure 100% relative humidity. Because of the extreme working conditions of several tapes, their test methods use elevated temperatures (see also Chapter 7).

Special tapes are designed for low-temperature application. Like freezer labels, freezer tapes are used at low temperatures (see also Section 4.1.2.2). They are applied as packaging tapes. Freezer tapes can be coated with an adhesive based on NR, SBR, regenerated rubber, rosin ester, or polyether. They are based on primed, soft PVC and must resist temperatures between −29 and +113°C. They can possess a complex construction also. For instance, a carrier for tapes for low-temperature application may have transverse cuts or holes.

4.2.2.5 Insulation Tapes

Insulating tapes are used for a variety of applications, mainly for electrical or heat insulation or as sealants. Insulation tapes must ensure thermal, electrical, sound, vibration, humidity, chemical, or other type of insulation. Depending on their application field, they are made with various carrier materials that have different thermal, electrical, and mechanical properties. Well-defined mechanical properties are necessary for insulation tapes that ensure elongation, shear, and peel resistance of the tape. In some cases, shear resistance should be measured after solvent exposure of the PSP. Thermal resistance and low-temperature conformability and flexibility are also required. For instance, electrically insulating tapes are based on PET, polyimide, PET/nonwoven, glass cloth, paper, PET/glass fiber, and other carriers. Wire-wound tapes, electrically insulating tapes, and thermally insulating tapes are the most important representatives of this product class.

Insulating tapes are generally carrier-based, adhesive-coated products. Carrierless (extruded) or adhesiveless (extruded) products are also available.

4.2.2.5.1 Electrical Insulation Tapes

As noted previously, electrical tapes must possess high dielectric strength and good thermal dissipation properties. Electrical insulation tapes (wire-wound tapes) are used for taping generator motors and coils and transformers, where the tape serves as an overwrap, layer insulation, or connection and lead-in tape. The carrier is woven glass cloth, impregnated with a high-temperature-resistant polyester resin, or woven polyester-glass cloth, impregnated with polyester resin. The latter variant is used when conformability is required. Resistance to delamination and tear resistance of impregnated fabric- or nonwoven-based tapes are also required. Oriented LDPE (80–110 μm) was also suggested for insulation tapes. For wire-wound tapes carrier deformability is needed. They must adhere to modified flame- and heat-resistant substrates. In general, the adhesive and the carrier for such tapes possess improved temperature and flame resistance.

The limits of temperature resistance vary depending on the tape application. For instance, an insulating tape with an acetate carrier used for wire-winding by a telephone manufacturer resists temperatures up to 130°C and must adhere to modified flame- and heat-resistant EPDM substrates. High-temperature-resistant insulating tapes on a PET basis can be used at 180°C and an electrical tension of 5000 V. Cable tapes must resist 1 month at 70°C. Such a cable tape is made with butadiene rubber (more than 90%

cis-butadiene content), an SBC (SIS or SBS), and a tackifier. This blend exhibits a T_g of −104°C. The peel resistance at −10°C is 790 g/15 mm and at +40°C it is 370 g/15 mm. The creep is measured at −5 and +40°C as well. The rolling ball tack changes from 18 to 28 cm between 3 and −5°C. According to Ref. [77], insulation tapes must resist temperatures between 5 and 60°C. Various tapes used in the automotive industry are often subjected to extremely high temperatures. In a closed automobile in the sun, the temperature can reach 100–120°C. Car undercarriage protection and insulation tapes must resist 20 h at 150°C. Their resistance to water or solvents is tested by immersion in liquid for periods up to 1 week.

Electrical insulating tapes must possess high dielectric strength and good thermal dissipation properties. The edges of electrical insulating tapes used for taping must be straight and unbroken to provide distinct boundaries of the finished areas when the taping is complete. Such tapes may be perforated.

Electrical insulating tapes are coated with an acrylic PSA [78]. Certain electrical insulation tapes must resist transformer oil; therefore, cross-linked acrylics are used for such products [79]. Nonflammable tapes can have an adhesive with 0–40% polychloroprene rubber [80]. Heat-cross-linkable, pressure-sensitive insulating tapes are manufactured using an adhesive based on an aromatic polyester coated on a textile carrier. Silicone polymers have excellent electrical properties, such as arc resistance, high electrical strength, low loss factors and resistance to current leakage, low surface energy, and excellent, low- and high-temperature performance (see also *Technology of Pressure-Sensitive Adhesives and Products*, Chapter 6). These characteristics are very important for electrical insulating tapes. Electrical tapes must be approved by Underwriter Laboratories.

4.2.2.5.2 Pipe Insulation Tapes

Pipe insulation (pipe wrapping) tapes are used for gas pipelines [81]. Cross-linked butyl rubber-based tapes for pipe insulation have been manufactured since 1963. Pipe insulation tapes are wounded by overlapping (50%) using special machines [82]. Pipe-wrapping tapes are plastic tapes reinforced with nylon cord and must adhere to metallic surfaces, wood, rubber, or ceramics, as well to oiled surfaces [83]. They are designed to offer puncture and tear resistance [84].

A mixture of polyisobutylenes (PIB) with molecular weights of 100,000 and 1500 can be used as a base recipe with rosin as a tackifier. For instance, a mixture of PIB (100 pts) with a molecular weight of 1000, low-molecular-weight PIB (120 pts) with a molecular weight of 1500, and rosin (150 pts) was proposed. Pipe insulation tapes with PVC carrier are coated with an adhesive based on NR, crepe (27 pts), ZnO (20 pts), rosin (22 pts), hydrated abietic acid (8 pts), polymerized trimethyl dihydrochinoline (2 pts), and lanoline (4 pts). Double-faced foam tapes can also be applied as pipe insulation tape for gas pipelines [85].

Pressure-sensitive insulation tapes are also manufactured as anticorrosion protection tapes [86]. Tapes applied as corrosion protection for steel pipes contain butyl rubber, cross-linked butyl rubber, regenerated rubber, tackifier (polybutene or resins), filler, and antioxidants. Low-molecular-weight PIBs (M_n less than 600,000) give better tack and peel [87]. Generally, blends of different molecular weights are used. Faktis improves the creep resistance of such tapes. Insulating tapes for gas and oil pipelines are manufactured by coextrusion of butyl rubber and a PE carrier. Bilayered heat-shrinkable insulating tapes for anticorrosion protection of petroleum and gas pipelines have been manufactured from photochemically cured LDPE with an EVAc copolymer as an adhesive sublayer [88]. The liner dimensions of the two-layer insulating tape decrease by 10–50% depending on the degree of curing (application temperature 180°C). The application conditions and physicomechanical properties of the coating were determined for heat-shrinkable tapes with 5% shrinkage at a curing degree of 30% and shrinkage force of 0.07 MPa.

4.2.2.5.3 Thermal Insulation Tapes

Double-side-coated foam tapes are designed for industrial gasketing applications. They are based on closed-cell PE foams and coated with a high tack, medium shear resistance adhesive. Air-sealing tape is made with butyl rubber on ethylene–propylene foam. Foamable insulation tapes also exist. For these products the PU foam is bonded *in situ* onto the (specially treated) carrier backside. An aluminum carrier and a tackified carboxylated butadiene rubber coated with 40 g/m² PSA are used for insulating tape in air-conditioning [89].

Special insulating tapes must adhere to refrigerated surfaces. For instance, a self-sticking tape for thermal insulation used to secure a tight connection between heat exchange members (metal foils) and PU foam-based thermal insulation materials is prepared by applying an adhesive that can adhere to refrigerated surfaces on an olefin copolymer carrier material (containing release) that is bonded to the PU [90].

Low-temperature-curable silicone PSAs have been formulated from gum-like silicones-containing alkenyl groups, a tackifying silicone resin, a curing agent, and a platinum catalyst [91] (see also *Technology of Pressure-Sensitive Adhesives and Products*, Chapter 6). Such special (temperature-, chemical-, and humidity-resistant) adhesives for heat-resistant aluminum tapes are coated with a thickness of 50 µm and cure in 5 min at 80°C. According to Ref. [92], heat-resistant insulating tape formulation can include a fluoropolymer. Such tape displays a rolling ball tack of 14 cm and peel resistance value of 45 g/20 mm. High-temperature-resistant polyimide films have been proposed as carriers for insulating tapes. Such films are flame and temperature resistant between −269 and +400°C. For instance, high-temperature (300°C)-resistant special tapes for the PCB industry are based on polyimide carriers [93]. Such polyimide materials have been recommended for transparent pressure-sensitive sheets coated with silicone-based adhesives, which resist 8 h at 200°C.

4.2.2.6 Medical Tapes

A special chapter of adhesive tape formulation concerns medical tapes. PSAs have been used since the late 1800s for medical tapes and dressing. Medical tapes have been developed for pharmaceutical companies, ostomy appliances, diagnostic apparatus, surgical grounding pads, transdermal drug delivery (TDD) systems, and wound-care products (see also Section 4.1.2.3).

4.2.2.6.1 Requirements for Medical Pressure-Sensitive Products

Table 4.3 summarizes the requirements for medical PSPs. Medical labels are used for wound care or in the operating room. Surgical labels must adhere well on the skin and not lose adhesiveness in hot or humid environments. No adhesive transfer should occur during removal of the adhesive from the skin. When the label is removed, no residue is left on the product to attract microorganisms. Medical labels can be defined as sterilization labels. For certain health care applications, pressure-sensitive labels are exposed to sterilization or autoclaving. Gas-sterilizable packages must be resistant to ethylene oxide. Such requirements must be taken into account by the choice of the carrier material. For instance, polyester laminated with PP can be used in medical applications that utilize hot-steam (134°C) sterilization. Typical examples of conformable carrier materials used for medical labels and tapes include nonwoven fabric, woven fabric, and medium- to low-tensile-modulus plastic films (PE, PVC, PU, PET, and ethyl cellulose). For conformability the films should have a tensile modulus of less than about 400,000 psi (in accordance with ASTM D-638 and D-882).

Adhesives for medical products must be physiologically compatible. Such adhesives are based on PU, silicones, acrylics, PVE, and styrene copolymers. According to Ref. [94], adhesives for medical tapes must fulfill the following requirements: they should not irritate the skin, they should adhere well but be easily removable, and they should be water resistant. For low skin irritation they should be breathable; they minimize skin irritation due to their ability to transmit air and moisture through the adhesive system. They should be cohesive, conformable, self-supporting, and nondegradable by body fluids. According to Ref. [95], skin tolerance, no physiological effects, resistance to skin humidity, and sterilizing capability without color changes are required. Certain applications require water-soluble or cross-linked adhesives. According to Ref. [96], the main performance characteristics for medical tapes are the conformability of PSA, its initial skin adhesion value, the skin adhesion value after 24–48 h, and adhesive transfer. A preferred skin adhesive will generally exhibit an initial peel resistance value of 50–100 g and final peel resistance value (after 48 h) of 150–300 g.

4.2.2.6.2 Adhesives for Medical Pressure-Sensitive Products

Key requirements for PSAs in biomedical applications are summarized by Subramony in Chapter 2. Like medical tapes and bioelectrodes (see also Section 4.2.2.6), medical labels have a special, conformable, removable, porous, skin-tolerant adhesive, with well-defined water solubility and electrical properties and a special, porous carrier material or a carrierless construction. The adhesive used for bioelectrodes applied to the skin contains water, which affords electrical conductivity. Electrical performances can be improved by adding electrolytes to the water. Adhesive balance, creep compliance (stretchiness), and elasticity are other properties required for medical PSAs. According to Chang (see *Fundamentals of Pressure Sensitivity*, Chapter 5), removable and medical-type PSAs fall within the application window quadrant for low G'-low G". This quadrant corresponds to low modulus and low dissipation. There is a difference between common removable and medical adhesives. The reference temperature for medical adhesives is the body temperature of 37°C, compared with 23°C in the removable case. This makes the bonding modulus of the medical adhesives even lower (i.e., more conformable) than that of removables because of the higher reference temperature. This is desirable for contact area considerations because of the rough, frequently varied, and contaminated nature of the skin. On the other hand, the debonding moduli (top right-hand corner of the window) are usually higher than that of the removables. Again, this is necessary to prevent lift or detachment because of frequent flexing of the skin, especially on curved areas such as knees and elbows.

"Historical" Recipes   The earliest medical labels and tapes were based on mixtures of NR, plasticized and tackified with wood resin derivatives and turpentine and pigmented with zinc oxide. The first adhesives for medical tapes were based on the reaction product of olive oil and lead oxide (lead oleate), together with tackifier resins and waxes. Lead oleate was used as an aseptic adhesive component for other medical tapes.

Later, NR tackified with rosin or PIB was recommended for adhesives for medical tapes. The earliest plaster-like, medical products were crude mixtures of masticated NR, tackified with rosin derivatives and turpentine, and filled with zinc oxide as pigment. Such formulation contains NR (100 pts), a terpene-phenol resin (80 pts), and a solvent (600 pts). Such compounds are hot pressed (see also *Technology of Pressure-Sensitive Adhesives and Products*, Chapter 10) in the fibrous carrier material; therefore, cloth served as the carrier material. Later, extensible, deformable paper was used. Now plastic films and plastic-based textile-materials are used as carriers. A broad range of macromolecular compounds are tested as adhesives.

In this application field, rubber- or PVE-based formulations compete with special acrylics, thermoplastic elastomers, PUs, or silicones. The irritation caused by the removal of the label was overcome by including certain amine salts in the adhesive. A typical formulation of rubber-based surgical tape is given in *Technology of Pressure-Sensitive Adhesives and Products*, Chapter 2.

Unfortunately, many of these compositions lose their adhesiveness in hot or humid environments and allow adhesive transfer during removal. Advances in macromolecular chemistry allowed the synthesis and characterization of pressure-sensitive hydrogels (see also *Technology of Pressure-Sensitive Adhesives and Products*, Chapter 8), which are biocompatible polymers and can be used as PSAs for various medical products (see also *Technology of Pressure-Sensitive Adhesives and Products*, Chapter 8).

Polyvinyl Ethers   PVEs do not produce skin irritation; therefore, they are suggested for medical application [97]. PVE is suggested for medical tapes alone or with polybutylene and titane dioxide as fillers. These compounds are warm coated without solvent (using coating by calendaring; see also *Technology of Pressure-Sensitive Adhesives and Products*, Chapter 10). By cooling the roll-pressed mass, a porous adhesive is achieved [98]. PVE processed as a hot melt can be used for double-side tapes with a textile carrier [99]. PVEs can also be coated by extrusion. The water vapor porosity of PVE is about the same as for skin, 259 $g/m^2/24$ h. Radiation curing has been proposed for the manufacture of medical tapes. This method is limited by the coating weight. Coating of UV-cured adhesives is practicable up to 70 g/m^2 [100]. UV-curable formulations can use difunctional or polyfunctional vinyl ethers for increased adhesive bond strengths after curing [101].

Polyisobutene and Synthetic Elastomers   PIB is used for medical tapes. According to Ref. [102], PIB (MW 80,000–100,000) is used for medical tapes because it does not adhere to the skin. This type of tackified PIB is plasticized at 80°C. Unlike many natural and synthetic elastomers, butyl rubber (BR; MW of 6000 to 20,000) does not need mastication to be dissolved. It is clear and tacky, it is not fragile after aging, and it can be used (together with PIB as tackifier) for medical tapes. BR (15 to 55 pts), untackified or tackified with PIB (7.5 to 40 pts) and cross-linked with zinc oxide (45 pts), has been proposed [103]. Such formulations give (180°C) peel values of 0.73 kg/25.4 mm to 1.81 kg/25.4 mm. Peel increases with the increase of the tackifier level, but at a tackifier/elastomer ratio of 40/15, legging appears. PIB or a blend of PIB (5–30% w/w) and BR, a radial SBC block copolymer (3.0–20% w/w), mineral oil (8.0–40% w/w), water-soluble hydrocolloids (15% w/w), tackifier (7.5–15% w/w), and a swellable cohesive strengthening agent are used to formulate a medical adhesive [104].

A mixture of *cis*-polybutadiene, *ABA* block copolymer, and tackifier resins was proposed for medical adhesive. Such an adhesive mixed with an anti-inflammatory paste is spread on a film carrier material for medical poultices.

Oil–gel-type adhesives based on SBCs are described by Hu and Paul in *Technology of Pressure-Sensitive Adhesives and Products*, Chapter 3. Table 3.9 in *Technology of Pressure-Sensitive Adhesives and Products* presents an example of an oil–gel adhesive and its skin-related performance.

Acrylics The past 2 decades have witnessed increased interest in bioadhesive polymers [105,106]. Bioadhesion involves the attachment of a natural or synthetic polymer to a biological substrate. Bioadhesion research investigated the adhesion of thin films and microparticles based on hydrophilic polymers like polyanhydrides and poly(acrylic acid) polymers [107,108]. There is a growing demand for bioadhesives that can be easily delivered and solidify *in situ* to form strong and durable bonds. Some of the potential applications for such biomaterials include tissue adhesives, bonding agents for implants, and drug delivery. It is also a necessity to prepare these adhesives in a toxicologically acceptable solvent that enables injection to the desired site and permits a conformal matching of the desired geometry [109]. Bioadhesives are discussed in detail by Feldstein et al. in *Technology of Pressure-Sensitive Adhesives and Products*, Chapter 7.

According to Ref. [110], commercial PSAs used for skin application are based on acrylic adhesives, which are easier to remove and cause less skin irritation than rubber–zinc oxide adhesives, which adhere well to the skin but may cause skin irritation because of their higher adhesion level. On the other hand, skin is a very rough surface with a hydrophobic exterior. Sweat and natural oils are exuded from its pores, which can lead to gradual debonding. Acrylic solutions and PUR solutions are both used. Both adhesives exhibit a continuous decrease of G' as a function of temperature; acrylics used for finger bandage show a more abrupt decrease. Their tan δ peak$_{max}$ is shifted to higher temperatures. Air- and moisture-permeable nonwovens or air-permeable tissue are used for medical tapes, coated with a porous, nonsensitizing acrylic adhesive for prolonged application of a medical device on human skin. Gels based on hydroxyethyl methacrylate have been used for medical applications for many decades, but they do not possess sufficient mechanical strength and cannot be mechanically processed [111].

Such materials minimize skin iritation due to their ability to transmit air and moisture through the adhesive system. The air porosity rate must attain a given value

(50 s/100 cc/in.^{2}) [112]. Acrylic adhesive compositions for medical tapes that do not leave adhesive residues on skin contain *tert* BA–maleic anhydride–VAc with a molecular weight of 2500 to 3000 and acrylic acid (AA)–BA copolymers [113].

Acrylic PSAs in medical tapes may display adhesion build-up over time or weakening of the cohesive strength due to migration of oils. These disadvantages are avoided by cross-linking with built-in dimethylaminomethacrylate or isocyanate. For a composition containing polar comonomers and built-in (carboxyl or hydroxyl group-containing) cross-linking comonomers, a coating weight of 34–68 g/m^2 has been suggested. Polyacrylate-based water-soluble PSAs are suggested for medical PSPs (operating tapes, labels, and bioelectrodes) [23]. An acrylate-based skin adhesive should have a creep compliance value of at least 1.2×10^{-3} [114]. Guvendiren et al. [115] prepared a modified acrylic triblock copolymer system that can be fully dissolved in toxicologically acceptable organic solvents. Hydrogels are formed by a solvent exchange mechanism when the solution is exposed to water naturally present within the body.

Neat acrylic block copolymers of methyl methacrylate (MMA)/BA/MMA block copolymer (MAM) type, based on MMA and BA, are described by Hu and Paul in *Technology of Pressure-Sensitive Adhesives and Products*, Chapter 3. According to Smit et al. [116], acrylates have a high moisture vapor transmission rate (MVTR). Hot-melt acrylates offer the advantage of low capital and operating costs and high line speed. Smit et al. [116] synthesized acrylic block copolymers (see also *Technology of Pressure-Sensitive Adhesives and Products*, Chapter 1), which can be formulated as a hot melt. They compare MVTR their with solution acrylics and SIS-based medical PSAs. In *Technology of Pressure-Sensitive Adhesives and Products*, Chapter 3, Hu and Paul compare MVTR values of two acrylic block copolymer based hot-melt adhesives with a solution of acrylic medical adhesive (PSA-M) and a SIS-based medical grade (see *Technology of Pressure-Sensitive Adhesives and Products*, Figure 3.21). Acrylic-based formulas have an MVTR value between 950–1100 g/m^2 day; SIS-based HMPSAs have an MVTR less than 100. Good long-term wear, MVTR, 4-day wear on a permeable backing, and 24-h adhesion levels were far superior for acrylic HMPSAs compared with SIS-based adhesives. No adhesive transfer was obeserved. Adhesion to permeable medical backing was excellent. Cytotoxicity was tested as well (see also *Technology of Pressure-Sensitive Adhesives and Products*, Chapter 3). For medical tapes, formulas with exceptional softness and fast wet-out have been obtained that provide excellent skin grab and long-term wear with painless removal. Such HMPSAs display improved heat resistance and thermooxidative stability.

Unfortunately, many of these compositions lose their adhesiveness in hot/humid environments and display adhesive transfer during removal. Adhesive polymers used for surgical adhesive tapes often exhibit a dynamic modulus that is too low for outstanding wear performance. A low storage and loss modulus result in a soft adhesive with adhesive transfer. Adhesives suitable for use on human skin should display a storage modulus of 1.0 to 2.0 N/cm^2, a dynamic loss modulus of 0.60 to 0.9 N/cm^2, and a modulus ratio of tan δ of 0.4 to 0.6 as determined at an oscillation frequency sweep of 1 rad/s at 25% strain rate and a body temperature of 36°C.

Polyurethane-Based Pressure-Sensitive Adhesives PU solutions are also used for medical PSAs. As noted in Ref. [117], different manner rubber formulations, which leave deposits on the skin, and acrylic formulations, which may irritate the skin, PU gels are inoffensive. According to Ref. [118], the PU-based PSA used for medical tapes includes a polyethylene polyol, polyisocyanate, resin, catalyst, and wetting agent. The type of polyethylene polyol and its functionality, structure, and hydroxy number are the main parameters. Isophorone diisocyanate, trimethylhexamethylene diisocyanate, hexamethylene diisocyanate, and naphthylene-1,5-diisocyanate are recommended as isocyanate components. Hydrocarbon resins, modified rosin, polyterpene, terpene-phenol, coumarone-indene, aldehyde- and keton-based resins, and β-pinene resins are suggested as tackifiers. Sn(II) octoate is used as a catalyst. The components of the formulation are mixed as follows: 100 pts polyethylene polyol, 50 pts resin, 1 pt catalyst, 7 pts diisocyanate, and 1 pt wetting agent. The resin is dissolved in the polyethylene polyol. This solution is blended with the isocyanate in an extruder, coated (closed blade), and cross-linked (120°C, 1.5–2.5 min). Special PUs have been suggested for medical tapes. Generally, the type of polyethylene glycol, its functionality (2–4), its structure (linear or branched), and its OH number influence its properties. For the polymer proposed in Ref. [117], with average functionality (F_I) of the polyisocyanates between 2 and 4 and the average functionality of the polyhydroxy compounds (F_P) between 3 and 6, the isocyanate number (K) is given by the following formula:

$$K = (300 \pm X)/(F_I F_P - 1) + 7 \qquad (4.1)$$

where $X \leq 120$ (preferably $X \leq 100$) and K is between 15 and 70.

Silicones Silicones have a long history of successful use in health care applications, first as silicone release agents for organic adhesives and more recently as silicone PSAs in applications ranging from wound management to biomedical devices to pharmaceuticals [70]. Although many of the properties developed for industrial PSAs are also important for medical PSAs, the latter have many additional requirements primarily driven by their intended and necessary use against human skin. Ulman and Thomas [119] discussed the use of silicone PSAs in health care applications. An excellent and exhaustive review by Lin et al. [70] concerning silicone-based PSAs includes their medical applications. The silicone chemistry set offers a range of materials and properties to help meet general requirements such as biocompatibility (e.g., biologically inert, nontoxic, nonirritating, and nonsensitizing), efficacy (e.g., suitability for use on skin and permeability to therapeutic substances), stability (e.g., good chemical inertness and retention of physicochemical properties at skin temperatures), and a wide range of formulation possibilities to meet specific application needs.

According to Ref. [70], the "standard" medical PSA consists of a condensed polydimethylsiloxane and methylsiloxane (MQ) resin network that can be further treated with a silane capping agent to yield an "endcapped amine-compatible" PSA. Some applications, such as drug delivery, require formulation with additives such as cosolvents, excipients, drugs, or skin-penetration enhancers to adjust the cold-flow characteristics of the adhesive. One method of raising resistance to cold flow is through the use of cohesive strengthening agents such as calcium stearate, magnesium stearate, or ethyl cellulose. Typical optimal loadings are in the 10 to 20% range. Silicone fluids act as plasticizers and lower the melting temperature of the adhesive solids matrix, allowing the adhesive to be delivered as a hot-melt formulation rather than as the more common dispersion in an organic solvent.

Alternatively, cross-linkable formulations can be used. These formulations can be designed to cure at room temperature or elevated temperature, as two-part cure systems, and as very soft, cohesive silicone adhesive gels. Silicone adhesives are generally coated and cured onto a release liner and then laminated to the backing substrate. Historically, it was thought that silicone adhesive gels had to be coated and cured directly onto the backing substrate to eliminate potential chemical interaction with silicone-based release liners. Recent developments demonstrated that transfer of the silicone adhesive gels from a release liner is possible when the backing substrate has been prepared with a titanate primer [120]. Titanate primers also act as adhesion promoters for silicone gels by increasing the anchorage of the gel to a given substrate. Gantner et al. [121] created silicone gel formulations for medical devices, including a hydroxy-substituted siloxane resin that does not require priming or surface treatments of the backing substrate to achieve adequate adhesion.

The use of silicone PSAs in the health care arena continues to expand and optimize in response to specific application needs. A number of publications describe compositions and constructions designed to provide specific benefits, such as efficient drug delivery, improved wound management/healing, greater friendliness to human skin, and resistance to biological fluids [122–132]. General trends in wound-care adhesive use have carried over into silicone adhesives. According to Posten [133], silicone gel sheeting was superior to other occlusive dressings in the treatment and management of scars. The market has witnessed a proliferation of scar therapies utilizing silicone gel sheeting. Several professional wound dressings are currently available that use silicone gels as the skin-contact adhesive. The need for medical dressings utilizing adhesives that can be applied to fragile skin types without causing irritation and can then be removed without traumatizing delicate skin may be highlighted by the aging population and the special needs of geriatric patients. However, this technology also finds utility in the treatment of neonatal and other chronic wound sufferers. Silicone gel dressings have low peel forces from skin and have been used effectively in these applications [134–138]. Shorter-than-desired wear times contributed to the limited acceptance of silicone gels in the wider market. The resin-containing silicone gel formulations reported by Gantner et al. [121] also have higher adhesion to skin, which should translate into longer wear times than previously achievable with traditional silicone gels, while maintaining the characteristic low peel forces from skin. Stempel et al. [124] combined a cross-linked gel and a silicone PSA to yield a blend displaying aggressive adhesion to steel and ostomy pouches. The trend toward skin-friendly alternatives received some attention in the drug delivery area. Developments in silicones and their use for health care products is discussed in detail by Shaow et al. in *Technology of Pressure-Sensitive Adhesives and Products*, Chapter 6.

Hydrogels A special class of biocompatible medical products was developed with pressure-sensitive hydrogels. The theoretical aspects of the design of raw materials for such medical PSPs are discussed by Feldstein et al. in *Technology of Pressure-Sensitive Adhesives and Products*, Chapter 7. As noted in Ref. [14], the special Corplex™ technology based on the molecular design of such PSAs allows the manufacture of (label-like or tape-like) products with varying adhesion, mechanical, and water-absorbing capabilities. Water-soluble, superabsorbent, medium absorbent, and weak absorbent PSPs with water absorbency of 10,000 to 100% and swelling ratio of 100 to 1 were developed. According to such characteristics, solid-state, film forming, liquid fill, dissolving film, and aqueous gel-like products can be manufactured and suggested for dry skin/wet mucosal and transdermal/transdermal mucosal or mucosal applications. Current commercial TDD systems use hydrophobic PSAs (see *Technology of Pressure-Sensitive Adhesives and Products*, Chapter 8). Bioadhesives should be hydrophylic. Typical bioadhesives include compositions of slightly cross-linked polyacrylic acid, as well as blends of hydrophylic cellulose derivatives and polyethylene glycol. Unfortunately, the cohesive strength of such highly swollen bioadhesives is generally low. Pressure-sensitive hydrogels according to the Corplex technology (i.e., on ladder-like or carcass-like simultaneously plasticized and cross-linked hydrogels) share properties of both hydrophobic and hydrophylic PSAs. Such adhesives may be also applied to skin, mucosa, or dental tissue as film-forming liquids. Thus, the adhesive can be formulated in transdermal and transmucosal drug delivery and wound care or tooth whitening [14].

Another possibility is the reinforcing of the biocompatible PSA with nanodimensional carriers, for example, liquid crystalline polymers and nanoparticles of NA-montmorillonite, according to the "nanocomposite concept" [15]. The fundamental aspects of the rheology of skin, correlated with the rheology of applied biocompatible PSA, are discussed by Kulichikhin et al. in Ref. [137].

The special new applications of medical products based on biocompatible pressure-sensitive hydrogels are described by Feldstein et al. in *Technology of Pressure-Sensitive Adhesives and Products*, Chapter 7.

4.2.2.6.3   Carrier Materials for Medical Pressure-Sensitive Products

Medical tapes are carrier-based products. Generally, they are manufactured by adhesive coating. An ideal soft tissue surrogate should possess a well-defined molecular weight and cross-link density to form identical gels from batch to batch and stable mechanical properties as a function of time and temperature that should mimic natural tissue mechanical properties. Previous research demonstrated that the elastic modulus of soft tissue spans a range from about 25 to 300 kPa [138]. Preferred carrier materials are those that permit transpiration and perspiration or the passage of tissue or wound exudate. They should have a moisture vapor transmission of at least 500 g/m^2 over 24 h at 38°C with a humidity differential of at least about 1000 g/m^2. Medical tapes require the adhesive, carrier, and tape construction as a whole to be porous. Because of the absorbtion, dosage, and storage function of such tapes, their special adhesives are coated on porous carrier materials with a high coating weight. Such adhesives do not contain volatiles. Therefore, as mentioned previously, medical tapes have been manufactured by roll pressing or coating of high-solid solutions and later by extrusion and radiation curing. The carrier must also meet air transmission criteria. A textile-based carrier, for example, cotton cloth, acts as a network that allows diffusion (of air, pharmaceutical agents, vital liquids, etc.).

Air- and moisture-permeable nonwovens (polyester nonwoven, embossed nonwoven, etc.) or air-permeable tissues are used for medical tapes for prolonged application of a medical device on human skin. Such products are developed for pharmaceuticals companies, ostonomy appliances, diagnostic apparatus, surgical grounding pads, TDD systems, and wound-care products.

The face stock material can repel, absorb, or contain chemicals, such as solid-state water-absorbent particles. For such application the face stock material serves as a dosage and storage function.

Conformability is also required for medical tapes. Extensible, deformable paper is required as a carrier for medical tapes. For such applications the direction of extensibility of the carrier is very important. For medical tapes cross-direction-elasticity may be given by a special nonwoven according to Ref. [139]. The wound plaster placed in the direction of elasticity on the carrier is narrower than the carrier. PVC has been introduced as a carrier for medical tapes as well. It can be used for self-adhesive medical tape or as an adhesive coating [117]. Soft PVC is used as a plaster (70–200 µm) with a silicone liner. Such films are dimensionally stable up to 200°C. Poly(vinyl isobutyl ether) and poly(vinyl alkyl ether) on cellophane are also used for medical tapes [140]. Polyolefins, for example, porous PE, are suggested as carriers for medical tapes to allow the diffusion of humidity and blood [141]. Special fillers allow the production of breathable polyolefin films. Cellulose acetate can be used as a porous carrier material as well. A special conformable carrier material contains a layer of a nonwoven web of randomly interfaced fibers, bonded to each other by a rewettable binder dispersed throughout, and at least one additional layer of the same composition. The fiber of the additional layer is laid directly on the first layer prior to addition of the second layer [142]. For instance, Rexam's Inspire line includes breathable components for wound care, for example, PU films, foams, adhesive, hydrogels, and composites [143]. A clean, fiberless tear or cutting section is required for medical tapes because of the contamination danger.

The preferred carrier materials are those that permit transpiration and perspiration or the passage of tissue or wound exudate. They should have a moisture vapor transmission of at least 500 g/m² over 24 h at 38°C (according to ASTM E 96-80), with a humidity differential of at least about 1000 g/m². A conventional polyethylene terephtalate film has an approximate value of 50 g/m² of moisture vapor transmission.

Nitrocellulose, PE, rubber, proteins, and silicones were proposed as a release layer for medical tapes. Embossed PVC film is recommended as a liner for plasters. For medical tapes a special release coating was proposed in Ref. [144]. Such formulations contain silicone rubber, a butoxytitanate–stannium octoate complex, oligomeric dimethyl-phenylpolysiloxane rubber, and triacetoxy methysilane. Polydimethylsiloxane–urea copolymer (a thermoplastic silicone elastomer) with elasticity and heat and cold resistance was developed for medical uses [145].

4.2.2.6.4 *Transdermal Applications*

Fluid-permeable adhesives are required for transdermal applications; such an adhesive must not irritate the skin. As discussed in Ref. [146], a fluid-permeable adhesive useful for applying transdermal therapeutic devices to human skin for periods up to 24 h is based on acrylic, urethane, or elastomer PSA mixed with a cross-linked polysiloxane. The therapeutic agent passes through the adhesive into the skin. Uninterrupted liquid flow through the adhesive must occur over a prolonged period (6–36 h) and at constant rate.

The skin adhesive must display an enhanced level of initial adhesion when applied to skin, but resists adhesion build-up over time. An antimicrobial agent such as iodine can be incorporated into the skin adhesive as a complex. The role of the adhesive as storage place for different end-use components may vary. The use of antiviral agents as fillers for the carrier or adhesive indicates new applications. Drug delivery and transdermal patches are discussed in detail by Subramony in Chapter 2. The requirements for such adhesive-dosage systems are described in detail in Chapters 5 and 6.

The versatility of silicones makes them a good adhesive choice for use in drug delivery applications [147,148]. Rapid growth in health care and self-applied treatment using medicines delivered via transdermal drug patches is undoubtedly expanding the use of silicone PSAs. The hydrophobic nature of silicone makes it an ideal carrier adhesive for nonpolar pharmaceutical agents, and some evidence implies that silicones demonstrate slightly faster release rates than acrylic adhesives [149]. Silicone PSAs have been used for some time in TDD therapies. Their use in these applications continues because of their desirable physicochemical properties and their ability to positively affect the release rates of many pharmaceutical and cosmeceutical actives. Silicone PSAs offer excellent solubility and permeability to lipophilic drugs, but in some cases it is desirable to deliver hydrophilic drugs. The adhesive can be modified by formulation with hydrophilic fillers, copolymers, or plasticizers or by modification of the network with silicone–organic copolymers. Silicone polyethers demonstrated special utility in facilitating the compatibility of powdered hydrophilic drugs with silicone adhesive matrices [150]. They also appear to positively impact the delivery of lipophilic drugs from these matrices. Bruner et al. [147] documented the use of silicone PSAs versus tacky gels to optimize drug delivery performance. Murphy et al. [123] added antiperspirant compounds to silicone PSA formulations and silicone gels to create a new composition that will inhibit perspiration during cosmetic and medical uses and thus retain the adhesive strength of the adhesive.

The use of methacrylic polyeletrolyte derivatives as diffusion matrices of transdermal therapeutic systems is discussed by Feldstein et al. in *Technology of Pressure-Sensitive Adhesives and Products*, Chapter 7.

A flow chart for the manufacture of transdermal therapeutic systems (based on HMP-SAs compounded in an extruder) is presented in Ref. [151]. Trenor et al. [152] examined how skin surfactants influence a model PIB PSA for TDD. These skin softeners can also plasticize the adhesive, thus lowering the T_g and reducing the peel strength. To evaluate the role skin surfactants play in affecting mechanical properties of PSAs used in TDD systems, Lee and McCarthy [153] synthesized butadiene maleic anhydride copolymer (BMA), which demonstrated superior bioadhesive properties and enabled the design of controlled-release drug delivery systems. They also studied the adsorption behavior of BMA and its reactivity with nucleophiles on Si wafers to elucidate the general mechanism for modification of anhydride-containing polymers.

4.2.2.6.5 Bioelectrodes

Special tapes are used for bioelectrodes because of their low electrical impedance and conformability to skin. Such products are manufactured by photopolymerization [154]. Compositions include AA polymerized in a water-soluble polyhydric alcohol and cross-linked with a multifunctional unsaturated monomer. A prepolymer is synthesized that is polymerized to the final product using UV radiation. This polymer contains water, which affords electrical conductivity. Electrically conductive PSAs can be manufatured as gels containing water, NaCl, and NaOH [155]. Such gels are useful in detecting voltages of living bodies. Such polymers based on maleic acid derivatives have a specifical resistivity of 5 kOhm (1 Hz).

4.2.2.6.6 Surgical Products

Surgical PSA sheet products include any product that has a flexible carrier material and a PSA. Labels, tapes, adhesive bandages, adhesive plasters, adhesive surgical sheets, adhesive corn plaster, and adhesive absorbent dressings were manufactured. First, natural adhesives mixed with natural additives were used for such products.

Surgical (operation) tapes are used for fastening cover materials during surgery. Cotton cloth and hydrophobic special textile materials coated with low-energy polyfluorocarbon resins are used as carrier materials, together with polyacrylate-based water-soluble PSAs. Such adhesive must be insensitive to moisture and insoluble in cold water. It must display adequate skin adhesion and skin tolerance. The water solubility of the product is reached at 65°C and a pH value of more than 9. Water solubility is given by a special composition containing vinyl carboxylic acid (10–80% w/w) neutralized with an alkanolamine [156]. Other compositions contain acrylic acid polymerized in a water-soluble polyhydric alcohol and are cross-linked with a multifunctional, unsaturated monomer. A prepolymer (precursor) can also be synthesized, which is polymerized to the final product by using UV radiation, according to Ref. [157]. This polymer contains water, which affords electrical conductivity. Electrical performances can be improved by adding electrolytes to the water. Such products can be used as biomedical electrodes applied to the skin. The adhesives used for surgical tapes often exhibit a dynamic modulus that is too low for outstanding wear performance. A low storage and loss modulus result in a soft adhesive with adhesive transfer. A creep compliance of 1.2×10^{-5} cm^2/dyn is preferred for a skin adhesive. So-called wet-stick adhesives (sterilized with ethylene oxide or γ-radiation) are used for surgical towels; they must adhere to skin, wet textiles, and nonwovens.

Reusable textile materials such as Goretex, PA, polyether sulfone/cotton, 100% cotton, or other hydrophobyzed carrier materials are suggested for operating tapes.

As dicussed previously, the adhesive can be hot pressed in the fibrous carrier. Such materials possess the advantage of preventing contact of small fibers and wounds.

Foam as a carrier (PU) alone or with a film is proposed for medical tapes for fractures; the adhesive is sprayed on the web [158].

4.2.2.7 Mounting Tapes

Mounting tapes have been developed for a number of special applications. Generally, they are double-side PSPs with a thick and conformable adhesive layer, and they work as transfer tapes (see also Chapter 1) for structural and semistructural bonding.

They can be used for the temporary or permanent mounting of parts of an assembly (e.g., fixing films in cartridge, mounting of glass panes, windowpanes, etc.). For instance, a high-temperature-resistant nonsilicone-type adhesive system based on rubber, coated on polytetrafluoroethylene or polyester backing, is used as an aircraft PSA for mounting. Such products are suggested for holding composite lay-up pieces in place during bonding [159]. A special mounting tape for instrument housing assembly is designed like a tear tape with a transfer adhesive [160]. The tape consists of a metallic foil strip of predetermined thickness and an acrylic transfer adhesive strip affixed to one side of the metallic foil, forming a tear band. Certain mounting tapes must be repositionable. Other mounting operations require high-peel, removable adhesives. In such cases postcross-linking can be carried out. Postapplication cross-linking is proposed to achieve easier delamination. According to Charbonneau and Groff [82,161], the procedure allows delicate electronic components to be removed from the tape, although they were nonremovable before cross-linking.

High-shear-resistant, low-tack applications are suggested for certain mounting tapes and double-side-coated foam tapes. For such mounting tapes, the target values of peel (180°, instantaneous), loop tack, and shear resistance (72°F) are 34.5/6.5 N/25 mm and 400 min, respectively. Manufacturers of foam mounting tapes or other double-face tapes must often utilize relatively thick films (6–8 dry mils or higher) of PSA. This is done to ensure that the adhesive film will be able to conform to a surface that is irregular or unpredictable. For such tapes a thick, 88-μm adhesive layer has been proposed in Ref. [162]. According to Haufe [56], the peel resistance of such foam tapes depends on the foam used and is about 3.6–9.1 foam tear N/25 mm.

Double-face mounting tapes can be manufactured as double-side-coated, carrier-based products or carrierless, foam-like webs (see also Chapter 1). Carrierless mounting tapes may have a monolayer or a multilayer, filled or unfilled, foamed or unfoamed construction. Generally, the manufacture of mounting tapes is based on (double-side) coating of a carrier material, which is postlaminated with a solid-state release liner. Mounting tapes with a carrier can be designed as "adhesiveless," that is, self-adhesive tapes (see also Chapter 1); which are extruded. Carrierless mounting tapes are transfer tapes. Such transfer tapes can be manufactured by coating, coating and foaming, or extrusion (see also *Technology of Pressure-Sensitive Adhesives and Products*, Chapter 10). Extensibility and deformability are general requirements for certain mounting tapes. A pleated carrier (which gives extensibility) is needed for some mounting tapes. Mounting tapes were manufactured with woven plastic carrier materials. Double-side mounting tapes may have a primer coating on the back side to allow the anchorage of a postfoamed layer. Double-side-coated tapes with neoprene, PVC, PU, or PE foam carrier were developed. When used as mounting tapes, such products must satisfy the requirements for stress relaxation, stress distribution, mechanical resistance, sealing and anticorrosion, reinforcing, and vibration/noise damping.

Certain double-face mounting tapes are built up with PET as a carrier material. They have a silicone adhesive on one side and a silicone release layer on the other. They are used in electronics. Temperature-resistant silicone-mounting tapes maintain their performance characteristics up to 288°C and adhere to a variety of surfaces, including metals, glass, paper, fabric, plastic, silicone rubber, silicone varnished glass cloth, and silicone glass laminates. They must exhibit good electrical properties and good resistance to chemicals, as well as to fungus, moisture, and weathering. Double-face foam tapes should also be fire and mildew resistant.

Water-borne acrylic PSAs without tackifiers exhibit better shear resistance and adequate low-temperature behavior, which are important in the manufacture of mounting tapes. A special cross-linkable, carboxylated, high T_g ($-25°C$) acrylate has been developed for mounting tapes and double-side-coated foam tapes. This compound contains tackifiers at a loading of 12–35% dry w/w [44].

Mounting tapes used in buildings must have the same expectancy as the building elements (i.e., 10 years). Such mounting tapes can replace other classic (mechanical) methods of fixing. For instance, a glass-mounting tape is specially designed to bond two sheets of glass around the edges for a composite glass with a strengthening layer. The old method used butyl tape to create a layer of air between two sheets of glass. Such tapes provide a long-lasting waterproof seal. Another double-coated adhesive tape for mounting provides a cleaner water-tight system and cuts labor costs. One side of the tape is applied to the window frame, and the glass pane is pressed against the other side of the tape. The pane is thus held securely when laths are nailed to the surrounding frame. In the first step of mounting, one-side-coated PE pressure-sensitive tape is applied. A double-layer pressure-sensitive tape intended for industrial and commercial use must bond to glass, concrete, steel, ceramic drywall, wood, tarpaulins, and covers [163].

Tissue tapes used for furniture manufacture are applied to temporarily hold seat upholstery together. The double-face tape holds leather in place prior to stitching and eliminates problems of leather creasing or slipping.

Pressure-sensitive tapes can be used to bond exterior trim to automobiles; fasteners are used in attaching automobile seat covers. Adhesives with excellent water resistance (resistance to humidity and condensed water) are required for technical tapes and pressure-sensitive assembly parts in the automotive industry [62]. Shear resistance values of about 600–1000 min (25 $\times$ 25 mm, 1000 g, on steel) and peel resistance values of 25–35 N/25 mm (20 min on steel) are required.

A PSA for double-face carpet tape, based on tackified SBS with 220 pts resin and 20 pts oil to 100 pts block copolymer (and a high coating weight of 40 g/m²), exhibits relatively low adhesive characteristics (a peel resistance of 16–21 N/25 mm, loop tack of 17–18 N/25 mm, rolling ball tack of 3–5 cm, and shear resistance on steel of more than 100 h) [164].

Mounting tapes are used in the fashion and textile industry for attaching textile parts to each another. They are protected from telescoping during storage with a woven on nonwoven supplementary layer [78].

Foam-like transfer tapes are used for structural bonding of aluminium or copper–titanium–zinc roof panels. Structural bonding tapes are now available that combine the instant adhesion obtained by a PSA tape with the structural properties of an engineering adhesive. The structural bonding tape can be treated as a structural adhesive and its high strain to failure contributes to joint performance, especially for long overlap lengths. Moreover, the tape could be stronger than other traditional structural adhesives in which strain concentration is encountered [165]. Special foam-mounting tapes are used for bonding of soft printing plates for flexoprinting. In this case, foam tapes can compensate for tolerances in the printing process and assure that dot gain is minimized. A foam-like pressure-sensitive transfer tape with good shear strength and weather resistance that is useful in the bonding of uneven surfaces has been prepared by forming foamed sheets from agitated acrylic emulsions, impregnating or laminating with adhesives, and drying. Such film is coated on a PET release film as a 1.2-mm adhesive sheet with a density of 0.75 g/cm^3. It displays (by bonding of acrylic polymer panels) an instantaneous shear strength of 4.5 kg/cm and demonstrates twice the resistance of a common PU film after weather-o-meter exposure [166].

4.2.2.8 Protective Tapes

Protective tapes are used as temporary or permanent surface-protective or reinforcing elements. Most of them are really protective films, which have a tape-like geometry. There are various application domains of such tapes, including face or back-side protection. There are various applications of protective masking tapes. For instance, special office masking tapes are applied to cover written text passages; the covered surface portion should provide a shadow-free copy. EVAc are used for HMPSA for weather-stripping tapes. A coating weight of 60 g/m^2 is applied for a bonding surface, with 50 g/m^2 on the opposite side, to bond the wipe-clean surface.

Masking tapes may have special carrier constructions to give them conformability and temperature resistance. Their adhesive is covered with a release liner. Some masking tapes must be flexible, stretchable, and contractible and capable of maintaining the contour and the curvature of the position in which they are applied. They are suitable for protecting a surface on a substrate that has irregular contours. For instance, a PSA tape for protection of printed circuits, based on acrylics and terpene–phenol resin, with a 100–200 g/m^2 siliconized crepe paper carrier, is made by first coating the adhesive mass on a siliconized PU release liner and then transferring it under pressure (10–30 N/cm^2) onto the final carrier material [167]. According to Lipson [168], a masking tape with PP carrier has a stiffened, longitudinal section extending from one edge, with an accordion-pleated structure to conform to small radii. The nonadhesive face of the tape is heat reflective. Masking tape based on foam was developed as well. For such products, smooth, white, cross-linked polyolefin foams with an average cell diameter of 0.50 mm and a yellowness index of 20.4 were prepared.

Decal premasking tapes are made from a latex-saturated paper that is coated with a PSA. In most cases, because low-adhesion adhesives are utilized, the decal premask/application tape does not require a release coating. Such papers are recommended where delamination resistance but clean removal is required; temperature resistance, high abrasion resistance, markability and translucency, dimension stability, and wet-rub resistance are needed.

Masking tapes must be removable. Because adhesion builds up due to use of highly polar comonomers (e.g., acrylic acid in acrylic PSAs), such components are not recommended for the synthesis of adhesives for masking tapes. Masking tapes on a paper basis (90 g/m^2) can be coated with 30 g/m^2 HMPSA. Generally, masking tapes are rigorously tested for deposit-free removability at elevated temperatures.

The manufacture technology for masking tapes is generally the same as that for protective films. However, some special tapes are manufactured using a complex technology. For instance, a laminated PSA tape with good chemical, heat, and water resistance, applied for soldering portions of a PCB, is manufactured using a craft paper carrier material impregnated with latex, treated with a primer or corona discharge, and coated on one side with a rubber–resin adhesive and on the other side with a PE film and release.

Thick masking tapes with "bubblepack" character and improved cuttability can be manufactured by classic coating technology, including an expandable filler in the adhesive. According to Ref. [169], a photo-cross-linked acrylic–urethane acrylate formulation used for holding semiconductor wafers during cutting and releasing the wafers after cutting exhibits a peel resistance of 1700 g/25 mm without cross-linking and 35 g/25 mm with cross-linking. Adding an expandable filler to the formulation (6 to 100 pts adhesive) reduces the peel resistance even more. Such a composition exhibits 1500 g/25 mm peel resistance before cross-linking and 15–20 g/m² after cross-linking. Such a releasable, foaming adhesive sheet can be prepared by coating the blowing agent containing PSA (30%) for a 30-μm dry layer [170]. Blowing up of the adhesive layer reduces contact surfaces. Such products exhibit lower bond strength, but they are only temporarily removable. Adhesive flow leads to rapid build-up of peel. Therefore, for adequate removability, both blowing up of the adhesive layer and cross-linking are needed.

According to Lin et al. [87], government regulation directives are changing, particularly in Europe, and driving a decrease in the use of lead-containing solder. Lead-free solder requires the use of higher temperatures in the soldering process and, therefore, requires the use of higher-operating-temperature adhesive tapes for masking electronic components during fabrication. Improvements in silicone PSA compositions are fulfilling the new performance needs and sustaining the growth of silicone PSAs in these application areas [171,172].

4.2.2.8.1 Car Masking Tape

According to Ref. [173], the first masking tapes allowed the two-color painting of cars. These tapes are used during spray-painting of vehicles. According to Stott [174], paper-based masking tapes were used in the 1920s for car varnishing. Crepe paper, coated with a cross-linkable silicone adhesive, has been applied up to 180–200°C. Later, masking tapes used for spray-varnishing in the automotive industry were partially replaced by adherent, non-pressure-sensitive cover films, which allow a less expensive, full-surface, easily removable protection [175] (see also Chapter 7).

Masking tapes used in the varnishing of cars must adhere to automotive paint, email, steel, aluminum, chrome, rubber, and glass and resist the high temperatures used for drying and curing varnishes. When tapes are delaminated from anodized aluminum, cohesive break may produce adhesive deposits. NR latex, tackified with PIB, is recommended for such tapes. PVE can also be used. For instance, a mixture of poly(vinyl-*n*-butylether) with different molecular weights (1000–200,000 and 200,000–2,000,000) is suggested for such tapes [176]. Butyl rubber (15–55 pts), untackified or tackified with PIB (7.5 to 40 pts) and cross-linked with zinc oxide (45 pts), has been proposed also [177]. Such formulations yield (180°) peel resistance values of 0.73 to 1.81 kg/25.4 mm. Peel resistance increases with increased tackifier level, but legging appears at a tackifier/elastomer ratio of 40/15.

4.2.2.8.2 Mirror Tapes

Mirror tapes are large-surface security-lamination films applied to the back side of fragile substrates (see also Section 4.3). Generally, security lamination of films for glass is carried out with security film, and antisplitting films. According to DIN52290, a 1-m^2 glass panel must resist three perpendicular consecutive impacts with a 4.1-kg steel ball [178].

4.2.2.9 Sealing Tapes

Sealing tapes are used as insulating and mounting elements because they have sealant-like functions. Pressure-sensitive sealing tapes are soft constructions, and their PSAs are tackified or plasticized formulations. They can be filled and cross-linked. Their quick stick values are low, and their peel resistance value (on aluminium) is about 120–140 oz/in. Special sealant tapes for gasket-type applications are made by extrusion onto a release film. Such tapes must possess high tack and be resistant to gasoline.

Sealing tapes based on a PVC carrier are printable. Sandwich printing of sealing tapes is possible (flexoprinting for the top side and gravure printing for the back side between the film and the adhesive) [38].

Classic sealing-tape formulations are based on BR. For instance, an air-seal tape is made with BR (on ethylene propylene foam). Such formulations may contain NR with polyisobutene (100 pts), hydrocarbon resin (50–70 pts), low-molecular-weight poly-butene (4–50 pts), and mineral oil. According to Risser [179], BR, plasticizer, carbon black, and terpene–phenol resins were tested for sealing tapes. Such tapes are applied by hand or using a sealing machine. These products are used for closing, manufacturer identification, antitheft protection, copyright protection, quality assurance, advertising, information, classification, and administration.

Polyacrylate rubbers can be used for pressure-sensitive sealant tapes and caulks [180]. A common tacky PSA can be added to the formulation as well (50 to 100 pts acrylic rubber). The composition can be cured using isocyanates or UV-curing agents (e.g., *p*-chlorobenzophenone). The recipe can include a high level of fillers. Such formulations can contain 200 pts of hydrated alumina as filler for every 100 pts of rubber, 50 pts of carbon black, and 50 pts of plasticizer. Polyisobutene can also be added.

Foam-sealing tapes are made of a foam-sealing tape layer and an interlayer strip rolled up together in a compressed form.

Automotive insert tapes are used as sealants. Such products are tapes based on a foam carrier and must exhibit aging stability, weatherabilty, sealing, and nonflammability. The tape roll is compressed before use (it is supplied in compressed 1/5 thickness).

The insulating glass sealant or tape must satisfy a number of requirements, the most important being optimal adhesion to glass and metals, aging resistance, high flexibility (even at low temperatures), and low water permeability.

Pharmaceutical sealing tapes are actually antitheft tapes. Some years ago, wet-gummed products were applied almost exclusively as sealing tapes for pharmaceutical use. Now pressure-sensitive tapes are applied. Such tapes are used as security closures for pharmaceuticals. An example of a useful application of a low-refractive-index mate-rial is in filling an air gap to help support and adhere the lenticular lens to the Fresnel lens of a rear projection screen for a television, a back-lit projection screen. A special UV-curable, water-soluble acrylic hydrogel was used by Chang and Holguin (see also Chapter 3).

4.2.2.10 Splicing Tapes

Splicing tapes have been used in the paper, converting, and printing industries. Paper-making and printing technology require splicing the end of one roll of paper to the beginning of another, as well as splicing parts of the roll after defective material has been cut out. Such splices must be made quickly and easily with an adhesive that rapidly attains maximum strength.

The adhesive and end-use requirements for splicing tapes are very rigid. They must resist temperatures up to 220–240°C, their carrier materials must be water dispersible, and the adhesive should be water soluble, but resistant to organic solvents. No migration is allowed. For high-speed flying splicing tapes, high tack and fast grabbing are required. They must exhibit adhesion to various papers, liner boards, foils, and films, as well as high shear resistance and temperature resistance while traveling through a drying oven. The application time (i.e., bonding time) for splicing tapes is very short (less than 60 s, according to Ref. [87]); their adhesive must bond at a running speed of 160 m/min. Under such conditions high instantaneous peel resistance and shear resistance are required. Reel changeover follows in less than 15 min and the joint length is less than 100 mm. The most difficult cases are characterized by high running speed, small reel diameter (for the next reel), maximum thickness of the web for the new, small-diameter reel, and short distance between the splicing point and the cutting line. Butt-splicing tapes, built up as a single-coated PET, must adhere to PP, PE, PA, and coated paper, resist up to 117°C, and possess high quick stick for high-speed splicing applications. The splice for flame-resistant tapes must be constructed in such a fashion that the roll can be unwound at a rate of 7.3 m/min.

Tack and heat stability, together with water solubility or dispersibility, are required for splicing tapes in papermaking and printing. Splicing tapes for paper manufacturing must be recyclable. The paper used as carrier material must be dispersible, and the adhesive must be water soluble, but resistant to organic solvents. Such formulations must adhere to the paper substrate at a running speed of 1200 m/min. The tear resistance of the tape must be as high as that of the paper to be bonded, and high instantaneous peel resistance and shear resistance are required. For splicing tapes the corresponding tack/peel resistance and shear resistance values are 2.2/1.5/360 [44]. The shear resistance value was measured at 250°F (50 μm adhesive thickness).

The main types of splicing tapes are one-side-coated tape, double-side-coated tape, and transfer tapes. In one-side splicing tapes the siliconized paper liner remains on the adhesive and must be recyclable. Carrier-free splicing tapes must be water-soluble. The liner for transfer tapes must allow the detachment of the adhesive, that is, it must display different degrees of release on its two sides. Water solubility or dispersibility of the paper carrier used for splicing tapes is a main requirement.

According to Czech [70], a water-soluble adhesive for splicing tapes can be manufactured using a polymer analogous reaction, polymerization, or polymerization and formulation. Water-soluble compositions used for splicing tapes contain polyvinylpyrrolidone with polyols or polyalkyglycol ethers as plasticizer, vinyl ether copolymers, neutralized acrylic acid–alkoxyalkyl acrylate copolymers, and water-soluble waxes or acrylics. According to Blake [156], water solubility is given by a special composition containing vinyl carboxylic acid neutralized with an alkanolamine. Other formulations contain acrylic acid polymerized in a water-soluble polyhydric alcohol and cross-linked with a multifunctional unsaturated monomer. Repulpable splicing tape specially adapted for splicing carbonless paper uses a water-dispersible PSA based on acrylate–AA copolymer, alkalies, and ethoxylated plasticizers. A polyamide–epichlorohydrine cross-linker may also be included [84]. As illustrated in Ref. [51], special acrylate-based, water-soluble adhesives for splicing tapes can change their adhesive properties as a function of atmospheric humidity. Such formulations must adhere to the paper substrate at a running speed of 1200 m/min and must be soluble in water at pH 3–9. Ethoxylated alkylphenols, ethoxylated alkylamines, and ethoxylated alkylammonium are used as water-soluble plasticizers (0–70% w/w).

Silicone pressure-sensitive tapes have been traditionally used as splicing tapes for silicone release liners. Increasing line speeds in silicone liner production shortened splicing time and increased adhesive performance needs in splicing operations. Improved double-side tapes, like those claimed by Adhesives Research, Inc. [181], continue to expand the use of silicone PSAs in this application.

Splicing tapes are manufactured by transfer coating [87] with 50 μm adhesive thickness [44].

4.2.2.11 Tear Tape

Self-adhesive tear tapes are designed to be used as an easy opening device for flexible packaging. They can be used to remove the top of pack, to form a lid, or in applications where a heavy-duty opening system is required. These tapes provide tamper evidence, indicating when the packaging is intact. They are effective on overwrap, shrink sleeves, flow packs, and blister packs for foodstuffs and pharmaceutical products. Some are provided with holograms, and some combine easy opening, tamper evidence, and inherent security using a motorized applicator [182]. The use of tear tapes in various end-use domains is based on their low mechanical resistance, that is, easy tear of their carrier material. This requires special carrier formulation, or postforming, of the carrier. For instance, a carrier for tapes for low-temperature application may have transverse cuts or holes. Easy-tear, breakable tapes allow easy tearing or splitting of the tape during handling. For instance, an adhesive tape used for fastening printer paper together is perforated along the center line in the long direction to allow easy tearing [183]. According to Ref. [184], tapes for easy handling comprise a continuous (film-adhesive layer-perforated film) laminate. Easy-tear breakable hank tapes are used in the same finished products as easy-tear paper tapes. These tapes are applied by hand. They are used for hanking and spot tapes when fast, easy removal is required. They are highly filled, with a low elongation and easy break force. PVC/PVAc compounds were proposed as base material for such tapes [185]. There is a difference between the tear behavior of paper-based and film-based tapes. Although a film-based tear tape can be used in the same finished product as a paper tape, its physical properties, application method, and resultant assembly are different. Because of the higher tear resistance of film-based tear tapes, care must be taken to ensure a maximum of two laps.

Cloth-based tapes with good dimensional stability and easy-tear properties are manufactured by laminating a synthetic resin layer on one side of a cloth with a warp of

10–25 yarns and weft of 10–20 yarns and a coating adhesive on the other side [186]. Metallic foils are used for tear-tape-like mounting tapes [160]. For instance, a special mounting tape for instrument housing assembly is designed like a tear type with transfer adhesive. The tape consists of a metallic foil strip of predetermined thickness and an acrylic transfer adhesive. For tear tapes, cross-linked acrylic formulations are suggested with *N*-methylol methacryl-amide as comonomer.

Easy-tear tapes include security tapes also (see also Section 4.1.2.7). According to UN Recommendation R 001 [187], packages of dangerous goods should be closed with tapes. Such tapes are called security tapes and must ensure antitheft protection, drop destroying, and tear resistance. For instance, a commercial product can use a multilayer construction (with a width of 24, 48, and 72 mm) and, if deteriorated, the logo *opened* appears [188].

4.2.2.12 Transfer Tapes

Like easy-tear tapes, transfer tapes possess a special construction that is valid for products applied in various end uses (e.g., medical, mounting, and security tapes). Transfer tapes are another class of uncoated and self-supporting monowebs. Transfer tapes have a temporary solid-state component that forms the release liner, supporting the adhesive core. The adhesive core may be a continuous, homogenous, or semicontinuous heterogeneous adhesive layer. The heterogeneity of the latter may be due to included solid, liquid, or gaseous particles (holes), that is, the adhesive layer can be a foam. Such carrierless tapes are prefabricated glues with more than sufficient dimensional stability to permit high-speed lamination. In actuality, such products are adhesive layers that have a higher mechanical resistance that allows their transfer during application, but does not allow their manufacture, storage, and handling (see also *Technology of Pressure-Sensitive Adhesives and Products*, Chapter 10, and Chapter 1 of this book). Therefore, before use they must have a laminate structure. An adhesive material in film form (without surface coating) can display many advantages during its application. Additional benefits are cleanliness; controlled, uniform thickness compared to a liquid system; and positionability. The principle of adhesive detachment from the inserted solid-state release liner is used for transfer tapes. Both surfaces of the carrier may have low adhesion coatings, one of which is more effective than the other. When the tape is used as a transfer tape, when unwound, the adhesive layer remains wholly adhered to the higher adhesion surface from which it can be subsequently removed.

Carrierless glues are also used for non-pressure-sensitive applications. For instance, a modified heat-curable epoxy-based film (25 µm) has been manufactured as a carrierless adhesive sheet for printed circuits. To achieve better mechanical performance, the adhesive can be reinforced. Another example is the repositionable acrylic adhesive used for liquid crystal thermometers, a fiber-reinforced free film. Acrylic foam was developed as a carrier-free adhesive construction for mounting tapes. Such structural adhesives are used in carrier-free adhesive tapes for bonding dissimilar substrates. Structural tape combines the properties of classic tapes and PSA. Virtually carrierless sealing tapes based on BR have been used in the automotive industry. Such carrier-free tapes are reinforced with a metallic wire included in the elastomer.

In some cases, the self-supporting adhesive layer is really a reinforced layer. Such constructions are called *tape prepreg* because the fiber-based reinforcing matrix is impregnated with the adhesive. In such tapes without backing, strength can be enhanced with a tissue-like scrim (cellulose or PA), included and coextensive with the adhesive layer. Such tapes behave like classic, adhesive-coated, carrier-based products. Foam-like carrierless tapes can be considered a development of tape prepregs.

Transfer tapes without a carrier can have a core of pure adhesive or include a reinforcing material. The reinforcing component can be a continuous web (e.g., a network) or a discontinuous filler-like material. For instance, thick PSA tapes (0.2 to 1.0 mm) have been prepared by UV-light-initiated photopolymerization of acrylics. In this case, the PSA matrix can contain glass microbubbles. The filled layers can be laminated together with the unfilled layers. Such acrylic formulations can be polymerized as a thick layer (up to about 60 mil) or the thick layer may be composed of a plurality of layers, each separately photopolymerized. The thickness of the layer is an important factor; a thicker layer requires a greater degree of exposure. Thick multilayered PSAs or PSPs are made in this way. If the thick layer is sandwiched between two thinner layers, it may be considered a carrier, although it also has pressure-sensitive properties. The support or carrier layer is 25–45 mil thick and it conforms well to substrates that are not flat themselves. The thick layer may include a filler, such as glass microbubbles, as disclosed in Ref. [189]. The thinner layers are about 1–5 mil thick. Thick, triple-layered adhesive tape can also be manufactured, with fumed silica as a filler material in the center layer. The viscoelastic properties of the layers are regulated using different photoinitiators and cross-linking monomer concentrations. A plastomer (polyvinylacetate) can be used as filler for the carrier layer. Such a carrier may be nonadhesive, but for certain applications the carrier itself possesses some pressure sensitivity. Curing by UV light is a preferred method to manufacture transfer tapes. EB curing allows simultaneous cross-linking of both sides of a two-side-coated release liner (transfer tapes).

For transfer tapes used at elevated temperatures, a solvent-based cross-linked acrylic is suggested. Formulations of transfer tapes that are conformable on uneven surfaces use fillers. Glass microbubbles can be incorporated as fillers to enhance immediate adhesion to rough and uneven surfaces. Such tapes are produced by polymerization *in situ* with UV radiation. Thick PSA tapes (0.2 to 1.0 mm) have been prepared by acrylic photopolymerization (UV). The filled layer can be laminated with an unfilled layer. The thinner layers range from about 1 to 5 mil thick.

According to Czech [190], a typical transfer carrier-free adhesive tape has a thickness of 30–200 g/m². The manufacture of UV-light-cured transfer tapes is described in Chapter 10.

4.2.2.13 Other Special Tapes

Many special tapes exist. Some have an ordinary construction and others are more sophisticated. Certain products function mechanically to fix and fasten elements; others have dosage, insulation, or other special functions.

Repair tapes are flexible plastic tapes reinforced with nylon cord that must adhere to metallic surfaces, wood, rubber, or ceramics. They are designed to offer puncture and tear resistance. Such tapes are supplied as continuous web or die-cut finite elements. They must be conformable and provide excellent bonding. Aluminum film coated with a tackified polyisobutene-based adhesive is applied as street marking tape. The release layer for this adhesive is PVA [191].

Nonflammable tape formulations must also include the carrier material and adhesive. Nonflammable tapes are coated with an adhesive with 0–40% polychloroprene rubber [192]. An adhesive for PVC-based tapes includes cold masticated nitrile rubber (100 pts), calcium silicate (40 pts), titanium dioxide (10 pts), oil (25 pts), coumarone indene resin (MP 45–55°C; 25 pts), chlorinated rubber (100 pts), and solvents (methyl ethyl ketone and toluene) [193].

Tapes used for automobile interiors must resist high temperatures. Common rubber–resin-based adhesives do not meet these requirements, so cross-linked acrylics are used. In some cases the composition contains glass microbubbles. For temperature-resistant tapes, PE, PET, and Teflon are suggested as carriers. A low coating weight is recommended for tropically resistant tapes. For such tapes, silica (particle diameter 0.01–0.03 μm) is proposed as a filler to increase the viscosity of the adhesive. Starch is also suggested as a water absorbent.

Cover tapes for bathrooms are made on a special PVC basis, with a bacteriostate (which inhibits bacterials growth) incorporated in the vinyl formulation. Such tapes are embossed to produce a secure, nonslip surface. They are coated with a repositionable adhesive that builds up adhesion from 26.7 to 36.2 oz/in. [194]. A 78-lb kraft release liner is used for such tapes.

Fluorescent adhesive tapes for highlighting must be removable from written surfaces. Such tapes are used on sheets bearing writing. These tapes are manufactured by coating plastic films with fluorescent inks and adhesives with low adhesion [195].

Special or standard tapes ("test tapes") are used to test the adhesion of varnish or printing ink [39]. Tapes can be used in testing of laminates (see also Chapter 7). To evaluate a laminate section by light microscopy, a microtome is used to produce perfect cuts. A rapid method for fixing (reinforcing) the sample is to laminate it with a tape [168]. The surface treatment is also tested according to ASTM 2141–68 using a PSA tape [196–199].

Plasma spray tape is used for the regeneration of worn surfaces with metal plasmas involving sandblasting, intense heat, and abrasion [54].

Lower and Jones [200] review a variety of chemistries that can be used to prepare hot-melt, moisture-curable silicone PSAs. The two primary preparation steps involve solvent removal and incorporation of silicon-bonded, hydrolyzable functionality, Si-X, with X being alkoxy, acetoxy, or oxime. For this HMPSA to be useful in a tape-type application, moisture must be excluded until the final application to avoid unintended postcuring in the tape prior to end use. This type of moisture-curable adhesive tape is limited to applications in which postapplication moisture cure is tolerated.

Silicone PSAs and tapes are useful for covers for analytical receptacles, such as microtiter plates and continuous multireservoir carriers used in bioanalytical applications [201]. The chemical and solvent resistance of silicone PSAs against chemicals typically used in bioanalytical applications makes them useful in these applications. Stretch–release adhesive tapes that bond strongly to substrates and then release from them by stretching have emerged. Stretch–release tapes are useful in attaching and mounting articles. The construction often contains adhesives composed of silicone and silicone-organic copolymer materials [202].

4.2.3 Tape Application Technology

Tapes are manufactured as a continuous web and applied as web-like or discontinuous items (segments of this web) by laminating on a solid-state surface. One-side-coated tapes are unwound; double-side-coated tapes are delaminated. Delamination conditions may vary. For instance, radiation-cured automotive tapes must present an initial breakaway peel and then the force needed to continue breaking the bond, the initial continuing peel, is measured. Initial continuing peel is lower (by about 30%) than breakaway peel [203].

Generally, quick-stick and high-temperature shear are required for PSA for tape application. Packaging tapes are applied either by hand or by automatic packaging machines for the fastening of goods. Special (e.g., medical, application) tapes are used manually. For instance, case-sealing units use adhesive tapes and close cartons [204]. In some cases (e.g., for special, hand-applied packaging tapes and masking tapes), an easy-tear carrier (see also Section 4.2.1.11) or high-strength carrier (closure tapes) is required. For some applications, transfer tapes are die-cut [44]. The use of plastic films as a carrier material for tapes allows their high-speed automatic application due to their flexibility and high mechanical strength.

Foam-like transfer tapes are applied under pressure [69]. A hydraulic press is used for bonding aluminum panels. According to Schroeder [205], the winding of tapes ensures the pressure necessary to adhere by lamination. Product surface, geometry, and application climate influence the end use of tapes. The application conditions for tackified adhesive-coated packaging tapes must differ from those for low-tack adhesive-coated or adhesiveless masking tapes. Soft, conformable, carrierless tapes need only low application forces. For medical tapes, deapplication is more important than application. Such examples illustrate the variable application conditions for tapes as a function of their end use.

Sealing tapes (without carrier) based on tackified BR have been applied with an extruder. Sealing tapes based on PVC are applied by hand or with a sealing machine [206]. A special application apparatus is suggested for folded tapes. Folding along the longitudinal axis has been proposed to prevent the tapes from rewinding. Thermal release tapes allow deapplication using heat.

Tape application machines also exist. These devices were first developed for hot-laminated tapes. They are used mainly for heavy insulation tapes. A self-adhesive tear tape, designed as an easy-opening device for flexible packaging, can be used with a motorized applicator. Transfer tapes are also often die-cut. Such tapes cannot be cleanly dispensed from a common adhesive transfer gun, which has no cutting blade; they tend to elongate during dispensing and leave excess adhesive both at the broken edge of the transferred strip and at the orifice of the gun.

4.2.4 Competitors of Pressure-Sensitive Tapes

Wet tapes are the main competitors to pressure-sensitive tapes. For some applications they are better than pressure-sensitive tapes. For instance, reinforced wet-glue tapes display 4% elongation and nonreinforced plastic tapes have elongation of 40–70%. Pressure-sensitive tapes on cellulose hydrate or PVC as a carrier material have been designed for box closures and are less heat and age resistant and less tamperproof than wet tapes. In comparison with wet tapes, PSA tapes are more sensitive to dirt. Wet tapes do not need an unwinding force. These characteristics were the main advantages that allowed a fast, exclusive application of wet tapes for security closures some years ago.

Heat-seal tapes compete with pressure-sensitive tapes in special application domains. They allow an application speed of 150 m/min.

4.3 Grades, End-Use Domains, and Application of Protective Webs

Protective webs are self-supporting, removable adhesive webs that build up a laminate with the product to be protected. The main characteristics of a protective web are its self-supporting character and its surface adhesivity. The self-supporting character is given by the use of a solid-state carrier material (film, paper, etc.). This material must display the mechanical resistance required during application of the web (bonding and debonding) and during the life of the laminate (protected material/protective web). Surface adhesivity is given by chemical composition and macromolecular characteristics and separates, in principle, such products in adhesive-coated and adhesiveless constructions. Table 4.5 illustrates how adhesive-coated and adhesiveless products are used for

TABLE 4.5 Main Protective Films, Their Characteristics, and Their Use

Application/Protection	Protection Type		Product Build-Up		Application Conditions		
	Static	Dynamic	Adhesive Coated	Adhesiveless	RT	Warm	Hot
Automotive storage	•	—	•	—	•	—	—
and transport	•	—	—	•	•	•	—
Building	•	—	•	—	•	—	—
Carpet	•	—	•	—	•	—	—
	—	•	•	—	•	—	—
Coil	—	•	•	—	•	—	—
	—	•	—	•	—	—	•
Deep-drawing							
Metal	—	•	•	—	•	—	—
Plastic	—	•	•	—	•	•	—
	—	•	—	•	—	•	•
Carpet	—	•	•	—	•	—	—
Environmental	•	—	•	—	•	—	—
	•	—	—	•	—	•	•
Furniture	•	—	•	—	•	—	—
	•	—	—	•	—	•	—
Glass impact	•	—	•	—	•		
	•	—	—	•	—	•	—
	—	•	•	—	•	—	—
Plastic plate and film	•	—	•	—	•	—	—
	•	—	—	•	—	•	—
	—	•	•	—	•	—	—
	—	•	—	•	—	•	—
Product security	•	•	•	—	•	—	—
	•	•	—	•	—	•	—

the same type of protection; however, adhesive protection webs have the advantages of free carrier choice, less sophisticated application conditions, and better adhesion.

Protective webs are conformable, laminated cover materials that must pass through the same processing or storage cycle as the protected item. Such materials are only temporary laminated and do not contribute to the value of the finished product. Therefore, the carrier materials for protective webs are generally common products, although they must be soft, mechanically resistant, thin, and deformable, with a balanced plasticity/elasticity, medium temperature resistance, and a low manufacturing cost. There are adhesive-coated and adhesive-less protective webs (see also Chapters 1 and 7). General quality requirements such as the ability to lay flat, narrow thickness tolerance, no wrinkles, and no gel particles are valid for both product classes. Carrier materials for protective films have lower mechanical performance characteristics than films used for labels or tapes. The manufacture of protective webs was described in detail by Benedek in Ref. [1]. Formulation of PSAs for protective webs was described in Refs. [6,22]. The end use of protective webs was discussed in detail by Benedek in Ref. [3].

Adhesive-coated protective webs can be classified according to adhesive nature, type of carrier material, adhesive properties, bonding/debonding nature, and application domain. Such parameters depend on the carrier type and thickness, PSA type and coating weight, and the substrate to be protected (see Table 4.6). Most protective webs are film-based.

Protective webs form a special class of PSPs, manufactured as self-wound webs and applied as a laminating component for web-like or discrete products. They can be used to protect finished surfaces from damage during manufacture, shipping, and handling

TABLE 4.6   Adhesive Characteristics and Application of Various Protective Films

		PSP Construction				Peel Resistance (N/25 mm)
		Carrier		PSA		
Product Use	Adherend Surface	Material	Thickness (μm)	Grade	Coat Weight (g/m²)	
Automotive						
Storage	Glossy, lacquered metal	PE/PP	50	AC	5–10	5–8
Processing	Rough, PP, ABS	PE/PP	100	AC	14	3–4
Plastic plate						
Storage	Glossy, PC, PMMA	PE	70	AC	4–5	1.2–1.5
Processing	Glossy, PA	PE/PP	70	AC	6–7	2–3
Carpet						
Storage	Textured, PP, PA, PET	PE	80	RR	10–12	12–14
Processing	Rough, PA, PP, PET, EVA	PE	25	RR	8–10	12–14
Coil						
Storage	Glossy, metal	PE	50	RR	2–3	0.3
Processing	Glossy, metal	PE	110	RR	4–5	1.25

and to mask areas of a surface from exposure during spraying operations by painting. Such products can be formed using a stencil and undergo the same processing steps as the protected item. Taking into account their protective function, they could also be considered packaging materials. In actuality, they cannot be defined as a packaging material because of their aesthetics. Some special tapes (e.g., masking tapes) may be considered special-grade (narrow web) protective films.

Adhesive-coated web-like protective PSPs may be used for metal- or nonmetal (e.g., glass, plastic, stone) surfaces. They can be classified according to their function (storage or processing protection), their place of application (face or back side protection), and the permanent (e.g., mirror tape, label overlaminating film) or temporary (masking films) character of the bond. Nonmetal surfaces may be film-like substrates, plates, profiles, and textiles (e.g., carpet).

As noted in Ref. [3], most adhesives for protective films must display adhesion that permits removal of the film. Therefore, the adhesive properties of protective films are characterized by removability. In working with protective-film-coated surfaces, their force of adhesion should ensure a nondestructible, permanent bonding (as in the case of permanent adhesives). On the other hand, delamination of the protective film after its use requires a very low peel force level, theoretically as low as possible. The real adhesion force of protective films is a compromise between the protection-related high-peel resistance and the low-peel resistance required for easy delamination. Although protective films display very low debonding resistance (measured as normed peel resistance), their delamination, in practice, is difficult because of their large surface area. The general principles of removability are discussed in Ref. [4] (see also *Technology of Pressure-Sensitive Adhesives and Products*, Chapter 8). For protection films this is achieved by an extreme reduction in coating weight (see also Table 4.6), but other formulation-related technical means are also used to reduce peel resistance. Earlier, softening of the adhesive was carried out through the use of PVC plastisols (plasticizer-gelled PVC compositions) to manufacture oil- and solvent-resistant, paper-back adhesives for covering sensitive substrate surfaces. In our case, extreme softening of the plastic carrier materials leads to adhesive-free self-adhesive films (see Chapter 7). The first formulating modality to achieve removability is the change in adhesion–cohesion balance. Because of required deposit-free removability, reduction of peel resistance values cannot be achieved by softening of the adhesive alone (see also *Technology of Pressure-Sensitive Adhesives and Products*, Chapter 8). Another possibility for detackifying is provided by cross-linking (see also *Technology of Pressure-Sensitive Adhesives and Products*, Chapter 8). Both methods must be combined for protection films. In this case, cross-linked adhesives are generally used to achieve removability. Various raw materials, elastomers (natural and synthetic rubber), and viscoelastomers (acrylics and PURs) were suggested for such PSA formulations. Although they are harder than rubber–resin-based compositions, acrylic adhesives are used as solutions as well as dispersions for protective films. Generally, such formulations are not tackified. Solvent-based acrylics are cross-linked recipes; acrylic dispersions can be used in certain cases as non-cross-linked formulations. Formulations based on rubber–resin adhesives (solutions) are the most important adhesives for protective films. Because of their softness and easy control of removability (by mastication, cross-linking, and tackification), for certain uses (e.g., deep drawable protection films) they cannot be replaced with other adhesives.

The choice between adhesiveless and adhesive-coated protective films is determined by the main characteristics of the products to be protected. The manufacture of adhesiveless protective films or protective films with embedded adhesive is discussed in Chapter 7. Such formulations lead to removable products with an adhesion of maximum 1.5 N/10 mm. Such peel resistance values can be achieved with special adhesives or cross-linked formulations and low coating weight (see also Table 4.6). Therefore, adhesive-coated protective films generally need a cross-linked adhesive with low coating weight. Protective films are used on very different substrates, depending on the nature of the products to be protected. As noted in Ref. [3], for protective films the adhesion of the PSA should be tailored to the surface to be protected.

Both the application and deapplication (separation) of the protective film depend on its adhesive properties. Therefore, protective films must display special adhesive characteristics manifested as an unbalanced, low-tack, and high-cohesion adhesion. They must be conformable to allow lamination on difficult adherend contours. Protective films must display plasticity to deform in parallel with the protected, processed adherend (e.g., deep drawn metal coil or plastic plate). They must exhibit permanent adhesion during such end use and removable low-force adhesion during deapplication. It is recommended that films possess enough adhesion to avoid film detachment, but not so high as to produce a deposit. Adhesive-coated protective films generally possess thicknesses of 40–120 μm. As discussed in *Technology of Pressure-Sensitive Adhesives and Products*, Chapter 1, down-gauging of the plastic carrier material is limited by its extensibility. In general, it is not possible to coat all thicknesses with every adhesive. Adhesives yielding higher peel resistance require higher gauge films.

The use of protective films is a relatively new application domain of PSPs. At the end of the 1980s, protective films were suggested for the postprotection of metal coils as an alternative to oiling or wax coating [207]. The main criteria given by the German Association of Surface-Finished Fine Coil (*Fachverband oberflächenveredeltes Feinblech*) for the choice of protective films are the nature, thickness, adhesion characteristics, forming, tear resistance, and light stability of the film, with the recommendation that "well-defined protective films should be applied outdoors for a given time only." Protective papers were developed first and later protective films.

As mentioned previously, the carrier for protection films is a common polymer used in the manufacture of packaging materials. PVC, polyolefins, and polyester are the most commonly applied materials. Modified poly(ethylmethacrylate) can be used as a protective film for various substrates such as PVC, acrylonitrile–butadiene–styrene (ABS) and polystyrene, wood, paper, or metal. It is weather resistant and UV absorbent. For self-adhesive protective films, tack-free plastomers (e.g., polyolefins) or viscoelastic compositions (based on EVAc or EBAc or tackifyed plastomers) are suggested (see Chapter 7) as a carrier material.

Protective films appeared on the market as improved variants of PSA-coated protective papers. Paper masking tapes were introduced in the 1920s in the automobile industry. Later, paper was replaced with the more conformable plastic film. During the 1950s, Japanese PVC-based products were the sole materials used as protective films. PVC has since been replaced with polyolefins. Classic protective films are pressure-sensitive self-adhesive films manufactured by a film conversion process and laminated on the product to be protected (see Section 4.4). Unfortunately, the range of such plastic films and their adhesivity is limited. From the technical point of view it was easier to manufacture the protective film in a two-step process in which common plastic films (mainly polyolefins) were off-line coated with a PSA.

Once applied, the film protects the product against mechanical damage during different working/logistic steps. Depending on the nature of the product, the nature of the applied stresses will differ too. The most common working operations are cutting, drawing, and punching. The film itself must resist shear, tensile, and compression stresses. On the other hand, like webless, unsupported protective coatings (see Section 4.2.5.1), the protective web must conform to the surface profile of the product. In conclusion, the choice of a solid-state, self-supporting carrier as a base material for protective films is influenced by the application conditions (laminating/delaminating), end use (product-working conditions), and economic considerations. Relatively thin plastic films tend to be used as common carrier materials for protective films. Their mechanical resistance is the result of their formulation and geometrics, which are given by the empirical experience of film and foil manufacturers with common packaging materials. It may be supposed that manufacture of protective films requires knowledge of the manufacture of removable film coatings and the mechanics of adhesive joints for plastics.

Generally, protective films do not need a separate release liner and in most cases need no release coating. However, for products with a higher coating weight or tacky adhesive, special, non-silicone-based release coatings are applied. The release layer for protective films displays a relatively low abhesivity. This is acceptable because of the low tack and low peel resistance of the adhesive. This is also necessary because of the increased mechanical requirements for the release coating. Economic considerations play a role in the choice of the release layer.

The choice of an adequate release coating for protective films (and tapes) is a complex problem that depends on the nature of the adhesive and the coating weight used. Different release materials impart different debonding forces and lead to different increases in release force with the coating weight. The best formulations display a "constant" release force for different coating weights. Carbamate-based release coatings possess this advantage. In a very different manner, the PA-fluoropolymer-based release formulation gives very low release forces at low coating weight, but high release forces (as high as that without release coating) at high adhesive coating weight.

As discussed previously, protective films have both static (packaging) and dynamic (processing-related) uses (see Tables 4.5 and 4.6). The requirements for these application fields are different. The classification of protective films according to their end use is related to the nature of the surface to be protected. As discussed in Section 4.3.4, metals (coils) and plastics (profiles, plates, films, and other plastic-coated materials) are the most important application domains of protective films.

Independent of their special use, the following general requirements exist for all protective films: adequate elongation at break, opacity, adequate coefficient of friction (film-to-film), release of backing, stability of adhesion, low unwinding resistance, high tensile strength, and no telescoping. In choosing film grades, their adhesive properties must be taken into account relative to the surface in their application domain.

4.3.1 Common Protective Films

Certain protective films are used as packaging materials only. The protected item does not undergo working processes. Often, films are used for protection during transport, for example, when stacking profiles; there is no paper between stacked items. Polycarbonate or poly(methyl methylacrylate) (PMMA) sheets are provided with protective films on both sides to avoid damage to the mirror-polished surfaces during transport (and processing). Adhesive-coated and adhesive-free protective films are used for coil coating. Time of storage on the sheet or on the roll should not exceed 1–1.5 years. The self-deformation of the coated plastic material in packaging uses must be taken into account as a dynamic force. Thermally induced extension on plates and profiles may produce tension and cause delamination of the protective film. PVC profiles display a coefficient of thermal expansion of $8 \times 10^{-5}/K^{-1}$. Plastics development reduces this risk. In the past decade the stability of plastic pipes at 80°C and 4 N/mm² force increased in the national norms from 170 to 1000 h [208]. Other protected films are processed together with the protected item and must resist mechanical and thermal stresses during the technological cycle.

4.3.1.1 Protection Films for Coil Coating

Coil coating is web coating in which calendered metal (steel, aluminum, copper, etc.) coils are coated with organic materials (lacquers, varnishes, or films). Liquid or dry coating materials can be used. Thin (10–20 μm) layers and thicker layers (60–100 μm for dry lacquers) are applied at a machine speed of 15–18 m/min. Steel plates 0.1–0.8 mm thick are used in automotive, building, furniture, packaging, and other industries. The protective film is laminated at the end of the varnishing process.

Until recently, protection films for coil coating were common products designed to ensure static protection during storage and transport. Because of outsourcing, the importance of coil coating has increased over recent years, which led to the "finish first, fabricate later" principle, that is, the pre-finishing of the metal plate surface before the coil is processed.

4.3.1.2 Films for Plastic Protection

Various plastic plates and other items are laminated together with protective films, during and after their manufacture, to protect them through storage and processing (see also Section 4.3.4). The main polymers used for plastic plates are PMMA, PC, polystyrene, styrene–acrylonitrile copolymers (SAN), PVC, and PP. Styrene-based safety glass for windows was also developed [209]. Butadiene–styrene copolymers for sheet extrusion replaced PET. A new range of compounds based on syndiotactic PS, a semi-crystalline, metallocene-catalyzed engineering polymer with a very high melting point,

TABLE 4.7 Application and Processing Characteristics of Protective Films Used for Plastic Plates

Protective Film		Application Conditions				
		Substrate		Laminating Conditions		
Adhesive Coated	Adhesiveless	Material	Temperature (°C)	Time (min)	Temperature (°C)	Peel Resistance (N/25 mm)
•	—	PMMA	70	15	150	1.25–2.00
—	•	PMMA	70	15	150	0.125–1.25
—	•	PC	70	15	170	0.125–1.25
—	•	PMMA	70	30	170	0.025–0.100

was developed. This material exhibits high continuous ST, hydrolysis resistance, dimensional stability, and good processing. The protective films for such plastic items are generally applied by in-plant lamination. A hot-roll laminator nip or a hot-roll press is used, and heat and pressure cause the film to conform and adhere to the surface. As illustrated in Table 4.7, the grade of the protective film used for a plastic plate and its laminating conditions depend on the plastic plate material.

Generally, adhesive-coated and self-adhesive protective films are used for the protection of plastic films and plates. Relatively hard, water-based acrylic adhesive-coated protective films have been suggested as adhesive-coated protective films for plastic plates and films.

The requirements concerning the dimensional stability and plasticity–elasticity balance of a protective film depend on the dimensional stability of the protected item. For instance, high-molecular-weight, cast PMMA displays mainly thermoelastic behavior in a broad temperature range [210]. Therefore, the protective film used must possess elastic properties. On the other hand, low-molecular-weight extruded PMMA displays a narrow elastic domain during high-temperature deformation. The transition from elastic to plastic deformation is not clear, and forming leads to dimensional changes. When the PMMA or PC items (plates) are first heated, manufacturing-technology-dependent shrinkage occurs.

To allow easier application, the plates have protective films of different colors on their fronts and backs, and plates on different chemical bases (e.g., PMMA or PC) or plates produced with different manufacturing methods are laminated with protective films of different colors.

If these plates are used outdoors under severe weathering conditions, the PE protective films must be peeled within 4 weeks to avoid the degradation (hardening) of the protective film and build-up of adhesion.

Another problem is the higroscopicity of plastic plates. They absorb humidity and reabsorb it after cooling under 100°C. The protective film scals the plate against water absorption; thus, film-protected plates do not require elimination of water before use.

Cast PMMA has a broader processing temperature range than extruded PMMA, which has a more exactly controlled thickness tolerance and lower shrinkage than cast materials. Therefore, extruded PMMA is easier to coat with a protective film. The product range for extruded grades is broader. Many surface qualities are manufactured into the product. Generally, cast PMMA is manufactured with gauges of 1.5–250 mm.

Extruded PMMA can have a thickness of 1.5–18 mm. A quality required for food contact (XII.BGA Recommendation and FDA Regulation 177.1010) is also available.

PMMA is used for illuminated advertisements, light domes, displays, pictures, lamps, vehicles, noise protection panels, solar panels, furniture, and sanitary items, among other things. The plates are processed in various ways.

Polycarbonate is manufactured as film (5–750 µm) and as plates. Polycarbonate films are laminated with the protective film using a laminating station (with rubber-coated cylinders) at 60°C and a lamination pressure of about 50 kPa/cm². Laminating pressure, cylinder hardness, cylinder cooling temperature, running speed, and web tension influence the final adhesion level and film shrinkage.

4.3.1.3 Automotive Protection Films

Automotive upper and undercarriage protection films exist. Protective films for storage and transport were developed to replace conventional systems (see Section 4.2.5.1 and Chapter 7) used to protect vehicle surfaces from scratches, iron dust, bird droppings, acid rain, etc. Such films must possess common pressure-sensitive properties with the exception of removability, which must be improved to be deposit-free after a long time weathering. Adhesive properties such as an initial peel of 4 N/25 mm, easy removability (hand peel-off), and no visible changes to the protected surface are required. Table 4.8 includes a list of the application requirements and test methods for automotive protective films.

TABLE 4.8 End-Use Requirements for Automotive Protective Films

Application Criteria					Test	
Resistance						
Temperature	Chemical		Aging	Air Pollutants	Principle	Conditions
High	—		—	—	Storage	10–15 h/50°C
						10–15 h/80°C
Low	—		—	—	Storage	10–15 h/−10°C
	Acid	—	—	—	Contact	1 h, RT, pH 3
	Alkali	—	—	—	Contact	4 h, 0.1 N NaOH, 80°C
	Salt	Calcium chloride	—	—	Contact	24 h, saturated solution
	Solvent	Motor oil	—	—	Contact	1 h, 70°C
		Washer fluid	—	—	Contact	4 h, 80°C
		Gasoline	—	—	Contact	0.2 h
		Humidity	—	—	Storage	125 h
		—	—	—	Sealing	Water leakage test
		External	—	—	Storage	1, 2, and 3 months
				—		Florida test
				Metal dust (iron)	Contact	4 h, 80°C
				Grain dust	Contact	4 h, 80°C
				Bird drop	Contact	4 h, 80°C
				Smoke	Contact	4 h, 80°C

Table 4.8 illustrates how these tests are related to the practical transport, storage, and end-use conditions of vehicles and includes mostly empirically noted conditions. The film should be stretched tight to minimize wrinkles and trapped air. It should be put on the substrate by pressing it slightly from the center area outward with a cloth, by hand, or with a sponge roller. Deapplication should be carried out by hand peel-off. Quality problems that may appear include air marks, a visible borderline, and adhesive deposits. The chemical composition of lacquers used for cars strongly influences the applicability (i.e., deposit-free removability) of such protective film. Soft lacquers, for example, lacquers based on low T_g aminoalkyd resins, are more sensitive than others. Highly cross-linked adhesives are suggested for automotive use.

According to automotive suppliers, protective films for cars should meet special requirements concerning outward appearance, adhesive properties, stainability, and paint protection. Outward appearance is controlled by visual evaluation and is concerned with color, thickness, and light permeability.

Adhesive properties are measured initially and after aging on the paint and by overlapping. Stainability is tested on automotive paint, but component stainability should also be tested, which means that the film should be tested on various automotive parts (glass, sunroof, rubber seal, emblems, stainless-steel moldings, headlight lens, etc.). As demonstrated in Table 4.8, paint protection must include resistance to acids, alkalies, rail dust, gasoline, and oil.

Automotive undercarriage protective coatings and films have been developed. Such films must display resistance to abrasion, light, weather, aging, washing, and gasoline. They must be heat deformable (150°C); therefore, certain films are based on a soft PVC carrier material.

4.3.1.4 Other Common Protection Films

Different films are used for the protection of textile webs. Their characteristics must be tailored to the manufacturing technology and build-up of the carpet, that is, its adhesivity and its postprocessing. Build-up includes fiber nature, length, and construction. Tufted carpets generally use PA pile fibers and polyester-based fabrics. Carpet processing depends on its end use. Automotive carpets are generally molded; thus, protective films for this use must ensure dynamic processing protection (see also Table 4.6). Common carpet-protecting films for normal use are based on solvent-based acrylic (SBAC) adhesives coated on an 80-µm PE carrier material. Special carpet-protective films must have antislip properties; they must fulfill the requirements of DIN51130 [211]. Like other adhesives used for automobiles, PSAs coated on such products must meet the requirements for volatile organic compounds according to DIN 75201 (Fogging test) and PV 3341 (VW) or PB VWL 709 (Daimler Chrysler AG) [212].

4.3.2 Special Protective Films

Special protective films laminated on items to be protected undergo the whole technological process of these items. Processing (forming) of coated surfaces requires good adhesion between the protected surface and the protective film. The film must be able to undergo processing. The main work processes are bending, punching, drilling, and deep drawing.

Cutting, drilling, shearing, routing, and heat bending are the most important working methods for protective films used for plastic plate protection, but vibration welding is also used for plastic plates. Polycarbonate plates may undergo sawing, shearing, punching (sometimes punching tools are heated to 140–180°C), drilling, turning, trapping, milling, grinding, and polishing [213]. During these operations, PE protective films should not come into contact with oily or greasy substances. They should not be used at temperatures greater than 70–80°C.

The thickness recommended for such a protective film depends on the thickness of the protected item (e.g., a thin, stainless-steel sheet of 3 mm needs a protective film at least 90 μm thick). For drilling and punching aluminum or stainless sheets, it is important to know that only a very few liquid processing aids can be used because PE films are oil permeable. The penetration of liquid between the sheet and film may produce residues. In this case, 70- to 90-μm films should be used.

The processing conditions are quite different for different plastics and plastic plates. As an example, drilling speed is the highest (2000 rpm) for PMMA in comparison with PC (1000 rpm) or SAN (200–300 rpm) [214]. The high processing speed must be considered, and the low thermal conductivity of the material must be taken into account as well. In comparison with steel, which has a coefficient of thermal conductivity of 47–58 W/mK, polycarbonate possesses a coefficient of thermal conductivity of just 0.21–0.23 W/mK. In some cases infrared radiant heating is used for warm-formed plastic plates. The halogen heating elements give their maximum radiation energy at 1.0-μm wavelength. However, at wavelengths of 0.5–1.5 μm both the processed plastic plate and the protective plastic film on it may absorb radiation energy.

Special protective films with improved deformability like deep-drawing films and security (mirror tape) films require sophisticated formulations that give a balance of plastic deformability and dimensional stability.

4.3.2.1 Deep-Drawing Films

Deep drawability is a general requirement of the protective films used as technological aids. Metals (coils) and plastics (films, plates, wovens, and nonwovens) are formed by deep drawing, but, as discussed previously, special carpets and furniture parts are drawn or embossed as well. Deep-drawable films can be classified according to the surface protected into films for processing of metals and films for processing of plastics. Metal coils are generally drawn at room temperature, whereas plastics are processed at elevated temperatures. In practice, processing of metal coils develops heat; on the other hand, the depth of processing is higher. Therefore, both protective films for metals and those used for plastic processing must possess thermal stability and a high degree of deformability.

Plastics can be processed by deep drawing at room temperature or at elevated temperatures. Among the common polymers, PC, PVC, and ABS can be processed at room temperature by deep drawing. PVC was used as the first carrier material for common protective films. Later, it was successfully tested for special applications such as deep drawing. Problems related to its recycling and environmental impact have recently arisen that forced its replacement with more acceptable, halogen-free, olefin-based polymers. The deep-drawing temperature of different PVC grades differs according to the polymerization procedure and film manufacturing process. Suspension-polymerized PVC has a lower and broader processing temperature range (125–165°C). High-temperature deep drawing of PE can be carried out at a temperature as low as 30°C. The deep drawing of PC is performed at 170–180°C. In such processes heat evolves by friction between the deep-drawing tool and the material processed. Therefore, water-based or organic lubricants are used. In the deep drawing of plastics, the different thermal expansion coefficients of different materials should be taken into account. For instance, the coefficient of thermal expansion for polycarbonate is higher (65×10^{-6} K^{-1}) than that of steel (11×10^{-6} K^{-1}).

Special protective films are appropriate for deep drawing. In this process wrinkles may form on the edge (PVC is better than PE). When the sheets are processed or embossed, wrinkles can be pressed into the surface. Warm deep drawing is one of the main processing operations for PMMA plates. The maximal pressing force by heat pressing may attain 200 kN. Form stability after processing depends on the product type. Cast PMMA achieves its form stability at 70°C, extruded PMMA at 60°C, and PC at 110°C. Polycarbonate requires a higher forming temperature (190–210°C), and its thermal processing temperature range is narrow.

Extruded PC or cast PMMA are processed as sanitary materials. The carrier film for such applications must be deep drawable and must support the processing conditions (160–170°C, 10–30 min) of the protected item. For such applications self-adhesive films (see also Chapter 7) or adhesive-coated films with a minimum thickness of 70–80 μm can be used. Generally, thin plates (less than 6–10 mm) are protected with SAF; for thicker plates, adhesive-coated protective films with a higher thickness (80–100 μm) have been proposed.

Extruded polycarbonate plates are laminated after processing and before cooling (at 60°C). Such plates are processed for 2–5 min at 180°C (e.g., for light domes, 2 min at 180°C). PC plates are hygroscopic and are generally dried before use. Drying is carried out at 130°C for 0.5–48 h, depending on the plate thickness (0.75 to 12 mm).

Taking into account the elevated temperature of their use, both the PSA and the carrier material must display special properties. The adhesive must possess good anchorage to resist shearing during embossing/drawing, but no adhesion build-up; the carrier (and PSA) must display minimum shrinkage. Therefore, short-term adhesion build-up and shrinkage are tested after storage at 130°C. Long-term adhesion build-up is tested after storage at room temperature and at 40°C after 2–8 weeks. The end adhesion should be situated at about 5 N/25 cm. If adhesiveless films are used, the strong influence of the laminating temperature should be taken into account. If adhesive-coated films are applied, attention should be paid to the dependence of the peel build-up on the coating weight.

Carpet-protecting films can be used for long-fiber tufted carpets (high-tack protection film), thermoformed, embossed carpet (high-tack deformable, deep-drawable protective film), and processing protection. Carpet films applied for protection during storage differ from automotive carpet protection films, which undergo the forming processes of automotive carpets.

Deep drawing of metals is more complex than that of plastics because metals are less ductile and require higher forces for cold drawing. According to the depth of the drawing, standard (depth of about 25 mm) or special deep drawing (more than 300 mm) is carried out. Advanced deep drawing may be carried out in one step or in multiple (one to four) steps. The forces acting on the film increase with the number of steps. Common deep-drawable film is used to protect metal plates (steel, copper, etc., with a thickness of 0.7–0.8 mm) that are deep drawn in one or more steps. This procedure is carried out at room temperature. (Friction may cause a slight increase in temperature.) The protective film coated on the substrate must display the same deformability as the adherend, that is, a plastic deformation. Elastic forces in the film should be limited to prevent delamination. The film must resist the higher temperature produced by friction and must also (partially) play the role of lubricating agent. As discussed in Ref. [215], the friction (F) during processing is the sum of the solid-state friction between metal and metal (F_S), friction in the lubricant layer (F_L) at the boundary between the coil and the tool, and the hydrodynamic and hydrostatic friction (F_H) in the thick lubricant layer:

$$F = F_S + F_L + F_H \tag{4.2}$$

For a coil covered with protective film, the PE film participates in the friction between the solid-state components. In this process a high contact pressure (15–70 N/mm²) is applied. The time-dependent interlayer phenomena during deep drawing are influenced by the compression stress caused by friction (τ_F), which depends on the temperature (T), normal stress (σ_N), relative speed (v_{rel}), and degree of forming (ε°) [216]:

$$\tau_F = f(T, \sigma_N, v_{rel}, \varepsilon^\circ) \tag{4.3}$$

The relative processing speed affects the temperature, and both depend on the drawing steps. The temperature influences the normal stress in the plastic component of the laminate (protective film).

Common deep-drawable protective films are based on a clear, 50- to 70-μm thin film of PVC or on polyolefin. No-slip agents are allowed in their composition. However, as mentioned previously, protective film should function as a lubricating agent. The carrier film for deep-drawable protective films should have good coatability, good slip, excellent deformability, and long-term deformability. Coatability refers to one-side wettability and adhesive anchorage. Good slip means that the bonded film should allow movement parallel to the drawing device. Deformability means that during embossing the film should follow the deformation of the metal plate. Metal plates are deep drawn up to 40–60 cm. Long-time deformability means that the film preserves its deformation after deep drawing and does not change form after the embossing device is removed. Such properties are given by soft PVC. Deformability is characteristic of LLDPE as well, but its elastic forces do not permit its use alone. Excellent deep drawability is given by EVAc.

According to Miura [217], the deep drawability of plastic films decreases as follows:

$$EVAc > PE > PP > PS \tag{4.4}$$

The processing temperature (°C) of the above polymers varies as follows:

$$PE(135) > PP(125) > PSt(105) > EVAc(90) \tag{4.5}$$

Unfortunately, the high blocking tendency allows the use of EVAc as coextrudate only. Special polymer compounds have been suggested. For instance, acrylic copolymers were suggested as additives for PP to improve its deep drawability [218]. Polyolefin-based deep-drawable carrier film compositions contain an extensible component like EVAc, PP, or LLDPE and other components that ensure adequate mechanical resistance, friction behavior, and machinability.

Generally, deep-drawable protective films are coated with a medium coating weight of $5.5-8.5$ g/m^2 of a water-based acrylic PSA or $4.5-5.5$ g/m^2 of a solvent-based rubber–resin PSA. Because of the higher adhesion of the rubber–resin-based formulation, such products are also release coated. For complex shapes and advanced deep drawing, rubber-based adhesives are preferred.

4.3.2.2 Protective Papers

Only a few special protective papers exist. Common (masking) protective papers, corrosion protective papers [219], and weather-resistant papers are manufactured [220]. Protective masking papers have a base paper of $49-147$ g/m^2 and can be classified as light weight, standard weight, medium weight, and heavy weight. For such products, surface and area protection requirements include all-surface protection applications such as polished metals, plastics, and wood; providing a surface on which layout drawings for holes, cutting, and bending or scoring can be made; and antimulation protection of components and finished products during assembly, shipping, and storage. Such papers are recommended when delamination resistance but clean removal is required and when temperature resistance, high abrasion resistance, markability and translucency, dimension stability, and wet rub resistance are needed.

4.3.2.3 Other Special Protection Films

Polarizing films are used for liquid crystal displays and must offer superior transparency and humidity and heat resistance. So-called E-masks are films that protect electronic components during the etching process.

Protective tapes containing silicone PSA are used extensively in the manufacture and assembly of PCB. Demand for small electronics and appliances is increasing, and silicone PSA tapes are ideal for protecting gold leaves in the processing of PCB during wave soldering. It is important to ensure that the silicone PSA leaves no residues on the substrate after removal when processing is complete. Consequently, proper composition materials must be selected to ensure this requirement can be met [75]. Growing applications for protective films in cellular phones and LCD monitors require new silicone PSAs with moderate tack and very low adhesion properties [221].

The transition of PCB board manufacturing from lead-based solders to non-lead-based solders placed new temperature requirements on the electronics industry. Whereas silicone adhesives were used early on in the masking, soldering, and plating operations involved in PCB manufacture, they were substantially replaced by silicone organic blends or high-temperature-stabilized organics. These replacement adhesives will now be replaced with silicone adhesives capable of withstanding the higher temperatures demanded by nonlead solders. To help meet these new requirements, Nakamura et al. [171] and Aoki [172] describe additives for silicone PSA formulations that enable cured silicone PSAs to withstand very high temperatures and still offer clean removability.

A recent application is for protective coverings on electronic components. These films comprise a number of layers, each providing properties to the protective coating. Satake [222] describes high-temperature applications for films. Further, Aoki [172] describes compositions that are useful in protective films for clean room technology (CRT) or optical displays.

4.3.3   Protective Film-Like Products

Another product class including protective, cover, and separation films has been developed to replace carrier-free protective coatings. The main protective film-like products include easy peel films, mirror films, separation films, and overlamination films.

Self-wound overlamination protection films are used for overlapping and permanent protection of labels or other printed items. Heat laminating has been (partially) replaced by pressure-sensitive protective laminating. Polyester and BOPP are used for pressure-sensitive overlaminating. Laminating films for book binding are used for reinforcement, repair, binding, and lamination [223].

Mirror tape (see also Section 4.2.2.8.1) is a permanent security film used to avoid dangerous destruction (cracking) of large glass surfaces. This film is laminated as a reinforcing layer on the back of mirrors or other impact-sensitive large-surface items. The film must have a high level of permanent adhesivity to avoid delamination of the glass by stress and dart-drop resistance against glass splits.

Separation films are used to protect web- or sheet-like metallic, plastic, or varnished items during storage. Because of the polar surface of the protected items (e.g., anodized aluminum or stainless steel) and the high laminating pressure, peel build-up occurs rapidly. Therefore, they must have a very low peel. They are based on rubber–resin formulations with a very low tackifier level (less than 10%) and a high degree of cross-linking.

4.3.4   Application Technology of Protective Films

As mentioned previously, protective films adhere to the surface to be protected. This is necessary to ensure a monoblock character of the product and retain its working (technological) performance. This is a main requirement toward protective films because, unlike classic packaging materials, protective films are used at the start and not the end of a technological procedure, that is, they are applied before the product to be protected undergoes its technological cycle. The contact with and the adhesion to the surface to be protected is achieved by laminating the web with the product. Within the range of protective materials, protective films form a separate class, characterized as *laminatable monoweb* construction. From the range of monoweb products, cover films (like protecting films) are functional, technological packaging materials, but they are used without laminating [175]. The application of a tape to the product to be protected also differs from the use of protective film. It is a low-pressure, discontinuous process. Other products like carrierless protective coatings are not pressure sensitive. They are tack-free polymer layers like varnishes with good adhesion on the coated surface; therefore, their removability is mainly chemically controlled.

Application and deapplication of the protective web refers to its lamination on the surface to be protected and its separation from that surface. During lamination and delamination the film is stressed by a tension; thus, its tensile strength is a very important property. Because they are used for temporary bonding only, application and deapplication are of equal importance for protective films. For instance, if the film is not detached completely, the different thermal dilatation of the metallic coil and plastic film may lead to stress cracking. In film application the chemical composition and processing quality of the surface play a special role. Generally, supplier checklists concerning the application conditions of a protective film should contain data about nature of the surface to be protected, the thickness of the web to be protected, the desired adhesive strength, and the future processing (forming) operations and UV resistance.

In a different manner from the laminating of other PSPs (labels, tapes, etc.), where low laminating pressure, high laminating speed, and (generally) normal temperature are used, the lamination of protective films may require a high laminating pressure and temperature (Table 4.2). On the other hand, the debonding (delaminating) speed of protective films is higher than that of other products. Therefore, the mechanical resistance requirements for the carrier material during application/deapplication of protective films are relatively high in comparison with those of labels and tapes. In this operation, the mechanical resistance must ensure the dimensional stability of the film necessary to allow standard debonding (peel) values (see also *Technology of Pressure-Sensitive Adhesives and Products*, Chapter 8).

Various types of substrates must be coated with protective films. Metallic or plastic surfaces are the most frequently encountered adherends. It should be taken into account that because of their higher elasticity, the processing of plastic surfaces for a given smoothness is more difficult [224]. The surface itself has a multilayered structure including a layer of dirt (3×10^{-6} mm), an adsorbtion layer (about 3×10^{-7} mm), and a reaction layer (1×10^{-4} mm), followed by the material itself [225].

Different surfaces must be protected within the same application field. A variety of lacquers are used for lacquering of various plastic automotive components (depending on their elasticity). For instance, for low-elasticity parts ($E > 3000$ N/mm²) one-component acrylic-melamine, epoxide, and two-component PU lacquers are applied. For parts with medium elasticity ($E = 1800$–3000 N/m²), one- or two-component PU acrylic- or polyester-based varnishes are suggested. For elastic parts, two-component PU lacquers have been proposed.

Customers refer to typical stainless finishes as dull, bright, or annealed. Steel and copper surfaces may be chemically etched to produce fibrous oxide growth on the surface. The level of roughness is achieved by transferring a pattern from the working (calendering) rolls to the material, which then goes to the tempering mill to be (eventually) tinned or coated with chromium or another metal.

A dull finish is developed by final rolling, followed by annealing and hot acid pickling. A low-gloss finish suitable for functional parts results when ultimate lubricant retention is required or severe forming with possible intermediate annealing occurs. Extensive polishing and buffing are required to acquire the desired finish quality.

A bright finish is achieved by rolling with high polish rolls, followed by conventional annealing and electrolytic pickling. A relatively glossy, gray surface results. This is a general-purpose finish that is used where a minimum of polishing and butting is employed to develop a bright luster. Typical uses include wheel covers, cookware, lids, etc.

For a bright annealed (BA) finish, the final rolling is done with highly polished rolls and annealing in a protective atmosphere. It is used for applications in which only a color coating is used to achieve the desired luster. A hardened finish is practicable only for certain martensitic, hardenable grades of steel. It is achieved by heat treating, annealing, and pickling. Additionally, there are many degrees of finish within each category [226]. For packaging steel, the measured roughness is expressed as the RA (average roughness) value (in micrometers). Roughness may vary from mirror finish (RA less than 0.17 µm) to matte finish (RA 0.70–1.00 µm). Generally, a steel surface is passivated. Passivation is a chemical or electrochemical process that increases the metal surface's resistance to oxidation and facilitates lacquering. Tin, chromium, and chromium oxide may be used for top coating as well.

Coil coating is the most important domain of metal coating. Various metals are used: unpainted and painted; powder and thermopainted; bright, glossy, and matte. Stainless-steel, bright annealed, cold-rolled, polished, or brushed, and aluminum, mill-finish, anodized, or brushed finishes have been protected. Coils are classified according to the metal, coating, and application. Uncoated pure metal surfaces and precoated varnished surfaced are laminated with protective films. Coated steel materials include plastisol-coated, epoxy-modified, polyester-coated, and hot-dip galvanized steel [227]. Alkyde-, PU-, or silicone-modified and fluorine-containing macromolecular compounds were coated on coils as well. Acrylics, melamine, PVC, lacquered, powder-coated, and thermolacquered surfaces may be used as substrates. Usually coils are oiled to ensure better handling and chemical protection. Plasticizers, like dioctyl sebacate or acetyl tributyl citrate, are used as the "oil." Their coating weight depends on the coil quality and use. For instance, tin-coated coils are coated with 0.5–11.2 mg/m² oil; uncoated packaging steel (black plate) has 25 mg/m² oil. For lacquering or printing 4.4 g/m² oil is used; for overseas shipments a larger quantity (10 mg/m²) is used.

The surface quality of the coil affects the choice of the PSA for the protective film. For instance, for stainless steel with a mill finish of polished or bright annealed surface quality, protective films with rubber–resin adhesive are recommended. It would be difficult to achieve adequate contact between a hard acrylic adhesive and this surface. Such an adhesive would make the edging difficult, and the film could loosen. Furthermore, it is useful to have higher adhesive strength when forming special steel. The "softness" of the rubber-based PSA formulations is used in other laminations as well. As is known from the practice of dry laminating, rubber-based PSAs are used for Zellglas/PE lamination, in which the mechanical properties of the components are quite different and a soft bonding layer is necessary.

The surface of aluminum with a mill finish is softer. For this surface an acrylic adhesive-coated protective film may be used. For aluminum coils various surface qualities (e.g., mill finish, polished, matte, and anodized) are manufactured. For electronic parts, brushed aluminum is used. Aluminum is difficult to laminate with protection film because the surface composition of aluminum changes during storage and the concentration of alkali-end alkaline-earth metals increases. An anodized aluminum surface can have very different oxide morphology. Generally, the surface of metals is rough and oxidized; the oxide layer contains water, and the water contains salts [228].

The main application fields of coated coils are the automotive industry, construction, and household machines. The varnishes for coil coating must resist processing steps such as deep drawing, drilling, and cutting. The nature of the varnishes may be very different. For instance, steel may be coated with epoxy-, polyester-, acrylic- or melamine-based paints [229]. Polyester-based coil-coating systems may contain nonvolatile ester alcohols [230,231]. Silicone acrylates are used for coil coating (surface panels for buildings) as well. Epoxyde and polyester, zinc, or acrylics and melamine are used for coating cans and coils [232]. High-solids paints, powder paints, and systems capable of being hardened by radiation were also developed.

Powders of cellulose, PA, epoxy, vinyl, acryl, and polyester may be used for metal coating. Electrostatic powder spraying is also used for coil coating and automotive parts. In Europe, the first powder-coating lines were developed in the early 1960s [233]; in the United States they were developed in 1971 [234]. The most used powder lacquers are based on epoxy, epoxy–polyester, polyester, and PU derivatives. There are cross-linked systems cured at 140–220°C. Epoxy–polyester systems are used particularly in the automotive industry, where acrylate powder lacquers are used as well. Powder-molded compounds, in which powder lacquering of the product occurs in the mold, are used for sanitary items and automotive parts.

Limitations exist for the use of adhesive-coated protective films. For instance, rubber–resin adhesive-coated protective films cannot be used for copper or brass (bright or mill finish); sulfur containing copper reacts with the rubber.

Other surfaces such as glass, plastic profiles, acrylic sheets, polycarbonate (PC) plates, textiles, carpets, linoleum, parquetry, marble, wood, and melaminated kitchen furniture also must be protected. Most are manufactured in different qualities (grades). Glass is covered with a PVA/plasticizer [235] (see also Section 4.1.1.2.2). Such plastic (ionomer)-coated glass surfaces are used as building panels. Polycarbonate plates are produced with bright, glossy, and matte surfaces (see later); melamines may be bright or matte; synthetic marble is produced with bright and matte surfaces. PVC plates are manufactured in soft and hard grades; they may also be coated (to increase their abrasion resistance and gloss) with a 0.003- to 0.004-in. layer of deck compound that displays adhesion other than common PVC [236].

Window frames are produced from hard high-impact PVC. Other polymers, like acrylinitrile–butadiene–styrene (ABS), PP, poly(ethylene-propylene) (PPE)–high-impact polystyrene (HIPS) blends, and PMMA, were tested as well. Window profiles from PVC were first marketed 40 years ago. In their formulation chlorinated PE or polyacryl ester are used as an impact modifier. Other components, for example, suspension PVC, filler (TiO_2, $CaCO_3$), and chlorinated PVC are added. Their compatibility strongly influences the surface quality of the profiles, that is, the laminability of protective films on such profiles. Lamination is hindered by short-time deformation of the profiles. After warm storage of the profiles, dimensional changes of 1.7% may occur. Screw rotation speed and the temperature in extruder also influence the shrinkage of the extruded product. Films to protect plastic window profiles must be age and weather resistant (months) and must have improved transparency to allow visual surface inspection. Generally, a 45-µm polyolefin film is used as a carrier for protective films used for plastic window profiles.

For furniture, UV-curable unsaturated polyester coatings and UV-curable polyester coatings are used. Polyester acrylates in tripropylene glycol diacrylate are UV cured on plastics for furniture. Decorative laminates are manufactured from paper impregnated with melamine resins. Modified poly-ethylmethacrylate may be used as a protective film for substrates such as PVC, ABS, and polystyrene, wood, paper, and metal.

PMMA-forming materials and semifinished products are used for autotomotive, building, light technical, advertising, household wares, and hygienic products. PMMA replaces polycarbonate in mobile phones, pocket PCs, and digital versatile discs. Semifinished products and cast and extruded PMMA plates were developed. According to Ref. [237], cast PMMA plates are manufactured with thicknesses of 2–25 and 30–100 mm; extruded plates have a thickness of 1.5–6 mm and extruded high-impact PMMA (HIPMMA) have a thickness of 2–6 mm. So-called acrylic glass is manufactured with a minimum thickness of 2–3 mm. HIPMMA plates have a gauge of 2–6 mm.

Semifinished products and cast and extruded PC plates were produced as well. Extruded PC plates have a gauge of 1–12 mm. According to Ref. [237], extruded PC plates are produced with a thickness of 1–2 mm. Their abrasion resistance is improved using special silicone alkoxy condensation products. Their surface quality may differ. As an example, polycarbonate plates can be patterned on both sides, with a pinspot pattern on one side and a prism pattern on the other side; both sides can be polished or patterned raindrop/haircell. Polycarbonate with a thickness of 0.25 mm is used for overlays for keyboards.

PE films become brittle at low temperatures and it becomes very difficult to remove them. At temperatures below zero, films should be warmed before removal. The recommended temperature for their delamination is 8–22°C. Protective films should be applied between +15 and +50°C. Table 4.5 summarizes the application conditions for the most common protective films. As demonstrated, very different application temperatures and pressures are used for protective films laminated on metallic or plastic surfaces. The application climate is strongly influenced by the nature of the substrate, substrate manufacturing procedures, and laminating equipment, as well as the construction (grade) of the protective film. For instance, high-temperature-extruded plates may be laminated with adhesive-coated protective films. In this case the adhesion of the film does not require an elevated laminating temperature; the temperature is imposed by the manufacturing procedure of the substrate to be protected.

Generally, protective films with higher adhesion (peel resistance) are required for complex processing (multiple deep drawing) and for matte surfaces. As mentioned previously, protective films used for deep drawing must also function as lubricating agents.

Protective films are generally applied by lamination, which involves the use of laminating equipment. The properties of the protective film depend on the chemical and physical properties of the adherend, its cleanness and surface temperature, and the elongation of the film during lamination (i.e., laminating method and equipment).

Large-volume objects and profiles are applied with application lines. Such lines are supplied by specialized firms. Automatic lines with stacking and cutting devices are used. The lines are generally built with an expander (banana) roller. The expander roller ensures a bubble- and wrinkle-free lamination. The brake should not be too strong, but strong enough to keep the films in a stretched position to avoid wrinkles. In some cases in which the film must be removed and reapplied, the relamination is carried out by hand. This is the case in small firms, where short (100–200 m) rolls are used. In many cases (coil coating, plastic plate manufacturing, extrusion of plastic profiles, etc.) the lamination equipment is part of the production equipment for the web to be protected. Here, the laminating speed and temperature are determined by the equipment. For instance, as mentioned previously, PVC window frames are produced at a speed of 3 m/min, so the lamination speed of the protective film must be the same.

4.3.5 Competitors of Pressure-Sensitive Protective Films

There are various coating-like and web-like competitors to pressure-sensitive protection films. Protective materials can be carrierless (coatings) or based on a solid-state carrier (protective webs). Protective materials differ according to their self-supporting or coating-like character (with carrier or without), their contact with the product to be protected (adhesion or overlapping only), and their removability from the product (with or without bond breaking). The carrier can be adhesive-coated or adhesiveless (self-adhesive). The web-like, self-adhesive competitors will be discussed in detail in Chapter 7. Coatings generally display adhesive contact with the surface to be protected. Protective webs may (e.g., protective films and separation films) or may not (e.g., packaging films and cover films) exhibit adhesive contact with the surface to be protected. This adhesive contact may be the result of an interaction between an adhesive and a substrate (e.g., protective film) or between two adhesiveless surfaces (e.g., cover film). For certain protective webs in adhesive contact with the surface to be protected, this contact is the result of lamination (e.g., lamination and separation films); in other cases (e.g., cover films), no lamination is carried out during application, and the contact between the protected surface and the protective web is not uniform.

The nature of the bond (temporary or permanent) may be different also. Common lacquers build up a permanent coating with the protected surface. The bond between laminating films used for surface protection (overlaminating films) is also permanent. Other protective materials (the majority of film-like products) are removable. Protective films were developed as a replacement for peelable varnishes and masking paper.

4.3.5.1 Protection Coatings

Peelable varnishes must fulfill the following requirements: coatability as a monolayer lacquer, weatherability, water and temperature resistance, chemical resistance, nonflammability, transparency, film-like rigidity and flexibility, and low but sufficient adhesion. According to Zorll [238], such formulations have an intermediate position between primers and varnishes concerning their adhesion and cohesion properties. Primers must display excellent anchorage on the surface to be coated, but medium cohesion (see also *Technology of Pressure-Sensitive Adhesives and Products*, Chapter 8). Varnishes must possess higher cohesion and mechanical resistance. The mechanical resistance and flexibility of removable protective lacquers must be as high as for varnishes, but their adhesion should be lower (their adhesion is about 1 N/mm^2).

Coatable films based on EVAc copolymers and waxes have been used as protection coatings [239,240]. Polyvinylbutyral has been applied for removable coatings [16]. For metal protection in applications where chemical and water resistance are required, immersion bath compositions based on thermoplasts, chlorinated paraffin, and esters of phthalic acid have been suggested.

Protective coatings for automotive use must fulfill the following requirements: chemical resistance (resistance to organic solvents, grease, and oils); versatility (suitable for all types of varnishes); weather resistance (outdoor exposure at least 6 months to pass the Florida Test on cars to resist harsh winter climates); minimum light transmission (lower than 0.04%); good environmental aspects (recycling, burning); and low costs compared to waxes (material, labor). The best known conventional protective systems for car protection are dispersions of low-molecular, nonadhesive, or slightly adhesive polymeric products like waxes or acrylic compounds. Such products build up a protective layer of about 15-μm thickness on the protected surface. These compounds are applied in a liquid state and are eliminated as a liquid system. Solvents are used for paraffin and alkalies (WB solutions) for acrylics. For temporary protection of automobiles during transport and manufacture, copolymers of C_{3-4} alkyl methacrylate and methacrylic acid have been proposed as protective coatings [241]. Such coatings sprayed on the surface to be protected yield a 6- to 10-μm film after drying in air (10–15 min) that is removable with a 10% NaOH solution. Such systems providing full surface protection are more effective than carrier-based films. Their disadvantages include working time and pollution.

Cross-linked elastomers have also been used as protective coating for metals. A patent describes the use of a polyisocyanate-cured butadiene polymer [242]. Later, protective papers replaced carrierless protective coatings. Water-removable protective papers have also been manufactured.

4.3.5.2 Non-Pressure-Sensitive Protection Films

Self-wound overlamination protection films are used for overlapping and permanent protection of labels or other printed items. They must display clarity, impact resistance, UV resistance, color stability, resistance to outdoor weathering, and printability. A matte finish, nonreflective character is needed. PP and PET have been suggested as carrier materials for such products: BOPP (25–50 μm, standard, and battery approved, normal and heat-resistant lacquered) and PET (25 μm) are generally used. For screen printers and lithoprinters, thermal laminating was used as the finishing process. For overlaminating of screen-printed labels a product build-up of 50/150, that is, 12.5 μm PET/75 μm heat-activated adhesive, has been recommended. This combination gives a greater gloss depth (in comparison with pressure-sensitive films), especially with embossed rough paper. UV-resistant thermolaminating films and inexpensive olefin copolymers have also been developed. Films based on cellulose have been applied by hot pressing on paper for protection. Ten to 15% of printed products are overlaminated. Thermal laminators vary in size and complexity. Tabletop models, machines with automatic feeders, and automatically synchronized cutoff models have been developed.

4.4 Other Pressure-Sensitive Products

As discussed previously, the main PSPs are common and special labels, tapes, and protection films. However, PSPs other than labels, tapes, and protective films with a lower production volume have also achieved economic importance. Plotter films, stickies, and decalcomanias are the main representatives of these specials products.

Plotter films are web-like PSPs that serve as raw material for the manufacture of information-carrying items. They are designed for use in graphics as signs, letters, etc. [243]. They are cut from special PSA-coated plastic films using computer programs that allow the design and mounting of letters or text fragments with dimensions of 10–1000 mm. The plotter film is used as a material for self-adhesive signs, letters, typefaces, or decorative elements. It is a bulk or surface-colored (coated) film with excellent dimensional stability, meaning there is minimal deformation during processing, storage, and application. Plotter film must display very good die-cuttability, sharp contours, no plasticity or elasticity during cutting, and excellent ability to lay flat and stiffness. They must also be weather resistant. Such carrier materials display elongation values of about 300% MD and 300% CD and modulus values of about 7 to 15 N/mm^2. High-quality products exhibit elongation values of 24–200% (DIN 53455) [244].

In Europe, the first (United States-made) pressure-sensitive PVC films were tested as plotter films in 1946 [245]. The first usable PVC films were manufactured in the 1950s. In 1984, computerized cutting plotters were launched.

PVC (cast and calendered) and polyolefin films are manufactured as carriers for plotter film. Plasticized PVC and vinyl chloride copolymers have been developed for plotter films. Films based on vinyl chloride copolymers can withstand 5–7 years of external use [246]. The finished product must exhibit excellent chemical and environmental resistance. Resistance to water, acid, alkalies, surface active agents, ethylene glycol, motor oil, methanol, gasoline, and certain other substances is required.

Such films are processed using special plotters for cutting. A procedure for the application of plotter films is described in Ref. [244]. Letter-cutting plotters have also been developed. Such cutting machines can use reels with a width of up to 1300 mm and may run at speeds of up to 400–600 mm/s and tolerances of 0.1 mm [245,247]. The pressure of the cutter is adjustable for different film thicknesses. A rollplotter can process up to 1320 × 25,000 mm web with a minimum sign dimension of 3 mm [248]. The working speed and acceleration of such machines are performance characteristics.

Plotter films are used together with a handling or fixing aid, that is, the application tape (see Section 4.2.2.1).

Such plotter films must be applied (laminated) at temperatures of 4–50°C. Wet and dry application methods are used. In a wet technique, warm water should be used for calendered film and cold water for the cast film. Their end-use temperature is in the range of −56 to +107°C. Special application papers (or films) can be used like a plotter film to produce logos or decorative elements. For such products the continuous, web-like part of the paper or film is used after the places to be lacquered or printed are cut off. Such stencil films or papers generally have a release liner because of their improved adhesivity. This is necessary to avoid ink penetration at the edges of the carrier.

Decalcomanias (transfers) are carrierless transfer tape-like products, but they can also be built up as labels. For such products the solid-state component is a technological (application) aid only. They can be manufactured using screen printing. First, a clear carrier lacquer is printed. Such a layer provides the mechanical resistance of the product. The image (in mirror print) is coated on this layer using screen (or offset) printing. The next layer is a transparent layer, followed by a screen-printed PSA [249]. In another procedure the PSA is coated on release paper. The PSA-coated liner is laminated together with a printed film.

Stickies are carrier-free PSPs designed to transfer graphical information. They use the solid-state carrier material as a release liner.

A number of pressure-sensitive imaging materials have been developed. Their number is growing with the increase in the number of new carrier materials, printing techniques, and packaging methods. For instance, in 1974 pressure-sensitive stamps were introduced on a trial basis [250]. They cost five times as much as conventional stamps. Pressure-sensitive envelopes have been developed as well. Message reply pads use low-peel PSAs [251].

PSAs can be applied as laminating adhesives. Decorative panels bearing PSA have been manufactured by pressing substrates (e.g., veneer) precoated with metallic alkoxides or chelates onto decorative sheets coated with PSA-containing reactive groups (e.g., N-methylol acrylamide).

Self-sticking carpet tiles comprise tiles with adhesive and nonsticky material on the back. Self-adhesive wall carpets have been manufactured. Self-adhesive wall covering consists of a layer of fabric with a visible surface, a barrier of paper that has one surface fixed to another fabric layer, a PSA coated on the barrier paper, and a release paper. Low-styrene-content (15%) SEBS block copolymers have been proposed for adhesives for wall covering [252].

Metallized film used for electromagnetic protection can be applied using PSAs. Such products are suggested for low-volume items with simple shapes and for optically low-quality applications [253].

PSAs or PSPs may be used as formulation or constructional components in other products. For instance, PSAs can be used for the attachment of plastic cards (see also Section 4.1.3.4) [254].

A cleaning device based on PSPs has also been developed [255]. Scent labels were produced by placing original perfume oils in a special two-layer label structure [256].

The PSA market for do-it-yourself applications has a higher growth rate than the average for PSAs. Sprayable HMPSAs are suitable for bonding laminating panels, thin plastics, fibrous paneling, and foam materials [257]. Pressure-sensitive aerosol adhesive is used to bond paper, cardboard, plastics, films, foil, felt, cloth, metal, glass, or wood. New application fields and requirements, new raw materials, and advances in manufacturing technology will force an increase in the number of special PSPs.

References

1. Benedek I., Manufacture of PSPs, in *Developments and Manufacture of Pressure-Sensitive Products*, Benedek I., Ed., Marcel Dekker, New York, 1999, Chapter 8.

2. Benedek I., Chemical Composition, in *Pressure-Sensitive Adhesives and Applications*, Benedek I., Ed., Marcel Dekker, New York, 2004, Chapter 5.

3. Benedek I., End Use of PSPs, in *Developments in Pressure-Sensitive Products*, Benedek I., Ed., Taylor & Francis, Boca Raton, 2006, Chapter 11.

4. Benedek I., Adhesive Performance Characteristics, *Pressure-Sensitive Adhesives and Applications*, Marcel Dekker, New York, 2004, Chapter 6.

5. Benedek I., Converting Properties of PSAs, *Pressure-Sensitive Adhesives and Applications*, Marcel Dekker, New York, 2004, Chapter 7.

6. Benedek I., Pressure-Sensitive Design and Formulation in Practice, *Pressure-Sensitive Design and Formulation, Application*, Benedek I., Ed., VSP, Utrecht, 2006, Chapter 6.

7. Benedek I., Manufacture of Pressure-Sensitive Labels, *Pressure-Sensitive Adhesives and Applications*, Marcel Dekker, New York, 2004, Chapter 9.

8. Benedek I., Manufacture of PSPs, in *Developments in Pressure-Sensitive Products*, Benedek I., Ed., Taylor & Francis, Boca Raton, 2006, Chapter 8.

9. Benedek I., Build-Up and Classification of PSPs, in *Developments in Pressure-Sensitive Products*, Benedek I., Ed., Taylor & Francis, Boca Raton, 2006, Chapter 2.

10. Waeyenbergh L., *19th Munich Adhesive and Finishing Seminar*, Munich, Germany, 1994, p. 138.

11. Feldstein M.M., Molecular Fundamentals of Pressure-Sensitive Adhesion, in *Developments in Pressure-Sensitive Products*, Benedek I. Ed., Taylor & Francis, Boca Raton, 2006, Chapter 4.

12. Borobodulina T.H., Feldstein M.M., Kotomin S.V., Kulichikhin V.G., and Cleary W.G., Viscoelasticity of PSA and bioadhesive hydrogels under compressive load, in *Proc. 25th Annual Meeting of Adh. Soc., and the Second World Congress on Adhesion and Related Phenomena*, Feb. 10–14, 2002, Orlando, FL, p. 147.

13. Feldstein M.M., Platé N.A., and Cleary G.W., Molecular Design of Hydrophylic Pressure-Sensitive Adhesives for Medical Applications, in *Developments in Pressure-Sensitive Products*, Benedek I., Ed., Taylor & Francis, Boca Raton, 2006, Chapter 9.

14. Feldstein M.M., Cleary G.W., and Singh P., Pressure-Sensitive Adhesives of Controlled Water Absorbing Capacity, in *Pressure-Sensitive Design and Formulation, Application*, Benedek I., Ed., VSP, Utrecht, 2006, Chapter 3.

15. Kulichikhin V., Antonov S., Makarova V., Semakov A., Tereshin A., and Singh P., Novel Hydrocolloid Formulations, Based on Nanocomposite Concept, in *Pressure-Sensitive Design, Theoretical Aspects*, Benedek I., Ed., VSP, Utrecht, 2006, Vol. 1, Chapter 7.

16. Benedek I., Principles of Pressure-Sensitive Design and Formulation, in *Pressure-Sensitive Design, Theoretical Aspects*, Benedck I., Ed., VSP, Utrecht, 2006, Chapter 4.

17. Lebez J., Development of Hot-melt Adhesives: Design of New co-terpolymers, in *Proc. 22nd Munich Adhesive and Finishing Seminar*, 1997, Munich, Germany, p. 71.

18. Benedek I., *Adhesion*, (12) 17 1987.

19. Benedek I., *Adhesion*, (3) 22 1987.

20. *Etiketten-Labels*, (5) 24 1995.

21. Hochwertige Getränkeausstattung mit Selbstklebeetiketten, Pago AG, Buchs, Switzerland 10, 2001.
22. Benedek I., The Role of Design and Formulation, in *Pressure-Sensitive Design, Theoretical Aspects*, Benedek I., Ed., VSP, Utrecht, 2006, Chapter 3.
23. Czech Z., Synthesis, Properties and Application of Water-Soluble Acrylic Pressure-Sensitive Adhesives, in *Pressure-Sensitive Design, Theoretical Aspects*, Benedek I., Ed., VSP, Utrecht, 2006, Chapter 6.
24. Benedek I., Rheology of PSAs, *Pressure-Sensitive Adhesives and Applications*, Marcel Dekker, New York, 2004, Chapter 2.
25. Czech Z., Removable and Repositionable Pressure-Sensitive Materials, in *Pressure-Sensitive Design and Formulation, Application*, Benedek I., Ed., VSP, Utrecht, 2006, Chapter 4.
26. *Verpackungs-Rundschau*, **53** 15 2002.
27. Tagsa, *Newsletter*, 1995, p. 1.
28. Nentwig J., *Converter, Flessibili- Carta-Cartone*, (1) 66 1991.
29. Post L.K., Development and Trends in High Performance Pressure Sensitive Acrylic Adhesives, in *Proc. 19th Munich Adhesive and Finishing Seminar*, 1994, Munich, Germany, p. 191.
30. Dostal J., *Verpackungs-Rundschau*, (5) 74 2004.
31. Dostal J., Innovative Chargenverfolgung, *Chemmanager*, (2) 1 2006.
32. Weiss H., RFID Industrie steckt im gefährlichem Teufelskreis, in *VDI Nachrichten*, (49) 25 2006.
33. Weiss H., RFID sorgt für regen Bibliotheksbesuch, *VDI Nachrichten*, (2) 9 2007.
34. Terfloth C., Reactive Hot-Melt Adhesives for Finishing Processes of Print and Packaging Materials, in *Proc. 31st Munich Adhesive and Finishing Symposium 2006*, Oct. 22–24, 2006, p. 243.
35. Bruckner W., RFID kommt im Mittelstand an, *VDI Nachrichten*, (2) 9 2007.
36. *Verpackungs-Berater*, (3)18 1997.
37. *Print and Produktion*, (5) 30 2004.
38. Pfeiffer H., *Druckspiegel*, **51**(6) 122 1996.
39. Hiroyoshi T., Kuribayashi H., and Usuda P. (Sumitomo Chem.Co.Ltd.), Japan Pat. 254,002/1988 *CAS*.
40. Lacave D., *Labels, Label*, (3/4) 54 1994.
41. *Labels, Label*, (3/4) 34 1994.
42. *Product Image & Security*, (2) 9 1997.
43. *Brit.Plast.*, (10) 42 2001.
44. Pierson D.G. and Wilczynski J.J., *Adhes. Age*, (8) 52 1990.
45. Müller J. and Eisele D., *ATP News-Letter*, 3.4.2004, ATP Adhesive Systems GmbH, Wuppertal, Germany.
46. *Caout. Plast.*, (9) 824 2004.
47. Dobmann A. and Planje J., *Papier Kunstst.-Verarb.*, (1) 37 (1986).
48. Dunckley P., *Adhesion*, (11) 19 1989.
49. *Kunststoff J.*, (9) 61 1986.
50. Benedek I., *Adhesion*, (4) 25 1987.
51. Czech Z., *European Adhes. Seal.*, (6) 4 1995.

52. *Label, Labels* (2) 14 1997.
53. *Package Print. Design. Int.*, (1/2) 8 1997.
54. Otter J.W. and Watts G.R., US Patent, 5,346,766, Avery International Corp., Pasadena, C.A., USA, 1994.
55. Sasaaki Y., Holguin D.L., and van Ham R., EP 0,252,717, Avery International Corp., Pasadena, C.A., USA, 1988.
56. *Label, Labels* (2) 98 1997.
57. Nitto Denko Products, Catalog Code 02,002, Tokyo, 20002, Japan, p. 6.
58. Haufe M., *ATP News-Letter*, 5.6.2004, ATP Adhesive Systems GmbH, Wuppertal, Germany.
59. Tomuschat P., UV-curable s'Silicones for Roll Labels, in *Proc. 31st Munich Adhesive and Finishing Symposium 2006*, Oct. 22–24, 2006, p. 282.
60. Dobmann A. and Viehofer A.G., in *Proc. of the 19th Munich Adhesive and Finishing Seminar*, 1994, Munich, Germany, p. 168.
61. LB2 Pago System 225, *Rundum Etikettierung im Drehstern*, Pago AG, Buchs, Switzerland 10, 2001.
62. Pago System 270DT, *Labeling of tubes*, Pago AG, Buchs, Switzerland 10, 2001.
63. *Labels, Label Int.*, (5/6) 24 1997.
64. Film-insert molding, in *British Plastics*, (10) 47 2001.
65. Ampoule labeling in the pharmaceitical industry, in *Verpackungs Rundschau*, **52** 12 2002.
66. Onusseit H., Hot Melts for Packaging-Requirements and Trends, in *Proc. 31st Munich Adhesive and Finishing Symposium 2006*, Oct. 22–24, 2006, p. 126.
67. Altenfeld F. and Breker D., Semi-Structurel Bonding with High Performance Pressure-Sensitive Tapes, 3M Deutschland, Neuss, 2nd Edition, Feb. 1993, p. 278.
68. Benedek I., Physical Basis of PSPs, in *Developments in Pressure-Sensitive Products*, Benedek I., Ed., Taylor & Francis, Boca Raton, 2006, Chapter 3.
69. ATP News-Letter, (2) 3 2004.
70. Lin S.B., Durfee L.D., Ekeland R.A., McVie J., and Schalau II, G.K., Recent advances in silicone pressure-sensitive adhesives, *J. Adhes. Sci. Technol.*, **21**(7) 605 2007.
71. *Neue Verpackung*, (3) 10 1995.
72. Shibata R. and Mixagawa H. (Hitachi Condenser Co. Ltd), Japan Pat., 6386785/ 18.04.1988, in *CAS, Adhesives*, **21** 5 1988.
73. Allmänna Svenska Elekriska AB, DBP 1276771.
74. Kilduff T.J. and Biggar A.M., US Pat., 3355545, in *Coating*, (7) 210 1969.
75. Ding Y. and Zhang W., *Zhongguo Jiaonianji*, **9**(6) 33 2000.
76. Do H.-S., Park Y.-J., and Kim H.-J., Adhesion Performance and UV-curing Behavior of UV-polymerizable Acrylic Pressure Sensitive Adhesives for Dicing Tape, in *Proc. of the 29th Ann. Meeting of the Adhesion Society Inc.*, Feb. 19–22, 2006, Jacksonville, FL, USA, p. 299.
77. Gerace M., *Adhes. Age*, (8) 1983.
78. Kendall Co., Canad Pat., 853145/1987.
79. Charbonneau R.R. and Groff G.L. (Minnesota Mining and Manuf. Co., St. Paul, MN, USA), EP 0106559B1/25.04.1984.

80. US Pat., 2925174, in Charbonneau R.R. and Groff G.L. (Minnesota Mining and Manuf. Co., St. Paul, MN, USA), EP 0106559B1/25.04.1984.
81. Bennett G., Geiß P.L., Klingen J., and Neeb T., *Adhäsion*, (7–8) 19 1996.
82. Stillwater D.C., Kostenmaki D.C., and Mazurek M.H. (Minnesota Mining and Manuf. Co., MN, USA), US Pat., 5344681/06.09.1994.
83. Suchy J., Hezina J., and Matejka J., Czech Pat. 247802/15.121987, in *CAS, Adhesives*, **19** 4 1988.
84. Blake F.D. (Minnesota Mining and Manuf., Co., St. Paul, MN, USA), EP 0141504 A1/15.05.1985.
85. *Coating*, (9) 33 1988.
86. Packard Electric, Engineering Specification, ES-M-2147.
87. Czech Z., *Adhäsion*, (11) 26 1994.
88. US Pat., 3096202, in Gleichenhagen P. and Wesselkamp I. (Beiersdorf AG, Hamburg, Germany), EP 0058 382B1/25.08.1982.
89. Haddock T.H. (Johnson & Johnson Products Inc., New Brunswick, NJ), EP 0130080B1/02.01.1985.
90. Allen Jr., D. and Flam E. (Bard Inc., Murray Hill, NJ) US Pat., 4650817/17.03.1987, in *Adhes. Age*, (5) 24 1987.
91. *Mod. Plast. Internat., Show Daily*, Oct. 23 & 24, 2004, p. 17.
92. *Adhes. Age*, (4) 6 1983.
93. Czech Z. and Sander D. (Lohmann GmbH, Neuwied, Germany), DE 44 33 005/16.09.1994.
94. Taubert K., *Adhäsion*, (10) 379 1970.
95. DDR Pat., 64111, in *Coating*, (1) 24 1969.
96. Sun R.L. and Kennedy J.F. (Johnson and Johnson Products, Inc., USA), US Pat., 4762888/09.08.1988, in *CAS, Hot Melt Adhesives*, **26** 1 1988.
97. Kishi T. (Sekisui Chem. Co. Ltd.), Japan Pat., 6369 879/29.03.1988, in *CAS Adhesives* **19** 6 1988.
98. Blackford B.B., Brit Pat., 8864365, in *Coating*, (6) 185 1969.
99. Müller H.W.J., *Adhäsion*, (5) 208 1981.
100. *Coating*, (6) 198 1993.
101. *Druckprint*, (4) 19 1988.
102. Adhesives Tapes Ltd., Brit. Pat., 861358, in *Coating*, (6) 185 1969.
103. Verseau J., *Coating*, (11) 309 1971.
104. Nagasuka A. and Kobari M. (Sekisui Chem. Co., Ltd.), Japan Pat., 6389585/20.04.1988, in *CAS Adhesives*, **19** 6 1988.
105. Peppas N.A. and Buri, P.A., *J. Control. Release*, (2) 257 1985.
106. Duchêne D., Touchard F., and Peppas N.A., *Drug Dev. Ind. Pharm.*, **14** 1912 1988.
107. Santos C.A., Freedman B.D., Ghosn S., Jacob J.S., Scarpulla M., and Mathiowitz E., *Biomaterials*, **24** 3571 2003.
108. Tamburic S. and Craig D.Q.M., *Eur. J. Pharm. Biopharm.* **44** 157 1997.
109. Tae G., Kornfield J.A., Hubbel J.A., *Biomaterials*, **26** 5259 2005.
110. Satas D., *Adhes. Age*, (8) 30 1988.
111. Wichterle O., Chemistry and Technology of Hydrogels especially for Applications in Ophtalmology, *Kaut. Gummi. Kunststoffe*, **41**(1) 85 1988.

112. Chand G., Brit Pat., 723226, in *Coating,* (6) 184 1969.
113. Gritskova I.A., Raskina L.P., Voronov S.A., Channova G.K., Avdeev D.N., Malyukova E.B., Vydrina T.K., Sandomirskaya N.D., and Solozhentseva O.M., *Otkrytiya, Izobret.,* (42) 87 1987, in *CAS, Coating & Inks and Related Products,* 11 11 1988.
114. Krampe S.E. and Moore C.L., EP 0,202,831 A2 (Minnesota Mining and Manuf. Co., St. Paul, MN, USA), 1986.
115. Guvendiren A.M., Lee B.P., Messersmith Ph.B., and Shull K.R., Synthesis and Adhesion Properties of DOPA Incorporated Acrylic Triblock Hydrogels, in *Proc. of the 29th Ann. Meeting of the Adhesion Society Inc.,* Feb. 19–22, 2006, Jacksonville, FL, USA, p. 277.
116. Smit E., Paul C.W., and Meisner C.L., Acrylic Block-Copolymer Hot-Melt PSAs, in *Proc. 31st Munich Adhesive and Finishing Symposium 2006,* Oct. 22–24, 2006, p. 295.
117. von Bittera M., Schäpel D., von Gizycki U., and Rupp R. (Bayer AG, Leverkusen, Germany), EP 0147588 B1/10.07.1985.
118. Wabro K., Milker R., and Krüger G., *Haftklebstoffe und Haftklebebänder,* Astorplast GmbH, Alfdorf, Germany, 1994, p. 48.
119. Ulman K. and Thomas X., in *Advances in Pressure Sensitive Adhesive Technology,* Satas D., Ed., Satas & Associates, Warwich, RI, pp. 133–157 (1995).
120. Gantner D. et al., PCT Patent Application WO2005051442 A1, in Lin S.B., Durfee L.D., Ekeland R.A., McVie J., and Schalau II G.K., Recent advances in silicone pressure-sensitive adhesives, *J. Adhes. Sci. Technol.,* 21(7) 605 2007.
121. Gantner D. et al., PCT Patent Application WO2005102403 A1, in Lin S.B., Durfee L.D., Ekeland R.A., McVie J., and Schalau II G.K., Recent advances in silicone pressure-sensitive adhesives, *J. Adhes. Sci. Technol.,* 21(7) 605 2007.
122. Lower L. and Jones, L., *Adhes. Age,* (2) 234 2002.
123. Murphy K., PCT Patent WO2006028612, in Lin S.B., Durfee L.D., Ekeland R.A., McVie J., and Schalau II G.K., Recent advances in silicone pressure-sensitive adhesives, *J. Adhes. Sci. Technol.,* 21(7) 605 2007.
124. Stempel E., US Patent 2005282977 A1, in Lin S.B., Durfee L.D., Ekeland R.A., McVie J., and Schalau II G.K., Recent advances in silicone pressure-sensitive adhesives, *J. Adhes. Sci. Technol.,* 21(7) 605 2007.
125. Inoue K., Ogawa K., Okada J., and Sugibayashi K., *J. Controlled Release,* 108(2–3) 306 2005.
126. Bott R., U. S. Patent 2004105874 A1, in Lin S.B., Durfee L.D., Ekeland R.A., McVie J., and Schalau II G.K., Recent advances in silicone pressure-sensitive adhesives, *J. Adhes. Sci. Technol.,* 21(7) 605 2007.
127. Bougherara C., PCT Patent WO2005021058 A2, in Lin S.B., Durfee L.D., Ekeland R.A., McVie J., and Schalau II G.K., Recent advances in silicone pressure-sensitive adhesives, *J. Adhes. Sci. Technol.,* 21(7) 605 2007.
128. Hoshi M., Japanese Patents JP 2004275329 A2 and 2004275213 A2, in Lin S.B., Durfee L.D., Ekeland R.A., McVie J., and Schalau II G.K., Recent advances in silicone pressure-sensitive adhesives, *J. Adhes. Sci. Technol.,* 21(7) 605 2007.
129. Fattman G., European Patent Application, EP 1424088, in Lin S.B., Durfee L.D., Ekeland R.A., McVie J., and Schalau II G.K., Recent advances in silicone pressure-sensitive adhesives, *J. Adhes. Sci. Technol.,* 21(7) 605 2007.

130. Kobayashi T., Japan Patent JP 2002200160 A2, in Lin S.B., Durfee L.D., Ekeland R.A., McVie J., and Schalau II G.K., Recent advances in silicone pressure-sensitive adhesives, *J. Adhes. Sci. Technol.*, **21**(7) 605 2007.
131. Cox, C., PCT Patent Application WO 2000053139, in Lin S.B., Durfee L.D., Ekeland R.A., McVie J., and Schalau II G.K., Recent advances in silicone pressure-sensitive adhesives, *J. Adhes. Sci. Technol.*, **21**(7) 605 2007.
132. Kosal J., U. S. Patent Application Publ. 2003065086 A1, in Lin S.B., Durfee L.D., Ekeland R.A., McVie J., and Schalau II G.K., Recent advances in silicone pressure-sensitive adhesives, *J. Adhes. Sci. Technol.*, **21**(7) 605 2007.
133. Posten J., *J. Wound Care*, (10) 9 2000.
134. Moffatt C.J., et al., European Wound Management Association Position Document on Pain at Wound Dressing Changes (2002), in Lin S.B., Durfee L.D., Ekeland R.A., McVie J., and Schalau II G.K., Recent advances in silicone pressure-sensitive adhesives, *J. Adhes. Sci. Technol.*, **21**(7) 605 2007.
135. Donovan D.A., Mehdi S.Y., and Eadie P.A., *J. Hand Surgery*, (12) 355 1999.
136. Platt A.J., *J. Internat. Soc. for Burn Injuries*, (11) 812 1996.
137. Kulichikhin V., Antonov S., Cleary G.V., and Singh P., Hydrocolloid Formulations, Based on Nanocomposite Concept, in *Proc. 11th Annual International Conference on Composites/Nano Engineering*, ICCE-Aug. 8–14, 2004, Hilton Head Island, USA.
138. Abe H., Hayashi K., and Sato M., Eds., *Data Book on Mechanical Properties of Living Cells, Tissues, and Organs*, Springer-Verlag, Tokyo, 1996.
139. Beiersdorf A.G., Hamburg, Germany, DBP 1667940, in *Coating*, (12) 363 1973.
140. DBP 1079252, in *Coating*, (6) 185 1969.
141. Lohmann K.G., US Pat., 3086531, in *Coating*, (6) 185 1969.
142. Riedel J.E. and Cheney, P.G. (Minnesota Mining and Manuf. Co., St. Paul Min., USA), US Pat., 4292360/29.09.1981, in *Adhes. Age*, (12) 58 1981.
143. *British Plastics*, (10) 63 2001.
144. Skuratovskaya T.N., Annikov O.V., Kvasko N.Z., Stolyarov V.I., Podvolotskaya M.D., Yushelevski Yu.A., and Filenko A.S., USSR Pat., 1395721/15.05.1988, in *CAS Cross-linking Reactions*, **22** 13 1988.
145. K2004, *Show Daily*, Oct. 23, 2004, p. 17.
146. Pennace J.R., Ciuchta C., Constantin D., and Loftus T., WO8703477A/18.06.1987.
147. Bruner S. and Freedman J., *Drug Delivery Technology*, **6**(2) 48 2006.
148. Reilly J., in *Proc. 31st Annual Meeting & Exposition of the Controlled Release Society*, June 12–16, 2004, p. 105.
149. Qvist M.H., et al., *Internat. J. Pharmaceutics*, **231**(2) 253 2002, in Lin S.B., Durfee L.D., Ekeland R.A., McVie J., and Schalau II G.K., Recent advances in silicone pressure-sensitive adhesives, *J. Adhes. Sci. Technol.*, **21**(7) 605 2007.
150. Raul, et al., PCT Patent Application WO2005092300 A1, in Lin S.B., Durfee L.D., Ekeland R.A., McVie J., and Schalau II G.K., Recent advances in silicone pressure-sensitive adhesives, *J. Adhes. Sci. Technol.*, **21**(7) 605 2007.
151. Yelin G., Hotmelt Extrusion-Quality and Flexibility through Continuous Production, in *Proc. 31st Munich Adhesive and Finishing Symposium 2006*, Oct. 22–24, 2006, p. 17.

152. Trenor S., Suggs A., and Love B., An examination of how skin surfactants influence a model PIB PSA for transdermal drug delivery, in *Proc. 24th Annual Meeting of Adh. Soc.*, Feb. 25, 2001, Williamsburg, VA, p. 144.
153. Lee J.-A. and McCarthy T.J., Synthesis of poly(butadiene-alt-maleic anhydride) copolymer and its absorbtion to silicon wafer substrates, in *Proc. of the 29th Ann. Meeting of the Adhesion Society Inc.*, Feb. 19–22, 2006, Jacksonville, FL, USA, p. 140.
154. Larimore F.C. and Sinclair R.A. (Minnesota Mining and Manuf. Co, St. Paul, MN, USA), EP0197662A1/15.10.1986.
155. Takashimizu K. and Suzuki A. (Advance Co. Ltd.), Japan Pat., 63 92683/23.04.1988, in *CAS Adhesives*, **19** 5 1988.
156. US Pat., Blake, 3.865770, in Larimore F.C. and Sinclair R.A. (Minnesota Mining and Manuf. Co, St. Paul, MN, USA), EP0197662A1/15.10.1986.
157. Sanderson F.T., *Adhes. Age*, (11) 26 1983.
158. Little A.D., Inc., US Pat., 3039893, in *Coating*, (6) 185 1969.
159. *Adhes. Age*, (9) 82 1984.
160. Bennett G.D. (Simmonds Precision, New York), US Pat., 925735/20.01.1987, in *Adhes. Age*, (5) 28 1988.
161. US Pat. 4286047, in Charbonneau R.R. and Groff G.L. (Minnesota Mining and Mnf. Co, St. Paul, MN, USA), EP 0106559 B1/25.04.84.
162. Hamada Y. and Takuman O. (Toray Silicone Co., Ltd.), EP 253601/20.01.1988.
163. *Adhes. Age*, (3) 8 1987.
164. Jacob L., New Development of Tackifiers for SBS Copolymers, in *Proc. 19th Munich Adhesive and Finishing Seminar*, 1994, Munich, Germany, p. 107.
165. Kadioglu F., Adams R.D., and Guilde F.J., The performance of a pressure-sensitive structural bonding tape as a structural adhesive, in *Proc. 23th Annual Meeting of Adh. Soc.*, Feb. 23–25, 2000, Williamsburg, VA, p. 246.
166. *Adhäsion*, (6) 14 1985.
167. Moldvai T. and Piatkowski N. (Inst. Cerc. Pielarie, Incaltaminte, Bucharest), Rom. Pat., 93124/30.12.1987, *CAS, Adhesives*, **19** 3 1988.
168. Lipson R.W. (Kwik Paint Products), US Pat., 5468533, in *Adhes. Age*, (5) 12 1996.
169. Kazuyoshi E., Hiroaki N., Katsuhisa T., Yoshita K., and Saito T. (FSK Inc.), Japan Pat., 6317981/25.01. 1988, in *CAS, Adhesives*, **12** 5 198.
170. Kurono T., Okashi N., and Tanaka N. (Nitto Electric Ind. Co. Ltd.), Japan Pat., 6333487/13.02.1988, in *CAS, Adhesives*, **12** 5 1988.
171. Nakamura A., PCT Patents WO2004111151 A3 and WO2006003853 A3, in Lin S.B., Durfee L.D., Ekeland R.A., McVie J., and G.K. Schalau II G.K., Recent advances in silicone pressure-sensitive adhesives, *J. Adhes. Sci. Technol.*, **21**(7) 605 2007.
172. Aoki S., European Patent EP1295927 A1 2003.
173. *Adhes. Age*, (9) 8 1986.
174. Stott D.M, *Surface Coatings*, (11) 296 1969.
175. Benedek I., Frank E., and Nicolaus G. (Poli-Film Verwaltungs GmbH, Wipperfürth, Germany), DE 4433626A1/21.09.1994.
176. *Hyvis/Napvis Polybutenes*, PB 301, BP Chemicals.
177. *Coating*, (3) 12 1985.

178. Cordorch C., *VDI Nachrichten* (33) 14 1994.
179. Risser A.J. (Cities Cervice Co.), US Pat., 3759780/18.09.1973.
180. Vipin M., *"Application of Acrylic Rubbers in PSA,"* in TECH 12, Advances in Pressure Sensitive Tape Technology, Technical Proceedings, Ithasca, IL., USA, May, 1989, p. 191.
181. Clemmens M., Wagner C.M., and Gabriele P.D., US Patent 2004/0219355 A1, in Lin S.B., Durfee L.D., Ekeland R.A., McVie J., and Schalau II G.K., Recent advances in silicone pressure-sensitive adhesives, *J. Adhes. Sci. Technol.*, 21(7) 605 2007.
182. *Openlines*, (8), Spring 93, P.P. Payne Ltd., England.
183. Nakahata H., Japan Pat., 07316510/05.12.1995, in *Adhes. Age*, (5) 11 1996.
184. Yamazaki N. (Matsushita Electric Ind. Co., Ltd.), Japan Pat., 63 86782/18. 04.1988, in *CAS, Adhesives*, 12 5 1988.
185. Becker H., *Adhäsion*, (3) 79 1971.
186. Sakai K., Inoue K. (Marubeni Co.), Japan Pat., 6348379/01.03.1988, in *CAS, Adhesives*, 15 3 1988.
187. *Papier Kunst. Verarb.*, (1) 20 1996.
188. 3M Presse information, Interpack-Scotch 3560-O/04.02, Scotch Security tape from 3M.
189. Vesley G.F., Paulson A.H., and Barber E.C., EP 0202938A2/26.11.1986.
190. Czech Z., Production of Carrier-Less Solvent-Free PSA Tapes, in *Proc. 31st Munich Adhesive and Finishing Symposium 2006*, Oct. 22–24, 2006, p. 253.
191. TNII Bumagi, SSSR Pat., 300561, in *Coating*, (7) 368 1969.
192. Carr A.F., *Coating*, (11) 334 1973.
193. Grabemann J. and Hauber R. (Hans Neschen GmbH & Co KG, Bückeburg, Germany), DE 42 31 607/01.09.1994.
194. *Adhes. Age*, (10) 125 1986.
195. Sala (International Carbon Solvent S.p.A.), EP 273997/13.07.1988, in *CAS, Adhesives*, 24 5 1988.
196. *Papier, Kunstst. Verarb.*, (19) 32 1990.
197. Menges G., Michaeli W., Ludwig R., and Scholl K., *Kunststoffe*, 80(11) 1245 1990.
198. Höflin E. and Breu H., *Adhäsion*, 10(6) 252 1966.
199. *Allg. Papier Rundsch.*, 29 798 1988.
200. Lower L. and Jones L., *Adhes. Age*, February 2002, in Lin S.B., Durfee L.D., Ekeland R.A., McVie J., and Schalau II G.K., Recent advances in silicone pressure-sensitive adhesives, *J. Adhes. Sci. Technol.*, 21(7) 605 2007.
201. Ko J., PCT Patent WO2000068336, in Lin S.B., Durfee L.D., Ekeland R.A., McVie J., and Schalau II G.K., Recent advances in silicone pressure-sensitive adhesives, *J. Adhes. Sci. Technol.*, 21(7) 605 2007.
202. Sheridan J.M., PCT Patent WO200204571, in Lin S.B., Durfee L.D., Ekeland R.A., McVie J., and Schalau II G.K., Recent advances in silicone pressure-sensitive adhesives, *J. Adhes. Sci. Technol.*, 21(7) 605 2007.
203. *Adhes. Age*, (12) 8 1986.
204. Kopp, Total Packaging Solutions, Catalogue 2002/20003, Kopp Verpackunssysteme, Reichenbach, Germany.
205. Schroeder K.F., *Adhäsion*, (5) 161 1971.

206. *Siegel Band*, Kalle Folien, Hoechst, Mi 1984, 38T 5.84
207. *Coating*, (8) 98 1987.
208. Gondro C., *Kunststoffe*, (10) 1082 1990.
209. *British Plast.*, (10) 48 2001.
210. Schlarb A.K. and Ehrenstein G.W., *Kunststoffe*, **83**(8) 597 1993.
211. Eisele D., *ATP News-Letter*, **2** 4 2004, ATP Adhesive Systems GmbH, Wuppertal, Germany.
212. Eisele D., *ATP News-Letter*, **2** 5 2004, ATP Adhesive Systems GmbH, Wuppertal, Germany.
213. *Polycarbonate*, Technical Booklet, 15/3/82, ERTA Tielt, Belgium.
214. Heinze M., *Kunststoffe*, **83**(8) 630 1993.
215. Doege E. and Hesberg U., *Blech, Rohre, Profile*, **39**(1) 25 1992.
216. Herrmann M., *Blech, Rohre, Profile*, **40**(2) 164 1993.
217. Miura T., *Giesserei*, **62**(17) 437 1975.
218. *Kunststoff Information*, (1169) 5 1994.
219. Ludlow Corporation, US Pat., 1157154, in *Coating*, (12) 353 1970.
220. Hamada K., Uchiyama U., and Takemura S., Japan Pat., 13765/69, in *Coating*, (12) 353 1970.
221. Kanar N., in *Proc. of 29th Pressure Sensitive Tape Council Technical Seminar* 2006, in Lin S.B., Durfee L.D., Ekeland R.A., McVie J., and Schalau II G.K., Recent advances in silicone pressure-sensitive adhesives, *J. Adhes. Sci. Technol.*, **21**(7) 605 2007.
222. Satake M., PCT Patent WO2003065086 A1. in Lin S.B., Durfee L.D., Ekeland R.A., McVie J., and Schalau II G.K., Recent advances in silicone pressure-sensitive adhesives, *J. Adhes. Sci. Technol.*, **21**(7) 605 2007.
223. *Coating*, (1) 29 1978.
224. *Adhes. Age*, (12) 38 1987.
225. Fauner G., *Adhäsion*, (4) 27 1985.
226. *Product Application Guide*, Poli Film-America, Cary, IL, USA, 1994.
227. *European Adhes. J.*, (6) 22 1995.
228. Bascom W.S.D. and Patrick R.L., *Adhes. Age*, (19) 25 1974.
229. Antonioli V., *La Revista del Colore*, (58) 79 1973.
230. Dynamit Nobel A.G., Netherlands Pat., 72.07134, in *Coating*, (5) 122 1974.
231. *Adhäsion*, (1/2) 28 1983.
232. Bond G.F. and Ralston J.N., *Ind. Finish. Surf. Coat.*, (305) 4 1973.
233. Seifert K.H., *Coating*, (2) 41 1972.
234. Brushwell W., *Farbe u. Lack*, (1) 34 1974.
235. Ball Brothers, Muncie, IN, USA, DBP 128585, in *Coating*, (7) 274 1969.
236. Saada M. and El-Moualled M., *Chimie et Ind.* (15) 1917 1970.
237. *Plastverarb.*, (9) 9 1994.
238. Zorll U., *Adhäsion*, (90) 236 (1975).
239. *Coating*, (6) 188 1969.
240. *Adhäsion*, (3) 83 1974.
241. Cucu O., Ilie C., Moga N., and Popescu M., Romanian Pat., 92, 466/1987, *CAS, Coating Inks, Related Products*, **11** 6 1988.
242. Hitco S.A., French Pat., 2, 045, 508, in *Coating*, (11) 336 1972.

243. Uhlemayr R. (O.T. Drescher GmbH, Rutesheim, Germany), Gebrauchsmuster, PS 3508114/22.05.1986.
244. *Selbstklebefolien maßgeschnitten*, Booklet, Grafityp, Houthalen, Belgium, October 1992.
245. *Grafityp Newsletter*, (11) 10 1997.
246. *Druckwelt*, (5) 51 1988.
247. Perigraf, Computer und Informationssysteme GmbH, Eibau, Germany, 2002.
248. Graphtec, RFC3100-60, Tischschneideplotter, Multiplot Grafiksysteme, Bad Emstal, Germany, 1997.
249. Hadert H., *Coating* (1) 11 1969.
250. Prane J.W., *Adhes. Age*, (1) 44 1989.
251. *Emulsion Adhesive E 959*, Sealock, Andover, Hampshire, U.K., 1995.
252. Ohata Y., Awano K., Atsuji M., and Hattori T. (Toa Gohsei Chemical Industry Co., Ltd., Aica Kogyo Co., Ltd.), Japan Pat., 63, 89345/20.04. 1988, in *CAS*, *Adhesives*, **21** 10 1988.
253. Weiss J., *Metall*, (5) 367 1994.
254. Weidauer J., *European Adhes., Seal.*, (5) 26 1995.
255. *Adhes. Age*, (3) 10 1987.
256. *Pack Report*, (3) 26 1997.
257. *European Adhes., Seal.*, (6) 16 1996.

5

Skin Contact Pressure-Sensitive Adhesives

Valery G.
  Kulichikhin
Sergey V. Antonov
Russian Academy
of Sciences

Natalya M.
  Zadymova
Lomonosov Moscow
State University,
Chemical Department

5.1 Introduction

The application of pressure-sensitive adhesives (PSAs) in skin contact products requires several important features that predetermine their composition, structure, processability, etc. These features are connected with biological properties of skin and its nature as a substrate (see also Chapter 6).

First, all components of skin contact PSAs should be safe and approved for contact with skin. Wearing the adhesive patch on the skin must not cause irritation or damage to tissues in any way. In the case of injured skin (wounds or burns), the PSA must not release any harmful substances into the tissue and should be easily removable without damage to the skin. The components of the adhesive should not leach out to the wound or remain in the wound after patch removal.

Second, skin has a complicated structure. Skin consists of three layers: the epidermis, which includes the stratum corneum, the dermis, and the hypodermis. Each layer has its own structure and hence its own mechanical characteristics. The combination of layers with different properties leads to nonlinear anisotropic viscoelastic nature of skin [1]. In addition, skin has evident relief as evidenced by wrinkles, etc. (see Chapter 6). As a result, the adhesive should have a good ability to flow (spreading upon application) to fill the dimples in the skin surface, thus establishing good contact between the adhesive and the skin.

Third, although hydrophobic adhesives may have good initial tack to skin, their wearing properties are poor. The flux of moisture (sweat, exudate) through the skin cannot penetrate the hydrophobic adhesive layer; therefore, moisture accumulates at the skin surface, which leads to detachment of the hydrophobic adhesive. This factor is even more crucial for wound dressings applied on open wounds that should absorb enormous quantities of exudate to maintain an optimal humidity level for healing [2–6]. Excessive moisture may cause not only detachment of the patch but also excessive maceration of the skin, which may result in damage to the healthy skin surrounding the wound upon removal of the dressing [7]. The patch should therefore either contain special components for proper management of incoming moisture (absorbents, etc.) or be capable of transporting the moisture through the adhesive layer to the breathable backing, with further evaporation of the moisture from the outer surface of the backing. In other words, hydrophilic components are necessary for moisture management (see also *Technology of Pressure-Sensitive Adhesives and Products*, Chapter 7). In many cases, breathable backings cannot be used and all incoming moisture must be absorbed and stored in the patch. Patches utilizing diffusion of moisture through the adhesive layer and a breathable backing should be thin, whereas a thick adhesive layer is necessary for moisture storage. If thick patches have insufficient shear strength, the adhesive could be squeezed out under a compressive load (e.g., under the weight of a patient). This process is sometimes called "cold flow" (see also *Fundamentals of Pressure Sensitivity*, Chapter 9). The initially good cohesive strength may deteriorate over time with moisture sorption as a result of poor cohesive strength of the solutions and gels of hydrophilic polymers.

Fourth, a prominent intersubject variability of skin properties exists among humans. The properties of skin also change with age [8,9].

For long-term wear, skin-contact adhesive should have good adhesion to both dry and wet skin. Good adhesion can be achieved using a combination of hydrophobic and hydrophilic components. Dressings of this type are usually (although maybe improperly) called "hydrocolloid dressings" [10].

Because they are multiphase systems, hydrocolloid dressings cannot be completely transparent. However, semitransparent (translucent) patches are often preferred for aesthetic reasons or to facilitate the observation of wound healing. Initially translucent patches may become opaque upon moisture sorption.

The overall performance of hydrocolloid formulations depends mainly on the following properties.

1. Adhesion to dry and wet skin
2. Moisture management
3. Shear resistance/cohesive strength

The relative importance of these parameters depends on the site of application. Let us consider, for example, foot care PSAs (Figure 5.1). The patch applied to the bottom

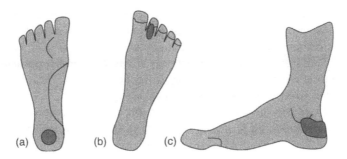

FIGURE 5.1 Sites of application for foot care products: (a) on the bottom surface, (b) on a toe, or (c) on the heel.

surface of the foot (Figure 5.1a) is exposed to the cycling compressive loading caused by the body's weight. The cold flow problem (and hence cohesive strength and elastic recovery) is of the utmost importance for this patch. The patch applied to the toe (Figure 5.1b) should withstand shearing forces from small movements of the feet in shoes during walking. On the other hand, the amount of moisture that should be absorbed in this position is relatively small. In the case of a patch applied to the heel (Figure 5.1c), moisture management and resistance to shearing forces that arise from small movements of the feet in shoes during walking are more important than resistance to compressive loads.

Hydrocolloid wound dressings are successfully used in healing moderately exuding wounds. They maintain the natural moisture level of the tissues without excessive maceration or dehydration. Moist healing enables faster reepithelialization compared with wounds kept under dry conditions [11].

The clinical efficacy and area of application of hydrocolloid wound dressings can be significantly expanded with the introduction of an antimicrobial agent [12]. The release rate of the antimicrobial agent and, hence, the antimicrobial activity of such dressings, is influenced significantly by the composition and structure of the hydrocolloid formulations.

The application of hydrocolloids in transdermal drug delivery products providing steady drug release is another point of interest. Hydrocolloid systems based on emulsions and microemulsions present some special advantages.

These issues will be discussed later. We begin with general approaches to the design of hydrocolloid products.

5.2 Structure of Hydrocolloid Dressings

As a rule, hydrocolloid formulations contain two parts: hydrophobic and hydrophilic. The hydrophobic part is responsible for dry tack, integrity, and mechanical properties, whereas the hydrophilic part provides wet tack, moisture transportation, and storage. If a hydrocolloid formulation contains active species (antimicrobial agents or drugs in transdermal delivery systems), these species can be associated with either hydrophobic or hydrophilic components, depending on their nature. The patch should be able to provide the necessary release rate of these species into the body upon application.

Both the hydrophobic and the hydrophilic parts of a formulation may contain immiscible components, thus increasing the total number of phases. The complicated heterophase structure should be arranged in a proper pattern so as to act efficiently. For example, moisture from the skin must have access to absorbing particles that can be otherwise locked in by the surrounding hydrophobic matrix. Morphology is very important in the performance of skin-contact adhesives.

The patch design is another important factor. The choice of breathable or occlusive backing, patch thickness, shape, etc., also influences the performance of the skin-contact product.

5.2.1 Hydrophobic Part

As noted previously, the hydrophobic part is responsible for the dry tack and integrity of the hydrocolloid dressing as a whole. This is explained by the fact that hydrophilic components easily lose their cohesive strength or even dissolve upon interaction with moisture, whereas the strength of the hydrophobic components remains practically unaffected. The hydrophobic components should therefore form a continuous phase and their content in the formulation is usually greater than 50%. This amount, however, can significantly decrease upon absorbing moisture due to swelling of the hydrophilic components and the adhesive may also lose its integrity. In early hydrocolloid formulations the content of the hydrophobic phase was lower than 50% [13]. Upon contact with exudate such compositions formed a soft gel that remained in the wound after removal of the hydrocolloid dressing. The gel had to be irrigated from the wound. The operation was both time-consuming and painful for the patient. Particles of the hydrophobic phase that were entrapped in the hydrophilic gel may have remained in the wound, causing irritation upon healing [7].

Generally speaking, an adhesive may be manufactured via solution or hot-melt technology. The drawbacks of the solution approach are well known: the unavoidable and time-consuming drying stage and the need for solvent recuperation and regeneration. These drawbacks are even more important for hydrocolloid formulations. As mentioned previously, the hydrocolloid patches are usually rather thick, which makes drying more difficult. The presence of residual solvents may be undesirable from a safety viewpoint. Furthermore, particles from the hydrophilic phase should be introduced into solution to be incorporated into the structure of the hydrocolloid formulation. The hydrophilic particles tend to precipitate in a suspension that may lead to variability in composition upon manufacturing. This problem can be solved by permanent agitation. Precipitation of the suspended particles may occur in the cast film upon drying, thus leading to the gradient of the properties along the thickness of the adhesive layer.

Some problems in solution technology can be overcome by the emulsion approach that will be discussed later in this chapter. Nevertheless, it can be concluded that the solution technology is not favorable for producing thick hydrocolloid patches.

Hot-melt technology involves mixing the polymers with other excipients in the melted state at elevated temperatures, followed by the extrusion or hot pressing of the prepared blend. Extrusion is a preferred method of manufacturing due to its high productivity. Sometimes it is possible to manufacture adhesives in a one-step process, where mixing proceeds in an extruder equipped with mixing elements. This method additionally increases the productivity and usually decreases the components' time of exposure to elevated temperatures.

The drawbacks of the hot-melt process are obvious: due to the high viscosity of polymer melts, mixing can be challenging; hot-melt PSAs have limited heat resistance; the method is inappropriate for producing very thin films; and components of the formulation and especially the active species need protection from thermal destruction.

Historically, low-molecular-weight (MW) polyisobutylene (PIB) was the first hydrophobic matrix to be applied in hydrocolloid PSAs since the 1960s [7]. An advantage of PIB is that it contains a chemically saturated aliphatic carbon–carbon backbone and therefore needs no stabilizer to reduce the degradation often seen in rubbery materials that have unsaturated backbones. However, PIB does not provide sufficient cohesive strength. Although PIB is chemically inert, it is suspected as the cause of abnormal foam cells (or giant cells) observed in the histology of healing dermal tissues. The shear behavior of PIB can be improved by reinforcing it with fillers, and particles of hydrophilic components can play this role [14] (see also *Technology of Pressure-Sensitive Adhesives and Products*, Chapter 4).

Curable rubbers can also be processed via hot-melt mixing/extrusion of the mixing/extrusion procedure is followed by curing, which, in turn, can be induced by action of chemical cross-linkers upon additional heating or ionizing radiation.

By controlling the degree of cross-linking with the amount of the cross-linking agent and the parameters of the process, it is possible to achieve a very good balance of adhesive and cohesive properties which is crucial for adhesives. To obtain homogeneous curing of the hydrophobic polymer, however, it is necessary to achieve uniform distribution of the cross-linking agent (usually taken in small amounts) in the matrix, which can be challenging. Temperature control of the extrusion and curing stages is very important for heat-induced curing: the curing must not start at the extrusion stage. Local overheating is therefore very dangerous for such systems. Mixing of high-viscous melts is always accompanied by the release of viscous heat, which is especially important for large batches. Migration of the cross-linking agent to the hydrophilic phase and its safety upon skin contact are also important.

Cross-linking in the presence of ionizing radiation seems attractive because ionizing radiation is widely used as a sterilization process. The main problem with this method is the variability of the radiation dose observed in practice [7].

Thermoplastic elastomers based on linear styrene block copolymers (SBCs) of the A-B-A type, described in detail in *Technology of Pressure-Sensitive Adhesives and Products*, Chapter 3, became very popular as main components of the hydrophobic part of hydrocolloid PSAs because of their favorable set of properties. In these copolymers, the A blocks are thermoplastic polystyrene end blocks and the B block is a rubbery midblock of polyisoprene (PI), polybutadiene, poly(ethylene/butylene), or poly(ethylene/propylene). They are meltable and can be easily processed in a molten state at reasonable temperatures. The two blocks are, however, thermodynamically incompatible and physical segregation of styrene and rubbery blocks proceeds upon cooling. Styrene domains that have a glass transition temperature of ~100°C [15] form a strong network that enables good mechanical properties, shear strength, and elasticity at exploiting temperatures. The properties of SBC at room temperature are similar to the properties of cured rubbers.

The SBC structure can be imagined as a set of styrene domains connected by rubbery midblocks. The size of the styrene domains estimated from X-ray data for a Kraton TR 1107 styrene–isoprene–styrene copolymer is approximately 20 nm [16].

Properties of SBCs depend mainly on three factors: nature of the midblock, styrene content, and MW (see also *Technology of Pressure-Sensitive Adhesives and Products*, Chapters 3 and 8). High-styrene grades demonstrate higher hardness, shear resistance, and modulus. Styrene–isoprene–styrene (SIS) copolymer, which is easily the most important styrene block copolymer used in hydrocolloid formulations, is usually softer than styrene–butadiene–styrene or styrene–ethylene–butylene–styrene, or styrene-ethylene-propylene-styrene (SEPS) [15] due to the higher entanglement MW of the isoprene midblock. The MW of the styrene end block is comparable with the corresponding entanglement MW; therefore, increasing the MW of the styrene end block improves the heat resistance of the adhesive.

Neat triblock SBCs, however, have rather poor tack to be utilized as PSAs. To improve their adhesive performance, they are blended with plasticizers and tackifiers.

Two types of plasticizers are commonly used in formulations with triblock SBCs: diblock SBCs of the same chemical composition as the initial triblock, and a low-MW polymer with a structure equivalent to that of the midblock of the triblock SBC. For example, the plasticizers used for SIS are styrene–isoprene (SI) diblock copolymer and oligomeric polyisoprene, respectively.

A molecule of the diblock copolymer has a styrene block that can be involved with the styrene-segregated joints and a free isoprene end that is not embedded in the styrene domain network. The molecule as a whole does not participate in the physical cross-linking and acts as a plasticizer. Molecules of the diblock SBC have the important advantage of having affinity to both styrene and rubbery blocks. Many commercially available SIS grades already contain SI as an additive [17,18].

As mentioned previously, low-MW polyisoprene is an example of the second type of plasticizers for SIS. Such plasticizers, built of the same monomeric units as rubbery midblocks of the triblock SBCs, have better miscibility with the midblocks.

Phase diagrams for SIS (Vector 4111)–SI (LVSI 101) and SIS (Vector 4111)–Isolene (low-MW polyisoprene) systems were presented in Ref. [14]. Both systems are characterized by upper critical solution temperature behavior. In similar experiments with other SIS grades, higher styrene content led to the extension of the heterophase region in the phase diagram.

Plasticizers with a chemical composition similar to that of triblock SBCs have one more important benefit in the area of skin contact application. Because they do not bring new chemical units into the formulation, they have a lower probability of causing skin irritation, etc.

Tackifiers are another type of additives used to improve the adhesion performance of SBCs. The addition of low-MW-compatible tackifying resins to SBCs usually improves peel resistance and probe tack [19,20], while decreasing cohesive-related characteristics (holding power and shear adhesion failure temperature [SAFT]) [21]. As low-MW substances, the tackifying resins have a higher chance of being harmful to the skin than polymeric components. Indeed, the pentaerythritol ester of hydrogenated rosin has a sensitizing potential for skin [22].

Typically, an SBC-based formulation consists of a triblock SBC (usually SIS), a plasticizer (diblock SBC or low-MW rubber), a tackifier, and a stabilizer (antioxidant). Although the content of the stabilizers is low (~2% or less), they play an important role in protecting the components of the formulation upon hot-melt processing.

Another way of improving the tack of SBC-based PSAs was described in Ref. [23]. Low-MW PIB, which is immiscible with SIS, has a lower surface energy. Upon processing and storage it tends to migrate to the surface of the adhesive patch, thus increasing its tack. This very thin PIB coating can contribute to filling skin dimples upon application of the dressing (wetting–spreading) that enables better contact between the adhesive and the substrate (see also *Fundamentals of Pressure Sensitivity*, Chapter 1).

5.2.2 Hydrophilic Part

As discussed previously, hydrophilic components are an unavoidable part of thick skin-contacting PSA dressings. Their main purpose is to handle the moisture that comes from the skin and to enable wet tack. Because these components either dissolve or swell in water, they eventually lose their cohesive strength. To maintain integrity of the dressing upon wearing, the components should be distributed in a continuous hydrophobic phase. The content of the hydrophilic components in the dressing should therefore be less than 50%. As already mentioned, however, the ratio between the hydrophilic and hydrophobic phases can change significantly upon absorbing moisture due to swelling and dissolution of hydrophilic components.

To understand the requirements of water handling capacity of hydrocolloid wound dressings, some data on exudate production are useful. According to Ref. [3], the evaporative water loss from burns is about 5 g/10 cm^2/24 h. Values of 1–13 g/10 cm^2/24 h were obtained for chronic wounds in Ref. [4]. In a study by Thomas et al. [2], the average exudate production from leg ulcers was about 5 g/10 cm^2/24 h, with a range of 4 to 12 g/10 cm^2/24 h. The hydrocolloid dressing itself influences the amount of the produced exudate: if a hydrocolloid dressing applied to a leg ulcer forms a secure waterproof seal on the surrounding skin, the exudate production is reduced by up to 50%. The proposed mechanism consists of pressure growth beneath the dressing caused by exudate production and swelling of sorbents. As this pressure approaches that in the capillaries, the exudate formation decreases.

The absorbed moisture can exist as unbound, loosely bound, and bound water. Bhaskar et al. [24] reported gelatin and pectin stored in a high humidity atmosphere contain only nonfreezable bound water, whereas Na-carboxymethyl cellulose (Na-CMC) exposed to these conditions contains loosely bound water as well. On the other hand, Na-CMC demonstrated the best water uptake in this study, although it was not able to lock the moisture reliably. These data indicate that a combination of sorbents can be more efficient than either sorbent alone.

Moisture sorbents can be divided into water soluble and water insoluble, meltable, and nonmeltable.

Water-soluble polymers, for example, Na-CMC and hydroxypropyl cellulose (HPC), usually form viscous tacky solutions that provide wet tack to tissues. If the concentration (and hence viscosity) of such solutions that form at the surface of the patch becomes too low, patch slippage will be observed. Water-soluble polymers therefore are not suitable for permanent storage of absorbed moisture.

Water-insoluble sorbents, for example, agar and pectin, are able to absorb and reliably lock enormous amounts of moisture. For example, the swell ratio of agar (i.e., the increase of its weight upon swelling) at 37°C is ~600% [14].

Meltable hydrophilic components can form a needed morphology during the hot-melt process. This will be discussed in detail below.

Because hydrophilic components dissolve or swell in moisture, there is always a chance that they can permeate through the wound or injured skin into the body during wearing or be left in the wound at removal. That is why the safety requirements for hydrophilic components are very important. Usually, components with a long record of safe use in the food industry or medicine are preferred.

The most widely used moisture sorbents are polysaccharides. They can be divided into the following groups.

- Cellulose and its derivatives (ethers and esters), for example, Na-CMC and HPC
- Starch and its derivatives
- Natural gums, for example, agar, carrageenan, alginic acid, and alginates
- Others (chitosan, pectin, etc.)

Gelatin is another natural substance that is widely used as a sorbent in hydrocolloid dressings. Synthetic hydrophilic polymers and their combinations, such as poly(N-vinyl pyrrolidone) (PVP), poly(acrylic acid) (PAA), and polyvinyl alcohol (PVA), can be also used in hydrocolloids dressings.

Moisture sorbents can be also divided by their charge into three groups: neutral (e.g., HPC, PVP), polyanions (e.g., pectin, agar, Na-CMC, alginates), and polycations (chitosan). It is said that application of charged absorbents leads to improved integrity of dressings [7].

Fluid handling properties of hydrocolloid dressings can be presented as a combination of moisture sorption with its transportation to the outer surface and evaporation from the outer surface of the backing. The contribution of the latter process to the overall performance depends on the moisture permeability of the adhesive and backing and on the thickness of the adhesive layer. According to the data presented in Ref. [25], in some cases fluid loss by transmission can be comparable with the amount of absorbed moisture. It is therefore desirable to create easy moisture transportation through the adhesive layer. Breathable backings are preferred in skin-contacting products unless their use is restricted for other reasons (e.g., in transdermal systems).

Thomas and Loveless [25] tested fluid handling properties of several commercially available hydrocolloid dressings. Typical values of moisture vapor transmission rate obtained in their study were ~1–2 g/10 cm²/48 h. Fluid retention of the majority of the tested hydrocolloid wound dressings after 72 h, as determined by immersing them in testing solutions, was in the range of 3–6 g/g.

Lanel et al. [26] studied the swelling kinetics of hydrocolloid dressings. They proposed an equation that describes the change of the gel thickness in time.

$$h(t) = h_0 = (h_f - h_0)\left\{1 - \frac{8}{\pi^2}\sum_{n=0}^{\infty}\left[\frac{1}{(2n+1)^2}\exp\left(-\frac{(2n+1)^2 t}{\tau}\right)\right]\right\} \tag{5.1}$$

In this equation, $h(t)$, h_0, and h_f are the current, initial, and final thickness of the dressing, respectively; t is current time and τ is the characteristic time of gel reflecting its chemical nature.

Equation 5.1 was successfully applied to describe the swelling of several hydrocolloid dressings. However, it has obvious drawbacks: it does not take into account partial dissolution of hydrophilic components, and parameters h_f and τ must be determined experimentally.

Channels for moisture transportation can be created within hydrophobic continuous phase by small particles taken in concentration over the percolation point. The percolation concentration depends on the particle size and their anisometry: smaller and anisometric particles can be used in much smaller concentrations. Kulichikhin et al. [14] used Na-montmorillonite particles with a layered structure and high specific surface area.

5.2.3 Morphology

As discussed previously, the content of the hydrophilic phase in a hydrocolloid formulation should preferably be less than 50%. Hydrophilic components should enable high moisture uptake and moisture transmission rate. If they are introduced as disperse fillers with isolated particles, the moisture transportation is controlled by moisture permeability of the hydrophobic phase, which is obviously low. To play their role efficiently, therefore, the hydrophilic components should form a continuous phase or contacting phases.

The optimal morphology of the hydrocolloid product should include a hydrophobic matrix, channels for moisture transportation from the skin-contacting surface to the backing, particles of insoluble moisture sorbents for permanent storage of moisture, and possibly reinforcing particles (Figure 5.2).

How can the desired morphology be created? We have already discussed the difficulties connected with the solvent cast process. The possibilities of controlling morphology in this case are very limited.

On the contrary, hot-melt process involves processing of high-viscous melts. Once created upon mixing/extruding, the morphology of the adhesive does not have enough time for relaxation and can be fixed while cooling down.

Let us discuss some components that can induce formation of specific morphological patterns.

As demonstrated in Ref. [14], due to its liquid crystalline (LC) structure, HPC is an interesting hydrophilic agent from the morphology viewpoint. At processing temperatures, HPC forms LC melt. LC melts can be easily oriented and elongate at shear [27]. Such elongated drops form liquid fibrils that can transform into solid fibers upon cooling. The ability of HPC to form fibers depends on its molecular weight: high-molecular-weight

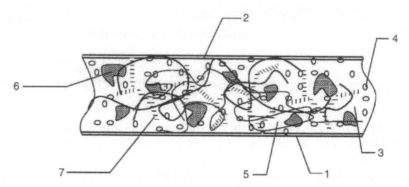

FIGURE 5.2 Sketch of the desired morphology of a hydrocolloid skin contact adhesive [23]. (1) Backing film, (2) release liner, (3) hydrophobic matrix, (4 and 5) droplets and fibers of cellulose derivatives, respectively, (6) particles of water-swellable polysaccharide, and (7) silicate bridges (channels).

grades tend to form fibers, whereas LM-weight grades form elongated drops. Thus, formed HPC fibers act as channels for moisture transportation.

Disperse particles can be also arranged in patterns. Kulichikhin et al. [14,28–30] observed the formation of both spherical and anisometric particle chains in viscoelastic polymer systems in shear fields. As discussed previously, anisometric and high-disperse particles are preferable due to their significantly lower percolation concentration.

5.2.4 Patch Design

Properties of skin-contact products such as moisture vapor transmission rate, moisture handling, and comfort upon wearing are influenced significantly by the patch design.

The shape of the patch or dressing should conform to the anatomical features of the intended site of application. Patch wrinkling due to improper shape should be avoided because such places are likely to become the starting points of detachment. Moreover, results demonstrated [31] that antimicrobial efficacy of antimicrobial wound dressings depends on their ability to conform to the contours of a wound. Patches with rounded corners and tapered edges are preferred because they provide a lower chance of accidental patch detachment.

If the adhesive layer has insufficient adhesion or if high adhesion may be harmful to tissues upon removal, the patch can be equipped with an adhesive overlay that would enable strong adhesion to healthy skin outside the wound, whereas the formulation that contacts the wound has lower adhesion. Similarly, an absorbent pad can be attached to the middle part of the patch with the hydrocolloid formulation acting as an adhesive overlay. This approach can be used to extend the application of hydrocolloid dressing to highly exuding wounds.

A layered skin-contacting product was described, for example, in Ref. [32]. The product would consist of two absorbent layers with different absorbent capacities: the layer closer to skin is less absorbent and works as a PSA, whereas the second layer is capable of absorbing and storing higher amounts of exudate. Such a layered design may help to maintain the integrity of the absorbent layers and minimize the amount of substances entering the wound.

Translucency is a typical requirement for hydrocolloid dressings. Translucency is preferred not only due to aesthetic reasons but also to allow observation of the healing process. Unfortunately, initially translucent dressings often become cloudy upon absorbing exudate. This phenomenon can be utilized. The change in color of Comfeel Plus® dressing manufactured by Coloplast is used as an indicator for dressing change [33].

The right choice of backing is a crucial point in designing skin-contact products. Insufficient moisture vapor transmission rate of the backing leads to accelerated accumulation of moisture in the adhesive and eventually decreases the wearing time. Therefore, breathable backings that are permeable for oxygen and water vapor but impermeable for liquid water are preferred for wound dressings. Such backings help to decrease skin maceration and prevent the growth of anaerobic bacteria. Backings can swell as a result of the moisture that goes across them, which can lead to wrinkling.

On the other hand, occlusive backings are usually used in transdermal drug delivery systems to prevent the drug from diffusing into the backing and to induce moisture accumulation in skin which is favourable for skin permeability.

The mechanical properties of backings are also important in dressing performance. Rigid backings worsen the conformability of the dressings and decrease comfort.

5.3 Advanced Skin-Contact Adhesives

Let us now discuss the application of skin-contact adhesives in two specific areas: antimicrobial wound dressings and transdermal drug delivery systems. Both applications are connected with the introduction of some active substances and the optimization of their release rate from the patch into a wound or across the skin. We will consider the specific features of skin-contact adhesives to be used in these fields, requirements for the active substances, and ways to control their release.

5.3.1 Antimicrobial Skin-Contact Adhesives

Introducing antimicrobial agents into wound dressings is a logical step toward improving their performance. Such dressings provide not only the optimum moist environment for heeling, but also prolonged sustained release of the antimicrobial agent into the wound. Results demonstrated [34] that the daily healing rate doubles upon application of an antimicrobial dressing compared with a plain alginate dressing.

What antimicrobial agents can be used in wound dressings? Antibiotics have demonstrated their high efficacy over the years. However, bacterial resistance to antibiotics has spread significantly in past decades, rendering many antibiotics ineffective [35,36]. The origin of bacterial resistance to antibiotics is connected with their large number in a population and high speed of reproduction. If a small subset of the population survives the treatment due to mutations and is allowed to multiply, the new population will probably inherit this ability. New antibiotics do not reliably solve the problem. Multidrug combinations with different mechanisms of action of individual drugs are becoming increasingly important in combating antibiotic-resistant bacteria [35,37].

The apparent resistance of bacteria to antibiotics has led to increased use of such agents as silver and iodine in antimicrobial dressings. Silver has been successfully used as an antimicrobial agent for thousands of years [38,39]. Its antimicrobial activity is provided by the ionic form [40]. The most attractive feature of silver as an antimicrobial agent is that it has a minimal chance of developing bacterial resistance. Furthermore, it has broad gram-negative, gram-positive, and antifungal activity, which is explained by the fact that silver ions demonstrate varying antimicrobial effects depending on their binding side: bacterial cell wall, proteins, DNA [41].

Iodine also has a long history of application as an antimicrobial agent. The PVP–iodine complex proved its usefulness in the treatment of infected and potentially infected wounds [42]. Iodine also has the advantage of a low chance of developing bacterial resistance [43].

In this article we will limit our consideration to silver as the most widespread antimicrobial agent used in wound dressings.

Silver can be introduced into dressings in several forms:

• Salts
• Complexes
• Metallic silver
• Silver oxide Ag_2O
• Silver nanoparticles

Silver salts release a silver cation into wounds upon contact with exudate. It can be assumed that their efficacy depends on the concentration of silver cations, which is, in turn, determined by the solubility product of given salts. Silver sulfadiazine, nitrate, sulfate, and phosphate should be mentioned among the salts to be used in antimicrobial dressings. Because wound exudate contains chloride ions, it is possible for the silver cation to precipitate in the form of poorly soluble silver chloride upon contact with exudates, which will decrease the antimicrobial efficacy of silver salts.

Silver forms a number of stable complexes with some anions, for example, chloride (in the excess of chloride ions) and thiosulfate [44]. Silver can also form "host–guest" complexes with low-MW polyethylene glycols [45]. Such complexes have the important advantage of protecting silver from interaction with other anions and from the influence of light and temperature upon processing as well as storage and application, which can lead to diminished antimicrobial activity. The efficacy of these complexes is determined by both their solubility and their stability constant.

The mechanism of release of soluble silver components into exudate is most likely as follows. In the first stage, the particles of silver compounds that contact with exudate are dissolved. Then exudate diffuses into the adhesive, preferably via moisture transportation channels, and dissolution of the particles of silver compounds that contact hydrophilic components of the adhesive begins. The driving force for the release of silver compounds to exudate at this stage is the difference in their concentration in the channel and in the exudate. The former of these values is controlled by the solubility of the substance in the exudate. The silver ions diffuse to the wound from this solution. Particles of soluble silver compounds isolated in the hydrophobic matrix are not likely to participate in the process.

Metallic silver and silver oxide cannot provide high concentrations of silver ions in solution. Therefore, these compounds should be localized on the surface of the dressing to be effective. The same can be said about silver nanoparticles. If properly used, they can provide outstanding antimicrobial properties [46].

Silver salts are more stable at acidic pH values. Basic pH causes the precipitation of silver oxide.

$$2AgNO_3 + 2NaOH = Ag_2O\downarrow + 2NaNO_3 + H_2O$$

A moist environment and poor photostability of silver salts can also cause their degradation, which is usually observed as patch discoloration and leads to a decrease in antimicrobial activity. Thermal stability of silver compounds is a challenging issue for hot-melt processing of hydrocolloid formulations.

The interaction of silver compounds with other components of the adhesive formulation is also important. For example, silver can form salts *in situ* with polyanions that are present in the hydrocolloid formulation. Formation of the silver salt of carboxymethyl cellulose was described in Ref. [47]. An exchange of Na^+ and Ag^+ ion in Na-montmorillonite treated with silver nitrate was discussed in Ref. [14]. Such interactions may be useful for providing constant silver release rates, but they can also lead to decreased silver salt solubility.

Because the release rate of silver from salts is determined by their solubility, combinations of silver salts with good and poor solubility (e.g., nitrate and phosphate) may be used to construct the desired release profiles.

5.3.2 Application of Skin-Contact Adhesives in Transdermal Products

The transdermal route of drug delivery is an alternate way to administer drugs, especially in cases when the traditional oral method is less effective [48,49]. Transdermal systems deliver the drug onto the skin surface at the intended rate. The drug molecules penetrate the skin to the blood vessels.

Transdermal drug delivery systems (TDS) offer many benefits compared with traditional methods of administration.

- Virtually constant concentration of the drug in blood without significant deviations in time and overdosing leads to a prolonged therapeutic effect and minimal side effects
- No loss of drug activity due to metabolism in the gastrointestinal tract and liver
- Convenience of application

The main drawbacks of TDS are as follows.

- Possible irritation or sensitization of skin due to interaction with active ingredients or other excipients may occur.
- There may be a significant lag time compared with drugs administered via injections.

- Because the driving force for drug penetration through the skin is the difference in concentrations on the skin surface and in the bulk, the drug is usually loaded into TDS in relatively high concentrations. Only a small part of the loaded drug, however, usually goes through skin during the wearing of the TDS. This increases the cost of TDS.
- The transdermal route can be used only for a limited number of drugs that have low dosage and acceptable flux rate across the skin, so as to maintain the required concentration in blood.

The physicochemical properties of a drug are decisive for its applicability in TDS. It is generally accepted that drug molecules with MW greater than 500 Da are skin impermeable [50]. Neutral molecules usually have higher skin permeability than ions. Hydrophilicity of the drug molecule also plays an important role due to the different nature of skin layers: the hydrophobic epidermis and hydrophilic dermis. It can be assumed, therefore, that a good drug molecule for transdermal administration should be the surfactant itself [51,52]. The thermal and photostability of the drug are essential for successful processing of the TDS. Finally, the drug must not cause any significant skin irritation or sensitization.

Two types of TDS design are currently used: drug-in-adhesive and reservoir. The first contains the drug, dissolved or distributed, in the adhesive that is applied to the skin. Reservoir systems have a more complicated design: they contain a reservoir with drug dissolved or solubilized in a liquid and a membrane between the reservoir and the skin. Such a system is more expensive and is used in cases where the drug-in-adhesive system is not capable of providing the required drug flux across the skin. An example of such a system is described in Ref. [53]. We will consider only drug-in-adhesive systems.

What are the main requirements for an adhesive to be used in a drug-in-adhesive TDS? The adhesive should maintain good adhesion over all prescribed wearing time without significant detaching or cold flow. The occlusive backings that are usually used in TDSs could make moisture management challenging. However, TDS are worn on healthy skin, usually at sites with a relatively low amount of releasing moisture. The adhesive must not contain components or impurities or contaminations that can interact with the drug, thus decreasing its activity. The solubility of the drug in the adhesive should be as high as possible to provide the maximum possible concentration of the drug at the skin surface and, hence, the maximum driving force for skin permeation. Unfortunately, a higher concentration often means more significant skin irritation. The structure of the drug-in-adhesive TDS should be stable over time. Oversaturated metastable systems undergo phase decomposition (e.g., drug or penetration enhancer crystallization) that may change their performance.

In the absence of experimental data, the skin permeability of a drug can be roughly estimated using the Potts–Guy equation [54].

$$\log K_p = \log \frac{D_0}{\partial} + f \log K_{oct} - \beta M_w \tag{5.2}$$

In this equation K_p is the permeability coefficient, D_0 is the diffusivity of the drug molecule, ∂ is the diffusion path length, K_{oct} is a drug partition coefficient in the octanol/water system, f and β are coefficients, and M_w is the drug's molecular weight.

K_{oct} in Equation 5.2 is the measure of hydrophobicity of the drug molecule and M_w is the measure of the individual molecular volume. Equation 5.2 illustrates that the skin permeability of drug molecules increases with K_{oct} and diffusivity and decreases with molecular weight.

Skin permeation of drugs can be improved by several techniques that can be divided into chemical and physical techniques [55,56]. The skin's barrier properties are primarily localized in the stratum corneum, which is a 10- to 15-μm-thick layer consisting of dead keratinocytes in a lipid matrix. Some physical methods of permeation enhancement, such as microblades, microneedles, and electroporation, involve skin treatment aimed at the formation of pores in the stratum corneum. Local heating increases the drug diffusion rate and blood flow. This technique has the advantage of not damaging skin. Iontophoresis involves application of an electromotive force to drive charged drug molecules through the skin layers. Its enhancement effect can be connected not only with electrorepulsion, but also with increasing permeability of the stratum corneum in the presence of electrical fields and electroosmosis. Iontophoretic systems require application of electrically conductive PSAs, which were described in Chapter 2. Mechanical vibrations and ultrasound are other methods proposed for improving skin permeability. It is understandable that the mentioned physical methods are inconvenient for self-administration and that they increase the cost of TDS and hence should be used only if other methods of skin permeation enhancement are not efficient.

Chemical enhancement includes the introduction of some additives or arrangement of a specific structure of a formulation so as to increase the flux of drugs through the skin barrier. Adjustment of pH upon drug release is one such approach. For example, in accordance with patent applications [57,58], the skin permeability of oxytocin, estradiol, etc., can be significantly improved by the incorporation of "hydroxide releasing agents," such as NaOH, Na_2CO_3, and MgO, into the transdermal delivery systems. The major concern regarding the use of such substances lies in the possible damage to skin from prolonged exposure to these agents.

Special skin penetration enhancers can also be very effective in improving skin permeability [55]. According to Ref. [59], a combination of chemical penetration enhancers can enable a 10-fold increase in the buspirone flux compared with the enhancer-free formulation. Typically, chemical penetration enhancers belong to one of the following groups: surfactants, fatty acids and their derivatives, and azone-like compounds. One of the major problems with chemical enhancers is that many of them are skin irritants. The potency of the chemical enhancers is associated in Ref. [60] with two possible mechanisms of their action: extraction of lipids from the stratum corneum and fluidization of lipid bilayers. The irritation behavior of chemical enhancers is correlated with the ratio of hydrogen bonding to polar interactions. It was assumed that competitive H-bonding of penetration enhancers can influence the native H-bonding in protein molecules, which can damage their structure.

Another chemical method of enhancing drug delivery across the skin is the application of emulsions [55]. We will consider emulsions in the next section.

5.4   Hydrocolloid Formulations Based on Emulsions

Emulsions consist of small drops of liquid (dispersed phase) distributed in another liquid (dispersion medium). The two liquids are immiscible. Typically, the size of drops in an emulsion lies within the range of 0.1–100 μm. Almost all emulsions contain water as a polar phase. The nonpolar phase is usually termed "oil," even if it is not oil in nature. Emulsions that have oil as a dispersed phase are called "direct," or oil/water (o/w), whereas those with "water" as a dispersed phase are "reverse," or water/oil (w/o).

So-called "double emulsions" also exist [61,62] in which the drops of the dispersed phase contain smaller drops of another liquid. Double emulsions can also be divided into two types: oil/water/oil (o/w/o) and water/oil/water (w/o/w) (Figure 5.3).

Emulsions are usually classified as lyophobic and lyophilic. In most cases, emulsions are lyophobic and have an excess of free energy; they are thermodynamically unstable and can exist for a long time only in presence of emulsifiers. Lyophilic emulsions are thermodynamically stable; they are formed spontaneously at temperatures close to the critical temperature of mixing for two phases. Such emulsions are stable only at elevated temperatures and usually are not practically important. Later we will discuss lyophobic systems.

It is important to distinguish between aggregation and sedimentation stability. Sedimentation stability is the ability of the emulsion to resist gravity and sedimentation or, vice versa, heaving of the dispersed phase drops. The aggregation stability is the ability of the emulsion to keep the constant dimensional distribution of drops in time. The aggregation stability of emulsions is kinetic in nature. Growth of the particle size may lead to a loss of sedimentation stability.

Several factors can help to stabilize lyophobic emulsions.

- Elasticity of films with adsorption layers of surfactants (Gibbs and Marangoni–Gibbs effects).
- Electrostatic repulsion of double electric layers in the case of application of ionogenic surfactants.
- Hydrodynamic resistance of the dispersion medium layer.
- A structural–mechanical barrier, a strong stabilization factor that includes formation at the interface of the adsorption layers of surfactants that lyophilize it. The structure and mechanical properties of such layers can protect the dispersed phase particles from coagulation and coalescence. These layers can be formed by fine particles of a solid stabilizer.

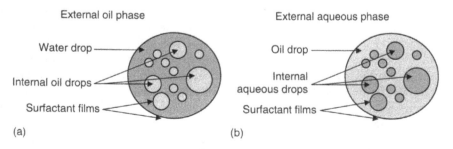

FIGURE 5.3   Schematic representation of o/w/o (a) and w/o/w (b) double emulsions.

The general rule for the choice of a stabilizer is that it should have an affinity to the dispersion medium—be soluble in it or be wettable by it (for solid stabilizers).

The type and properties of emulsions depend on the nature and ratio of the liquid phases, amount and nature of the emulsifier, method and conditions of emulsion preparation, etc. A change in composition or external actions can cause phase inversion, that is, transformation of a direct emulsion into reverse or vice versa. In emulsions with low concentrations, drops move freely and independently. Upon increased dispersed phase concentration up to density packing (74% for emulsions with equal drop sizes), the viscosity of the systems increases abruptly and they behave like gels. The shape of the drops of such emulsions, which were initially spherical, becomes nearly polyhedronic. The content of the dispersed phase in emulsions can be as high as 99% (vol); the dispersion medium in such emulsions exists as thin films, similar to films between bubbles in concentrated foams.

Microemulsions are lyophilic, thermodynamically stable colloid systems that are formed spontaneously upon mixing two liquids with limited mutual solubility in the presence of a micellar surfactant. As a rule, the system includes an electrolyte and a nonmicellar surfactant that is also called the "cosurfactant." The radius of the dispersed phase drops is usually 5–100 nm [63]. Winsor [64] proposed a classification of microemulsions (Figure 5.4). The first type (I) is direct microemulsions (o/w) that are in equilibrium with an excess of the oil phase. Type II represents reverse (w/o) microemulsions in equilibrium with an excess of the water phase. Type III is bicontinuous microemulsions in equilibrium with excesses of oil and water phases. Emulsions of type IV have the appearance of single-phase systems, although they are microheterogeneous (o/w or w/o). Changes in temperature and composition may lead to changes in the type of microemulsions.

Potential applications of emulsions and microemulsions in drug delivery systems were studied in the past 2 decades [65–72]. Double emulsions are mentioned [65] as liquid reservoirs capable of protecting drugs from the degradation caused by oxidation, light, and enzymes. The application of double emulsions for slow, sustained drug release was demonstrated in Refs. [66,67]. Results presented in Ref. [70] indicated that nifedipine flux from microemulsions through mice skin was 15–37 times higher compared with that from gel formulations. Solubilization of the drug in microemulsions helps to prevent its degradation. Similar results were obtained in Ref. [71] for lidocaine. It can be concluded that systems based on emulsions and microemulsions have great potential of application in TDS.

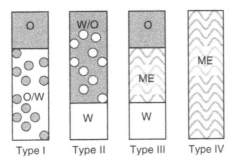

FIGURE 5.4 Winsor's classification of microemulsions.

The potential application of emulsions for preparation of dry films that can be loaded with drugs or other active species was also experimentally studied by our group. The emulsions under investigation contained a hydrophobic polymer enabling good tack, a hydrophilic polymer (hydroattractant), nonionogenic surfactants, and, in some cases, fine alumosilicate powder (Na-montmorillonite, clay) as an emulsion stabilizer.

By changing the ratio of polar and nonpolar phases, it is possible to obtain direct, reverse, and double emulsions. This feature can be useful in developing a universal platform for drugs of different hydrophilicity. The hydrophobic polymer was SIS triblock copolymer Vector 4111, manufactured by DexCo (Houston, TX). HPC was chosen as a hydrophilic polymer. Tween 80 and Tween 20 surfactants and Cloisite Na^+ and Cloisite 15A (Southern Clay Products, Gonzales, TX) were used to stabilize prepared emulsions. Cloisite Na^+ is a fine powder of natural montmorillonite; Cloisite 15A is a montmorillonite with a hydrophobized surface.

The emulsions were prepared by mechanical or ultrasonic dispersion. The films were prepared by solvent casting followed by a drying method.

Direct emulsions of SIS and HPC were prepared by ultrasonic dispersion of their solutions in toluene (or cyclohexane) and water, respectively. The volume fraction of the oil phase varied from 0.2 to 0.8. The water phase contained 5–10% HPC and 5–10% Tween 80, whereas the oil phase contained 5–10% SIS and 0–5% Tween 80. In several cases clay was used as a stabilizer. All these emulsions were of the direct type (o/w) with a drop size of 50–300 nm. SIS/HPC emulsions are unstable in the absence of surfactants. The introduction of clay particles (hydrophilic to the water phase and hydrophobic to the oil phase), however, allowed us to obtain emulsions that remained stable for several weeks.

The rheological properties of selected emulsions were studied by means of a Haake RheoStress rotational viscometer (Thermo Electron Corporation, Waltham, MA) equipped with a cone-and-plate operating unit. The measurements were carried out in three modes: CR (constant shear rate), CS (creep and recovery at constant shear stress), and oscillatory regime.

Typical results of the rheological experiments are presented in Figures 5.5–5.15. All studied HPC or SIS solutions with different additives demonstrated Newtonian rheological behavior. The viscosity of HPC solutions increased with the solution concentration faster than that of SIS.

Diluted emulsions with a volume fraction of the dispersed phase lower than 0.1 have Newtonian behavior [73]. However, the studied direct emulsions with moderate oil phase content demonstrated a significant viscosity anomaly with yield stress behavior (Figure 5.5). The viscosity decreased with shear stress (Figure 5.6) and increased with HPC content (Figure 5.7).

Yield stress values were calculated from the flow curves using the Casson equation [74]. The yield stress of the studied emulsions lies within the range of 5–40 Pa and increases with polymer content (Figure 5.8).

Viscoelastic properties of the emulsions were analyzed in a creep-recovery test. The measurements were carried out in the linear viscoelasticity region. The Burgers model [75] was successfully used to describe the experimental results. The model is a series connection of Maxwell (with elastic G_0 and viscous η_0 elements connected in series) and

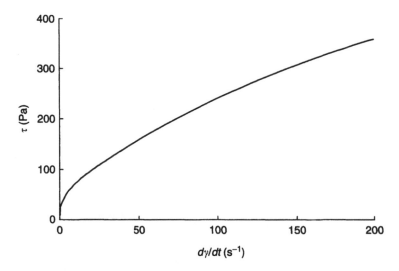

FIGURE 5.5 Shear stress versus shear rate for o/w emulsion (5% SIS + 3% Tween 80 in toluene/ 10% HPC + 5% Tween 80 in water). The volume fraction of the dispersed phase is 0.33.

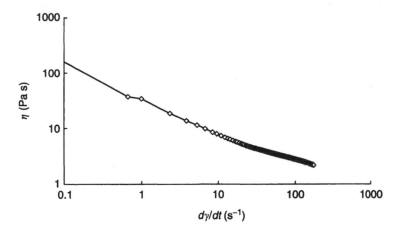

FIGURE 5.6 Viscosity versus shear rate for o/w emulsion (5% SIS + 3% Tween 80 in toluene/10% HPC + 5% Tween 80 in water). The volume fraction of the dispersed phase is 0.33.

Kelvin–Voigt models (parallel connection of elastic G_1 and viscous η_1 elements). The equation for deformation in the creep stage can be written as follows.

$$\gamma(t) = \frac{\tau_0}{G_0} + \frac{\tau_0}{G_1}\left(1 - e^{-t/\lambda_1}\right) + \frac{\tau_0}{\eta_0}t \tag{5.3}$$

In this equation, λ_1 is the retardation time, τ_0 is initial stress, G_1 and G_2 are elastic moduli of the elastic elements, η_1 and η_2 are viscosities of the viscous elements, and t is current time.

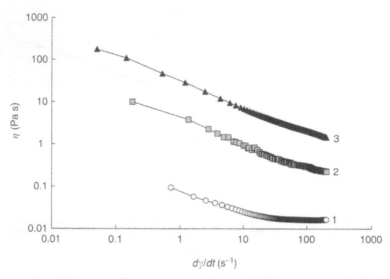

FIGURE 5.7   Viscosity versus shear rate for emulsions of the following compositions: 1. (5% SIS + 1.325% Tween 80 solution in toluene)/(0% HPC + 5% Tween 80 solution in water), volume fraction of the oil phase is 0.5; 2. (5% SIS + 1.325% Tween 80 solution in toluene)/(5% HPC + 5% Tween 80 solution in water), volume fraction of the oil phase is 0.5; 3. (5% SIS + 1.325% Tween 80 solution in toluene)/(10% HPC + 5% Tween 80 solution in water), volume fraction of the oil phase is 0.5.

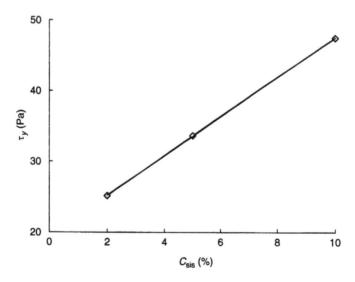

FIGURE 5.8   The effect of SIS concentration on the yield stress τ_y for o/w emulsions with volume fraction of the dispersed phase set at 0.33 and aqueous phase composition (10% HPC + 5% Tween 80).

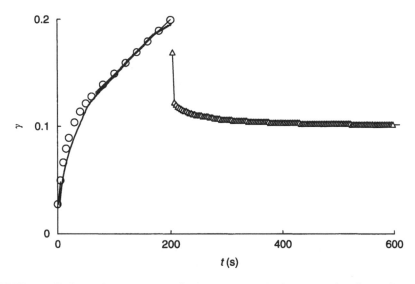

FIGURE 5.9 Deformation at constant shear stress $\tau = 1$ Pa (creep stage) and 0 Pa (recovery stage) for o/w emulsion (5% SIS + 2.6% Tween 80 in toluene)/(5% HPC + 5% Tween 80 in water). The volume fraction of the dispersed phase is 0.5. Points are calculated in accordance with Burgers model.

Figure 5.9 demonstrates an example of the experimental data obtained in the creep test. The calculated parameters of the Burgers model in this case are as follows: $G_0 = 36.2$ Pa, $G_1 = 14.0$ Pa, $\eta_0 = 2000$ Pa·s, $\eta_0 = 217.4$ Pa·s, $\lambda_1 = 15.6$ s.

Similar results were obtained for clay-stabilized emulsions. These systems can be characterized by viscosity anomaly (Figures 5.10 and 5.11), yield stress (Figure 5.12), and viscoelastic behavior following Burgers model (Figure 5.13).

Dynamical mechanical analysis of emulsions also was performed in the linear viscoelastic region. Typical relationships obtained for elastic G' and loss G'' moduli and complex viscosity η^* are illustrated in Figure 5.14 for an o/w emulsion with composition (5% SIS + 5% Tween 80)/(5% HPC + 5% Tween 80). As illustrated in Figure 5.14, the elastic modulus is greater than the loss modulus in the whole explored frequency range. This may indicate strong structure formation in the studied emulsions. Results obtained for clay-containing emulsions (Figure 5.15) also confirm this statement.

Prepared emulsions were used to obtain heterogeneous films by casting. The scheme of the process is presented in Figure 5.16. One-side siliconized polyethylene terephthalate (PET) tape (Loparex 7300A) was used as a backing. This film has hydrophilic and hydrophobic (silicon-coated) sides. Direct emulsions with a volumetric fraction of the oil phase below 0.35 could not form stable films on the siliconized side of the tape. Emulsions with an oil phase content greater than 0.5 were able to form stable films on both (hydrophilic and hydrophobic) sides of the PET film.

The morphology of the prepared films was studied by means of optical and atomic force microscopy (Figures 5.17 and 5.18). As discussed previously, preparation of heterogeneous films by casting via emulsions has a very important advantage: sedimentation stability of

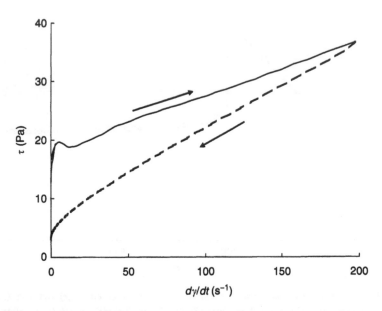

FIGURE 5.10   Hysteresis of the flow curve of clay-stabilized o/w emulsion (1.5% Cloisite 15A + 5% SIS in toluene)/(1.5% Cloisite Na$^+$ + 10% HPC in water). The volume fraction of the dispersed phase is 0.57.

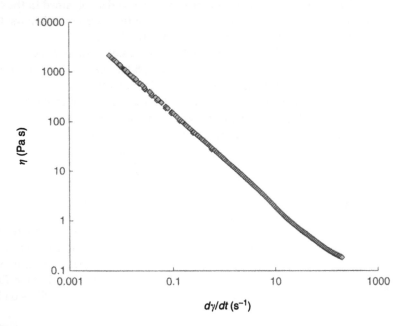

FIGURE 5.11   Viscosity versus shear rate for clay-stabilized o/w emulsion (1.5% Cloisite 15A + 5% SIS in toluene)/(1.5% Cloisite Na$^+$ + 10% HPC in water). The volume fraction of the dispersed phase is 0.57.

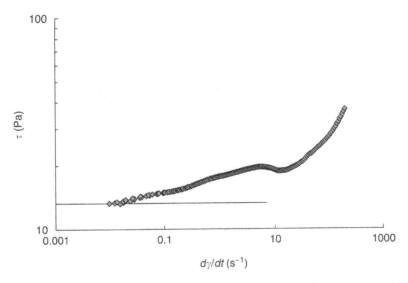

FIGURE 5.12 Flow curve for clay-stabilized o/w emulsion (1.5% Cloisite 15A + 5% SIS in toluene)/(1.5% Cloisite Na+ + 10% HPC in water). The volume fraction of the dispersed phase is 0.57. CR mode. τ_y = 13.3 Pa.

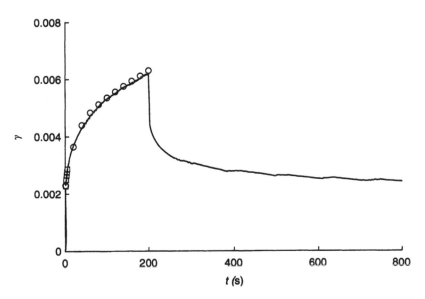

FIGURE 5.13 Deformation at constant shear stress τ = 0.5 Pa (creep stage) and 0 Pa (recovery stage) for o/w emulsion (1.5% Cloisite 15A + 5% SIS in toluene)/(1.5% Cloisite Na+ + 10% HPC in water). The volume fraction of the dispersed phase is 0.57. Points are calculated in accordance with Burgers model.

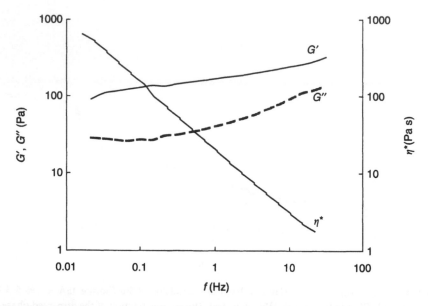

FIGURE 5.14 Elastic G' and storage G'' moduli and complex viscosity η^* for o/w emulsion (5% SIS + 3% Tween 80 in toluene)/(5% HPC 0 +5% Tween 80 in water). The volume fraction of the dispersed phase is 0.33. Oscillatory mode; shear stress amplitude is 1 Pa.

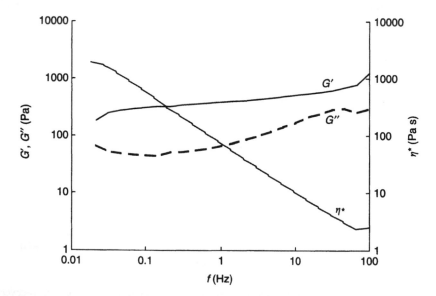

FIGURE 5.15 Elastic G' and storage G'' moduli and complex viscosity η^* for clay-stabilized o/w emulsion (1.5% Cloisite 15A + 5% SIS in toluene)/(1.5% Cloisite Na$^+$ + 10% HPC in water). The volume fraction of the dispersed phase is 0.57. Oscillatory mode, shear stress amplitude is 0.5 Pa.

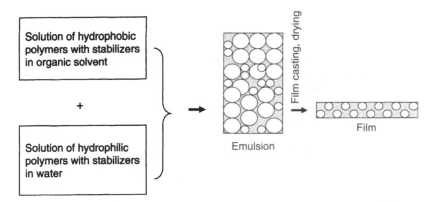

FIGURE 5.16 The scheme of preparation of films based on emulsions.

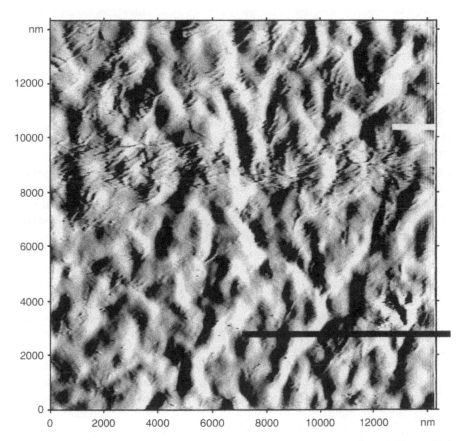

FIGURE 5.17 Atomic force microscopy image of o/w emulsion (2% SIS + 3% Tween 80 in toluene)/(10% HPC + 5% Tween 80 in water). The volume fraction of the dispersed phase is 0.33.

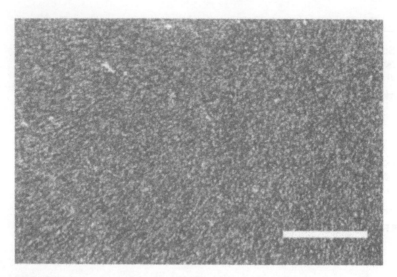

FIGURE 5.18 Optical micrograph of film (on glass) from o/w emulsion (1.5% Cloisite 15A + 5% SIS in toluene)/(1.5% Cloisite Na$^+$ + 10% HPC in water). Volume fraction of the dispersed phase is 0.57. Nonpolarized light. Scale bar, 100 μm.

the emulsions can be achieved through the proper choice of surfactants or other stabilizers. This enables a uniform distribution of components both in the cast solution and in the prepared film without concentration gradients in the cross-section of the film.

An additional advantage of emulsion-based films lies in the possibility of using different drugs that vary in hydrophilic–lipophilic balance.

The capability of heterophase solid platforms to release into water lysozyme–(3,2,1,17)-N-acetylmuramid-glucan hydrolasa, a water-soluble enzyme containing one peptide chain and 129 residues of amino acids, was also studied experimentally. This enzyme has good antibacterial activity. The release of lysozyme into water was estimated using the calibrated dependence of enzyme activity (EA) on its concentration. The lysozyme activity was measured using the hydrolysis rate of the *Micrococcus lysodeikticus* biomass suspended in a phosphate buffer solution (pH 6.86). The method is based on the change of optical density (*D*) of the initial suspension under action of lysozyme released from the film. The EA value was determined from kinetic measurements of optical density extrapolated on the initial stage of hydrolysis ($\lim_{t \to 0} dD/dt$). A definite time after immersion of the film into water, the lysozme-containing probe was mixed with the biomass volume and kinetics of optical density change was measured. In this manner the amount of released enzyme was estimated in micrograms per square centimeter per hour.

Figure 5.19 illustrates the release of lysozyme from the film prepared on the basis of complex emulsion containing 20% SIS in toluene, 10% HPC in water, 10% surfactants, and 3% enzyme. Lysozyme loading was ~300 μg/cm^2. The release of enzyme to water can be characterized by rapid onset (40% after 1 h), followed by a stagnant release rate. Such a release profile is good for antimicrobial systems: rapid onset can enable fast killing of bacteria and the slow stage can maintain the sterility of the medium. This result confirms the applicability of emulsion-based films as carriers of active species.

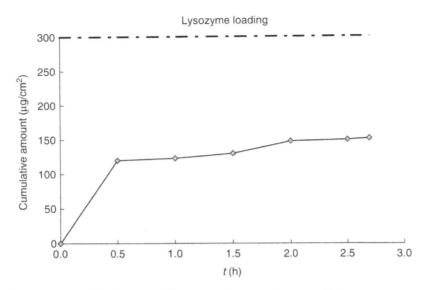

FIGURE 5.19 Cumulative amount of lysozyme released at 37°C from a film prepared from o/w emulsion. Lysozyme content is 3%. The volume fraction of the dispersed phase is 0.5.

5.5 Conclusion

Skin-contacting PSAs are used as protective and wound care dressings, as carriers of active species (antimicrobial and transdermal products), etc. Specific features of skin-contacting PSAs are connected with the nature of the substrate (skin), its mechanical properties, and its biological functions. All components of skin-contacting products should be safe and approved for contact with skin.

Skin-contacting products are usually multicomponent systems with a complicated structure that is created to manage the sweat/exudate flux. A proper product design is needed for optimum performance of the product.

Hydrocolloid wound dressings have become very popular due to the acceleration of wound healing by maintaining the optimum moisture level in the wound. To be successful, dressings should have outstanding fluid-handling properties. Their area of application can be further extended with the introduction of antimicrobial agents.

Transdermal drug delivery systems have several important advantages over traditional routs of drug administration. Methods of increasing of the drug flux across skin include the use of chemical penetration enhancers and different physical factors.

Preparation of skin-contacting PSAs from stabilized emulsions may not only lead to improved processability but can also have benefits from the viewpoint of delivery of both hydrophilic and hydrophobic drugs and other active species.

Comparison of skin-contacting PSAs prepared via melts, solutions, and emulsions may provide a good topic for discussion about the potential advantages and disadvantages of said preparation methods.

Acknowledgment

This work was partially supported by grants from the Russian Foundation of Basic Research 05-03-08028, 07-03-00998, and 07-03-00515.

References

1. Khatyr F., Imberdis C., Vescovo P., Varchon D., Lagarde J.-M. 2004. Model of the viscoelastic behaviour of skin in vivo and study of anisotropy. *Skin Res. Technol.* 10:96–103.
2. Thomas S., Fear M., Humphreys J. et al. 1996. The effect of dressings on the production of exudate from venous leg ulcers. *Wounds.* 8:145–149.
3. Lamke L.O., Nilsson G.E., Reichner H.L. 1977. The evaporative water loss from burns and water vapour permeability of grafts and artificial membranes used in the treatment of burns. *Burns.* 3:159–165.
4. Dealey C., Cameron J., Arrowsmith M. 2006. A study comparing two objective methods of quantifying the production of wound exudate. *J. Wound Care.* 15:149–153.
5. Winter G.D. 1962. Formation of scab and the rate of epithelialization of superficial wounds in the skin of the young domestic pig. *Nature.* 193:293–294.
6. Thomas S., Fram P., Phillips P. 2007. The importance of Compression on dressing performance. World Wide Wounds. (November) http://www.worldwidewounds. com/2007/November/Thomas-Fram-Phillips/Thomas-Fram-Phillips-Compression-WRAP.html.
7. Lipman R. 1999. Hydrocolloid PSAs: New formulation strategies. *Medical Device & Diagnostics Industry.* Jun. 1999, p. 132.
8. Skin Problems in the Elderly. 2001. *Wounds.* 13:59–65.
9. Escoffier C., de Rigal J., Rochefort A., Vasselet R., Lèvêque J.-L., Agache P.G. 1989. Age-related mechanical properties of human skin: an in vivo study. *J. Invest. Dermatol.* 93:353–357.
10. Heenan A. 1998. Frequently Asked Questions: Hydrocolloid Dressings. *World Wide Wounds.* (April) http://www.worldwidewounds.com/1998/april/Hydrocolloid-FAQ/hydrocolloid-questions.html
11. Hinnman C.D., Maibach H.I. 1963. Effect of air exposure and occlusion on experimental human skin wounds. *Nature.* 200:377–378.
12. Ovington L.G. 2007. Advances in wound dressings. *Clin. Dermatol.* 25:33–38.
13. Chen J.L. 2006. Bandage for Adhering to Moist Surfaces. U.S. Pat. 3,339,546 assigned to E.R. Squibb & Sons.
14. Kulichikhin V., Antonov S., Makarova V., Semakov A., Tereshin A., Singh P. 2006. Novel hydrocolloid formulations based on nanocomposites concept. In *Pressure-Sensitive Design, Theoretical Aspects. Volume 1*, ed. I. Benedek, Leiden—Boston, VSP, pp. 351–401.
15. Park Y.J. 2006. Hot-melt pressure-sensitive adhesives. In *Pressure-Sensitive Design and Formulation, Application. Volume 2*, ed. I. Benedek, Leiden—Boston, VSP, pp. 141–179.

16. Siemann U., Ruland W. 1982. Determination of the width of the domain boundaries in polymer two-phase systems by x-ray small-angle scattering. *Colloid. & Polymer Sci.* 280:999–1010.

17. Vector SIS grades. Dexco Polymers LP. http://www.dexcopolymers.com/prodselector.pdf

18. Kraton Polymers and Compounds. Typical Properties Guide. Kraton Polymers. http://www.kraton.com/content/includes/Kraton%20Typical%20Properties%20Guide.pdf

19. Lim D.-H., Do H.-S., Kim H.-J. 2006. PSA performances and viscoelastic properties of SIS-based PSA blends with H-DCPD tackifiers. *J. Appl. Polym. Sci.* 102:2839–2846.

20. Kim D.-J., Kim H.-J., Yoon G.-H. 2005. Effect of substrate and tackifier on peel strength of SIS (styrene-isoprene-styrene)-based HMPSAs. *Int. J. Adhes. & Adhes.* 25:288–295.

21. Kim D.-J., 1995. Ph. D. Thesis. Performance of SIS-based hot-melt pressure sensitive adhesives. Seoul National University.

22. Sasseville D., Tennstedt D., Lachapelle J.M. 1997. Allergic contact dermatitis from hydrocolloid dressings. *Am. J. Contact Dermat.* 8:236–238.

23. Kulichikhin V., Parandoosh S., Feldstein M., Antonov S., Cleary G.W. 2007. Composition for cushions, wound dressings and other skin-contacting products. U.S. Pat. 7,217,853 assigned to Corium International, Inc.

24. Bhaskar G., Ford J.L., Hollingsbee D.A. 1998. Thermal analysis of the water uptake by hydrocolloids. *Thermochimica Acta.* 322:153–165.

25. Thomas S., Loveless P. 1997. A comparative study of the properties of twelve hydrocolloid dressings. *World Wide Wounds.* (July) http://www.worldwidewounds.com/1997/july/Thomas-Hydronet/hydronet.html.

26. Lanel B., Barthès-Biesel D., Regnier C., Chauvè T. 1997. Swelling of hydrocolloid dressings. *Biorheology.* 34:139–153.

27. Kulchikhin V., Vasileva O., Litvinov I., Antipov E., Parsamyan I., Plate N. 1991. Rheology and morphology of polymer blends containing LC component in melt and solid state. *J. Appl. Polymer Sci.* 42:363–372.

28. Kulichikhin V., Plotnikova E., Tereshin A., Subbotin A. 2000. Rheological and interfacial phenomena in homophase and heterophase anisotropic polymers. *Polym. Sci. C.,* 42:2235–2264.

29. Kulichikhin V., Plotnikova E., Subbotin A., Plate N. 2001. Specific rheology-morphology relationships for some blends containing LCPs. *Rheol. Acta.* 40:49–59.

30. Kulichikhin V., Semakov A., Plotnikova E., Subbotin A. 2005. New approach to explanation of "Spurt" effect, advances in rheology and its application. *Proc. of the 4th Pacific Rim Conference on Rheology.* Schanghai. pp. 282–283.

31. Jones S., Bowler P.G., Walker M. 2005. Antimicrobial activity of silver-containing dressings is influenced by dressing conformability with a wound surface. *Wounds.* 17:263–270.

32. Burton S.A. 2005. Multi-layer absorbent wound dressing. U.S. Pat. 6,903,243 assigned to 3M Innovative Properties Company.

33. Coloplast Group. 2006. Comfeel Plus. Instructions for use. http://www.woundcare.coloplast.com/EEndCom/CWC/Homepage.nsf/(VIEWDOCSBYID)/B0DD1A7C839981B9C125716F0042372D.
34. Meaume S., Vallet D., Morere M.N., Teot L. 2005. Evaluation of a silver releasing hydroalginate dressing in chronic wounds with signs of local infection. *J. Wound Care.* 14:411–419.
35. Chait R., Craney A., Kishony R. 2007. Antibiotic interactions that select against resistance. *Nature.* 446:668–671.
36. Levy S.B., Marshall B. 2004. Antibacterial resistance worldwide: causes, challenges and responses. *Nature Med.* 10:122–129.
37. Kollef M.H., Micek S.T., Dellinger R.P. 2005. Strategies to prevent antimicrobial resistance in the intensive care unit. *Crit. Care Med.* 33:1845–1853.
38. Burrell R.E. 2003. A scientific perspective on the use of topical silver preparations. *Ostomy Wound Manage.* 49:19–24.
39. Russell A.D., Hugo W.B. 1994. Anti-microbial activity and action of silver. *Prog Med Chem.* 31:351–370.
40. Parsons D., Bowler P.G., Myles V., Jones S. 2005. Silver antimicrobial dressings in wound management: a comparison of antibacterial, physical, and chemical characteristics. *Wounds.* 17:222–232.
41. Ovington L.G. 2004. The truth about silver. *Ostomy Wound Manage.* 50:1–10.
42. Vogt P.M., Hauser J., Roßbach O., Rosse B., Fleischer W., Steinau H.-U., Reimer K. 2001. Polyvinyl pyrrolidone—iodine liposome hydrogel improves epithelialisation by combining moisture and antisepsis. A new concept in wound therapy. *Wound Repair Regen.* 9:116–122.
43. Reimer K., Fleischer W., Brogmann B., Schreier H., Burkhard P., Lanzendorfer A., Gumbel H., Hoekstra H., Behrens-Baumann W. 1997. Povidine-iodine liposomes – an overview. *Dermatology.* 195:93–99.
44. Capelli C. 2000. Silver-based antimicrobial compositions. U.S. pat. 6,093,414 assigned to C. Capelli.
45. Capelli C. 1997. Antimicrobial compositions useful for medical applications. U.S. pat. 5,662,913.
46. Yin H.Q., Langford R., Burrell R.E. 1999 Comparative evaluation of the antimicrobial activity of ACTICOAT antimicrobial barrier dressing. *J. Burn Care Rehabil.* 20:195–200.
47. Yoshimura S., Minami H. 1998. Antimicrobial agent. U.S. Pat. 5,709,870 assigned to Rengo Co., Ltd.
48. Chong S., Fung H.L. 1989. Transdermal drug delivery systems: pharmacokinetics, clinical efficacy, and tolerance development. In *Transdermal drug delivery: Developmental issues and research initiatives,* ed. J. Hadgraft, R.H. Guy, New York, Marcel Dekker, p. 135.
49. Audet M.C., et al. 2001. Evaluation of contraceptive efficacy and cycle control of a transdermal contraceptive patch vs. an oral contraceptive: a randomized controlled trial. *JAMA.* 285:2347–2354.

50. Bos J.D. and Marcus M.H.M. Meinardi. 2000. The 500 Dalton rule for the skin penetration of chemical compounds and drugs. *Exp. Dermatol.* 9:165–169.
51. Attwood D. 2006. Micellar drugs. In *Encyclopedia of Surface and Colloid Science*, ed. P. Somasundaran. vol. 5. New York and London, Taylor & Francis, pp. 3675–3689.
52. Attwood D., Florence A.T. 1982. Surface activity and colloidal properties of drugs and naturally occurring substances. In *Surfactant Systems*. London, Chapman and Hall, pp. 124–228.
53. Hansen J., Mollgaard B. 1992. Transdermal system. U.S. Pat. 5,120,546 assigned to Pharmacia AB.
54. Potts R.O., Guy R.H. 1992. Predicting skin permeability. *Pharm. Res.* 9:663–669.
55. Tiwary A.K., Sapra B., Jain S. 2007. Innovations in transdermal drug delivery: formulations and techniques. *Recent Patents on Drug Delivery & Formulation* 1:23–36.
56. Cross S.E., Roberts M.S. 2004. Physical enhancement of transdermal drug application: is delivery technology keeping up with pharmaceutical development? *Curr. Drug. Deliv.* 1:81–92.
57. Luo E.C., Hsu T.M. 2003. Transdermal administration of steroid drugs using hydroxide-releasing agents as permeation enhancers. U.S. Pat. 6,562,370 assigned to Dermatrends, Inc.
58. Luo E.C., Jacobson E.C., Hsu T.M. 2003. Topical and transdermal administration of peptidyl drugs with hydroxide-releasing agents as skin permeation enhancers. U.S. Pat. 6,565,879 assigned to Dermatrends, Inc.
59. Beste R.D., Hamlin R.D. 2001. Skin permeation enhancer compositions comprising a monoglyceride and ethyl palmitate. U.S. Pat. 6,267,984 assigned to ALZA Corporation.
60. Karande P., Jain A., Ergun K., Kispersky V., Mitragotri, S. 2005. Design principles of chemical penetration enhancers for transdermal drug delivery. *PNAS.* 102:4688–4693.
61. Sjoblom J. (ed.) 2001. *Encyclopedic Handbook of Emulsion Technology*. New York, Marcel Dekker, pp. 377–407.
62. Garti N., Lutz R. 2004. Recent progress in double emulsions. In *Emulsions: Structure Stability and Interactions*. Interface Sci. Techn. ed. D.N. Petsev. vol. 4. Elsevier Acad. Press, pp. 557–605.
63. Friberg S.E., Botorel P. 1987. *Microemulsions: Structure and Dynamics*. Boca Raton. CRC Press.
64. Winsor P.A. 1948. Hydrotropy, solubilization and related emulsification processes. *Trans. Faraday Soc.* 44:376–398.
65. Yoshida K., Sekine T., Matsuzaki F., Yanaki T., Yamaguchi M. 1999. Stability of vitamin A in oil-in-water-in-oil-type multiple emulsions. *J. Am. Oil Chem. Soc.* 76:195–200.
66. Pays K., Giermanska-Kahn J., Pouligny B., Bibette J., Leal-Calderon F. 2001. Coalescence in surfactant-stabilized double emulsions. *Langmuir.* 17:7758–7769.

67. Okochi H., Nakano M. 2000. Preparation and evaluation of w/o/w type emulsions containing vancomycin. *Adv. Drug Deliv. Rev.* 45:5–26.
68. Osborne D.W., Ward A.J.I, O'Neill K.J. 1991. Microemulsions as topical drug delivery vehicles: in vitro transdermal studies of a model hydrophilic drug. *J. Pharm. Pharmacol.* 43:451–454.
69. Kemken J., Ziegler A., Müller B.W. 1992. Influence of Supersaturation on the pharmacodynamic effect of bupranolol after dermal administration using microemulsions as vehicle. *Pharmaceut. Res.* 9:554–558.
70. Boltri L., Morel S., Trotta M., Gasco M.R. 1994. In vitro transdermal permeation of nifedipine from thickened microemulsions. *J. Pharm. Belg.* 49:315–320.
71. Lee P.J., Langer R., Shastri V.P. 2003. Novel microemulsion enhancer formulation for simultaneous transdermal delivery of hydrophilic and hydrophobic drugs. *Pharmaceut. Res.* 20:264–269.
72. Bagwe R.P., Kanicky J.R., Palla B.J., Patanjali P.K., Shah D.O. 2001. Improved drug delivery using microemulsions: Rationale, recent progress, and new horizons. *Crit. Rev. Ther. Drug Carrier Syst.* 18:77–140.
73. Barnes H.A. 2004. The rheology of emulsions. In *Emulsions: Structure Stability and Interactions.* Interf. Sci. Techn. ed. D.N. Petsev. vol. 4. Elsevier Acad. Press, pp. 557–605.
74. Casson N. 1959. *Rheology of disperse systems.* ed. C.C. Mill, London, Pergamon Press, p. 84.
75. Burgers J.M. 1935. Mechanical considerations—model systems—phenomenological theories of relaxation and of viscosity. In: *First Report on Viscosity and Plasticity,* ed. J.M. Burgers, New York, Nordemann Publishing Company.

6

Factors Governing Long-Term Wear of Skin-Contact Adhesives

Anatoly E. Chalykh
Alexey V. Shapagin
Anna A. Shcherbina
Russian Academy of Sciences

Parminder Singh
Corium, USA

6.1 Introduction

The problem of maintaining adhesion contact stability between an adhesive and skin during patch wear has particular significance in the design and application of transdermal drug delivery systems (TDS). Transdermal patches are designed for extended wear (~1–7 days) application to human skin. In this chapter we first describe the mechanisms of pressure-sensitive adhesive (PSA)–skin interactions and, based on experimental results, offer a quantitative approach for the prediction of skin wear time of various types of PSAs.

Our investigations illustrate that for quantitative characterization of the durability of the PSA–skin adhesive joint, it is convenient to use a critical time, t_{cr}, over which the peel strength of the skin–PSA bond (P) achieves some minimum critical value, P_{cr}, when spontaneous patch detaching occurs. These factors, t_{cr} and P_{cr}, are of great fundamental and practical importance because they define the application kinetics of TDS.

As a first approximation, the critical time characterizing the stability of a PSA–skin adhesive joint should be equal or longer than the expected time, t^*, of the patch application. The t^* value can be estimated based upon the independently measured drug release rate of TDS [1,2]. However, *a priori* estimation of the t_{cr} value is still an open issue and experiments for t_{cr} estimation are not defined.

This chapter is intended to provide a discussion of the problem. The starting principles of this investigation were as follows:

1. Overall kinetics of patch wearing up to its spontaneous separation from skin surface should be measured *in vivo* in human volunteers.
2. Each peel test should be performed under at least three different detaching rates to take into account the contribution of work of skin deformation in the work of patch peel off.
3. As test patches we used commercially available materials and model PSAs (see Experimental) with occlusive backings that were free of any drugs, excipients, and low-molecular-weight migrating components to exclude the effects of additives on adhesive performance and possible skin irritation.
4. The analysis of measurement results should focus on a formal mathematical description to obtain an analytical equation, allowing the calculation of critical patch-wearing time.

6.2 Skin as a Substrate

Any substrate as a component of an adhesion joint can be characterized by several parameters. First, chemical composition and the structure of the surface govern the value of thermodynamic work of adhesion. The second parameter is the viscoelasticity of the substrate, which contributes to the work of adhesive joint failure [3,4]. The third characteristic is surface relief. We can state, to a first approximation, that substrate relief relates to the structure of the substrate. However, this relationship is not always evident because methods of surface pretreatment before adhesive joint formation greatly affect the morphology of the surface layer [3].

Human skin is a multilayer structure that consists of dead cells (corneocytes), live cells (keratinocytes), and intercellular space formed by lipid layers. Skin anatomy includes the epidermis (outer skin layer) with the outermost stratum corneum, the dermis, and the hypodermis (inner skin layer). A simplified construction of the epidermis is presented in Figure 6.1.

The stratum corneum is the outermost protective skin layer contacting the applied patch, as illustrated in Figure 6.1. From a morphological point of view, the stratum corneum can be considered a two-phase system consisting of a disperse phase composed of keratinized corneocytes that are rich in proteins and a dispersion medium composed of intercellular lipids that are organized into bilayer membranes (Figure 6.2). Traditionally, such a structure is compared to the "brick–mortar" model [5,6].

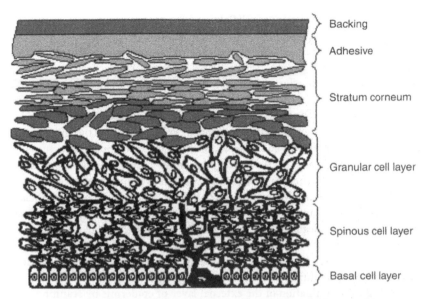

FIGURE 6.1   Schematic structure of the TDS patch adhesive joint with human skin.

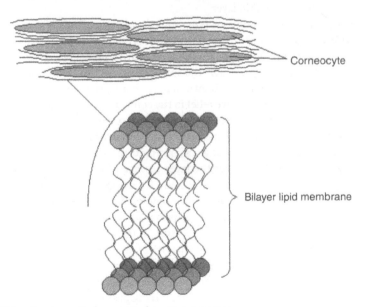

FIGURE 6.2   Structure of a fragment of stratum corneum.

The stratum corneum is an effective barrier of transepidermal water loss from internal skin layers and prevents the penetration of exogenous substances. This layer is impermeable to the components of the adhesive matrix. The stratum corneum is about 10–30 μm thick, whereas the entire epidermis is about 100 μm thick, and dermis thickness ranges from 1 to 1.5 mm. From a morphological point of view the stratum corneum may be visualized as a tile-like gathering of cell flakes, which are derived from the keratinocytes over their life cycle (25–28 days).

Besides protein (keratin), cell flakes contain a mixture of hygroscopic substances such as free amino acids, derivatives of α-pyrrolidone carbonic acid, lactic acid and urea, and their salts. This blend can effectively absorb and accumulate water (up to 60% wt) even at low relative humidity in the surrounding atmosphere. Corneous flakes are fastened together by intercorneocyte lipids (Figure 6.2) consisting of cholesterol, fatty acids, and ceramides, which form liquid–crystalline lamellar layers. Owing to the latter configuration, multiple lipid interlayers do not separate and stay as a complete structure [7]. The same substances are present in the hydrolipid mantle of a very thin film that is located on the surface of the stratum corneum, which protects it against microorganisms [8].

One of the major distinctive features of the skin compared with other artificial substrates is that the structural elements of the skin transform continuously with time due to differentiation into dead and live cells. Within the epidermis layer a desquamation of corneous flakes (cells) occurs continuously under the action of lipases arising from granular cell layers (see Figure 6.1). Thus, young cells migrating from the underlying sublayers renew the stratum corneum permanently.

Morphological examination of the external layer of epidermis by scanning electron and scanning x-ray probe microscopy revealed flat corneous plates lying on top of one another to form the surface skin layer.

Between the plates, the continuous pores and slits are clearly observed (Figures 6.3a and 6.3b) with a diameter of 50–70 μm. Pore depth is comparable with a flake thickness of 0.5–1 μm. Local pores of complicated shape occur in places where several corneocytes contact each other (Figure 6.3c). The surface layer of the forearm skin of human subjects (35–45 years old) contains about 100 pores per square centimeter.

Alongside the macroscopic surface relief in the contact zone of epidermis cells (Figure 6.3), the surface of the keratin plates has its own micro relief, illustrated in Figure 6.4. It is evident that the amplitude of imperfections at the corneocyte surface is within 400 nm, and these ledges are located along a scanning direction with 500 nm periodicity. Thus, well-developed relief underlies the large surface of the stratum corneum. These skin characteristics should be taken into account for interpretation of the initial stage of adhesion joint formation, when adhesive flow onto epidermis surface defects occurs.

The dermis layer of the skin provides the mechanical properties of elasticity, strength, and flexibility. Within an accepted classification of transition zones in adhesive joints (see *Fundamentals of Pressure Sensitivity*, Chapter 3), the skin–patch system can be considered a flexible joint with an elastic substrate. As experiments have demonstrated, high skin compliance and a relatively low value of apparent elasticity modulus greatly affect the mechanism of adhesive joint failure during peeling. Under a low deformation rate of model joints (lower than 5 mm/min) from skin substrate, the geometric dimensions of a zone of adhesive plastic deformation are comparable with the size of deformation zone when the same adhesive is peeled off a rigid substrate (steel or glass plates); the strain zone distortion due to dermis deformation is comparatively negligible.

With the increased peel rate, a significant change in the type of failure of the skin–patch adhesive joint occurs. As illustrated in Figure 6.5, the size of the skin deformation zone increases, and at 20 mm/min it achieves a few centimeters in lengthwise

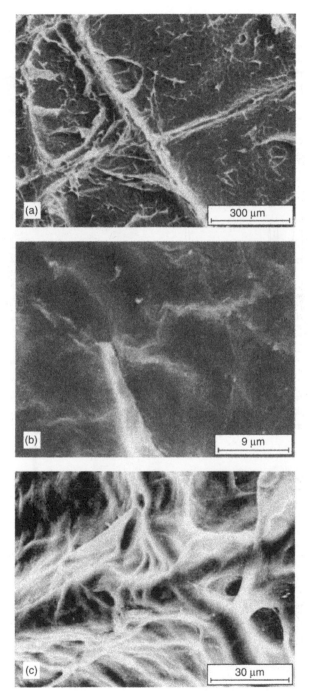

FIGURE 6.3 SEM morphology pictures of silicon replica prepared at the surface of human forearm skin.

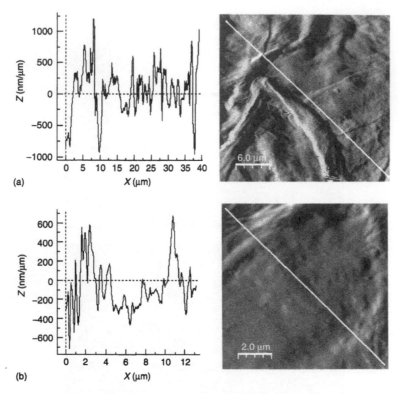

FIGURE 6.4 Atom Force Microscopy (AFM) picture of silicon replica prepared at the surface of forearm human skin. The white line indicates the direction of scanning; Z denotes the height of relief; and X is scanning coordinate.

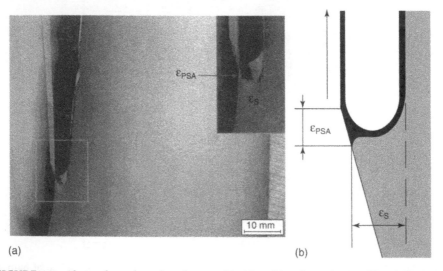

FIGURE 6.5 Photo of a peel test from human skin (a) and its schematic view (b); tensile strain is designated by symbol ε.

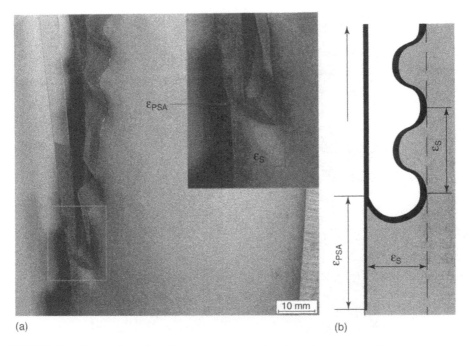

(a)   (b)

FIGURE 6.6 Photo of a peel test from human skin (a) and its schematic view (b).

direction and 1–1.5 cm in the transverse direction. A further increase in the detaching rate (more than 50 mm/min) results in the formation of a very large skin deformation zone (Figure 6.6). In this case, the skin deformation zone consists of the following areas:

- An area of already debonded joint (deformation within this area is related to the presence of residual adhesive that is still in contact with the epidermis)
- An area related to the deformation process within the zone of the peeling front (also illustrated in Figure 6.5)
- An area including the rest of the joint (Figures 6.6 and 6.7). In this region periodic waves and wrinkles occur along the whole undebonded joint as a result of skin deformation. The stress in such elements is in excess of a critical value that leads to a failure of the adhesive on the skin.

Evidently, it is impossible to quantitatively describe a stress pattern for such a system, particularly in the immediate proximity of a peeling front. It is also difficult to estimate the contribution of skin deformation in total energy of adhesive joint debonding.

Our studies indicate that this difficulty can be overcome with some changes in the test scheme. It may be enough to prevent skin deformation growth near a peeling front, for example, by pushing a finger on this place as illustrated in Figure 6.8.

Using this technique, it is possible to neglect the skin deformation independently of test conditions. Using this scheme, the geometric dimensions of the zone of plastic deformation of an adhesive against a skin substrate are comparable with the parameters

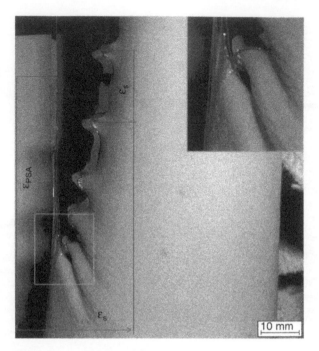

FIGURE 6.7   Photo of a peel test from human skin.

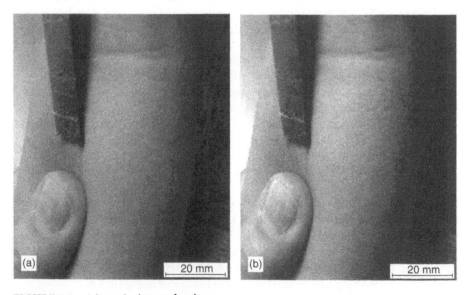

FIGURE 6.8   Adjusted schema of peel test.

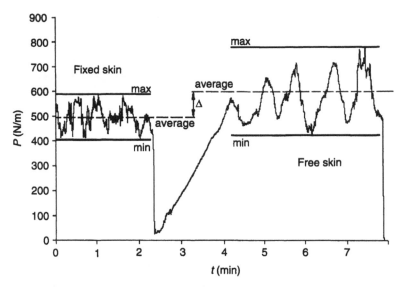

FIGURE 6.9 Peel curves for two peel test conditions: fixed and free substrate (skin).

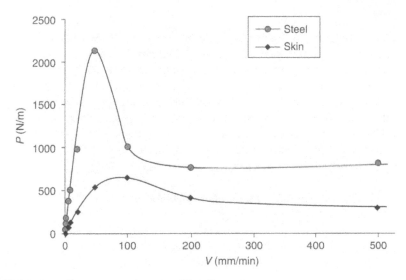

FIGURE 6.10 Adhesive strength versus debonding rate.

of a deformation zone when peeling the same adhesive off a rigid substrate. Using these test conditions we obtain reproducible peeling curves with a slight deviation (~5–10%) of the local values of peeling stress from the average value (Figure 6.9).

Using a modified test scheme, the debonding rate dependence of the work of adhesive joint failure was measured for the acrylic adhesive Duro-Tak 87–900A from skin and from steel, illustrated in Figures 6.10 and 6.11. The peel strength (P) and debonding rate

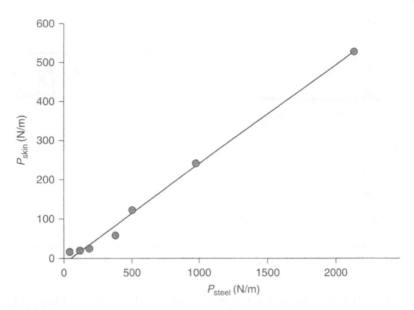

FIGURE 6.11   Comparison of skin and steel peel adhesion.

(V, mm/min) curves are of similar shape, with a linear correlation between peel from steel (P_{steel}) and peel from skin (P_{skin}) ($P_{skin} = 0.2546 * P_{steel} - 13.139$), with $R^2 = 0.9949$.

The human skin is characterized by low surface energy of 28–29 mJ/m² [5], great surface relief, low elasticity modulus, high compliance, and at the same time, a dynamic nature of both morphology and humidity. Prediction of adhesive performance requires quantitative experimental information about the changing kinetics of the strength of skin–PSA joints *in vivo* and the data on the mechanism and failure type of these adhesive joints at different stages of the wearing process.

6.3   Some Ideas on How the Studies Should Be Organized

Analysis of the data regarding adhesion interactions in any adhesive–substrate systems, including PSA–skin adhesive joints, demonstrates that there is a general tendency to the change in adhesive joint strength with time, starting from the moment when the adhesive contact is first formed through the moment of its failure during wear (t). A schematic presentation of the function $P(t)$ is given in Figure 6.12. As the first approximation, one can divide this curve in three stages: I, II, and III.

During the first stage of adhesive joint formation (I), the strength of PSA–substrate contact usually increases asymptotically, approaching an upper limit ($P(t) \to P_{max}$). The time of approaching (t_1) the limit P_{max} may be influenced by several factors, but as will be illustrated, it basically depends on the substrate relief and adhesive viscosity. In real systems P_{max} and the equilibrium value of the strength of the adhesive bond P_∞ (at $t \to \infty$)

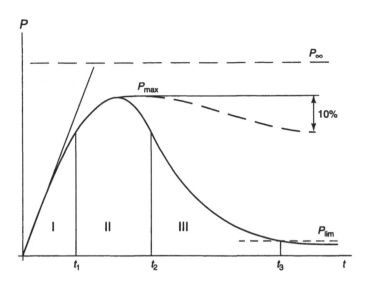

FIGURE 6.12 Schematic view of the kinetics of adhesive strength variation under the conditions of adhesive joint formation and long-term-wear.

often do not coincide because the process of adhesive contact formation is limited by a certain time range.

In the second stage (II), the adhesive joint is relatively stable. In the ideal case the objective is to reach the constant strength of the adhesive bond at the conditions of patch wearing ($P_{max} \approx$ constant) or allow a small deviation of P_{max} (not more than 10%). During this stage, $t_1 < t < t_2$, three possibilities exist: all the characteristics of the adhesive bond do not change (first, a strength of interfacial forces and viscoelastic properties of adhesive and substrate), their changes compensate each other, for instance, due to adhesive–substrate interdiffusion (see *Fundamentals of Pressure Sensitivity*, Chapter 3), or they change slightly due to the effect of external medium-like moisture sorption or the action of aggressive components.

In the third stage (III), $t_2 < t < t_3$, the change in the adhesive joint strength proceeds gradually, although sometimes it may happen dramatically. For composite adhesives, t_2 is considered a limit of application time. However, for transdermal systems the situation is simpler, because in most cases a definite limiting value of strength is preset, $P_{lim} < P_{max}$, which determines the length of wear time t_3.

Our goal in this research was to study and characterize every stage of the formation and exploitation (wear) of the PSA–skin adhesive joint and to develop some approaches for prediction and regulation of the strength of adhesive joints for transdermal systems.

6.4 Experimental

Commercial drug-free antiseptic patches were used for the wear study (see Table 6.1). Duro-Tak 87–900A PSA was used as a model industrial medical grade acrylic adhesive. Model patches were produced in the laboratory by casting 40% wt solution of adhesive

TABLE 6.1 Commercial Patches Employed in This Study

No.	Name	Size	Information
	Commercial Patches		
1	Johnson & Johnson Band-aid "Antiseptic Clear"	7 × 2 cm	Lot 18468_180 Exp 2009_07
2	Johnson & Johnson Band-aid "Antiseptic Flex"	7 × 2 cm	Lot 16669_165 Exp 2009_05
3	Johnson & Johnson Band-aid "Antiseptic Washproof"	7 × 2 cm	Lot 17269_171 Exp 2009_07
4	Pharmadoct Eurosirel "Econom"	6 × 50 cm	Lot − Exp 2010_04
5	Pharmadoct Eurosirel "Mini"	38 × 38 mm	Lot − Exp 2010_04
6	Paul Hartmann Cosmos "Classic"	6 × 50 cm	Lot 62803381 Exp 2011_06
7	Paul Hartmann Cosmos "Aqua"	30 × 40 mm	Lot 62005381 Exp 2011_06
	Adhesive model system		
8	Duro-Tak 87-900A (150 μm) on Scotch Pak 9732 backing film	10 × 80 mm	Lab preparation

in ethyl acetate on Scotch Pak 9732. Prepared films were dried at 60°C for 30–40 min and then in a vacuum (10^{-3} mm Hg) for 24 h. The residual content of ethyl acetate was controlled by thermogravimetric analysis (TG 209F1 Iris, Netzsch, Germany) and it was not more than 1.5–1.8% by wt. The final size of the model patches was 10 × 80 mm, and the adhesive layer was 100–120 μm thick.

The wear study was carried out in 10 volunteers (male and female)[*] ranging in age from 23 to 50 years old. The patch application site was predominantly the inside part of the forearm. Wearing time varied from 20 min to 14 days.

Adhesion performance was estimated by 180° peel using an Instron 1121 Tensile Tester (Instron Ltd, High Wycombe, Bucks, England) at a debonding rate of 5, 10, and 50 mm/min. Failure type was detected using contact-angle measurement of the skin surface and scanning electron microscopy of a patch surface after debonding. A scanning electron microscope JSM-U3 (Jeol, Japan) and x-ray electron probe (WINEDS High Performance X-Ray Microanalysis, Germany) were also used for determination of skin morphology. The atomic force microscope BerMad 2000 (Nanotec Electronica, Madrid, Spain) was used to estimate the fine surface structure of the skin.

---

[*] Based on tentative wearing data, 10 volunteers were chosen who did not demonstrate any skin irritation to the studied patches and model adhesive.

6.5 Kinetics of PSA–Skin Adhesive Bond Formation

6.5.1 Background

Among the modern theories of adhesion, the rheological theory occupies a prominent position. In the framework of this theory, the kinetics of the increase in the strength of an adhesive contact at isobaric–isothermal conditions is related to the kinetics of the establishing the area $S(t)$ of a "true contact" between the adhesive and the substrate.

Several analytical expressions have been introduced describing the kinetics of the adhesive contact formation. Wu and Bright [9,10] supposed that the driving force of this process is the coefficient of spreading λ_{12} of the phase of adhesive 1 upon the substrate surface 2,

$$\lambda_{12} = \gamma_2 - \gamma_1 - \gamma_{12}, \tag{6.1}$$

whereas the size of a defect (nonwetted area at the interface) d is correlated with λ_{12} via the equation

$$d = d_0 \left[1 - (\lambda_{12}/\gamma_2) \right]^n, \tag{6.2}$$

where d_0 and n are constants, γ_i is surface tension of the i-component, and γ_{12} is the interfacial tension between the substrate and adhesive. If we assume that the size of a defect to be related with a nonfilled area between the contacting phases by the equation $S_{\text{def}} = kd^2$ and the full surface of the substrate is S, then the true contact area may be written as $S_{\text{true}} = S - kd^2$, where k is a proportionality coefficient.

Then adhesion P follows,

$$P = P_0 \left(S - kd_0^2 \left[1 - (\lambda_{12}/\gamma_2) \right]^{2n} \right) \tag{6.3}$$

where P_0 is the specific adhesion. The rate of wetting of the interface defects (micro vacancies at the interface) may be described by the exponential equation

$$d \cong d_\infty (1 - \alpha^{-t/\tau})^{-2} \tag{6.4}$$

and the change in peel strength with time by the following expression:

$$(P_\infty - P_i)/(P_\infty - P) = \exp(t/\tau) \tag{6.5}$$

where d is the size of defects at time t, d_∞ is the same value at the infinite time, α is a constant, P is the adhesion at t, P_∞ is adhesion after an infinitely long time, P_i is initial adhesion ($t = 0$), and τ is relaxation time.

A different approach was developed by Gul' [11,12]: the kinetics of formation of the contact surface during the bonding of an adhesive with a substrate is described by the following Osborne equation:

$$S(t) \cong k\sqrt{\frac{pt}{\eta}}$$

and the adhesion strength is

$$P(t) \cong P_i + P_0 k\sqrt{\frac{pt}{\eta}} \tag{6.6}$$

where p is the bonding pressure, t is time, and η is the viscosity of the adhesive. The value of the constant in Equation 6.6 depends on the geometrical parameters of the substrate relief. Thus, micro defects of cylindrical capillary shape $k = 3.17$, for closed pores of triangle shape equal to $2h/cos\alpha$, where h is an average of the triangle height and α is the half of the angle at the top of triangle.

If we assume that at t_∞ the front of an adhesive simultaneously fills all the irregularities in the relief of the substrate surface, Equation 6.6 may then be rearranged in the following form:

$$(P - P_i)/(P_\infty - P_i) = (t/t_\infty)^{1/2}. \tag{6.7}$$

Thus, according to Equation 6.5, the rate of the adhesion strength increases with time and obeys first-order kinetics, whereas according to Equation 6.7 it obeys the exponential law with a fractional exponent.

6.5.2 Experimental Results

Figure 6.13 demonstrates typical curves characterizing the change over time in the peel strength during skin–patch adhesive joint formation for several commercial patches and Duro-Tak 87–900A PSA. Evidently, all kinetic curves $P(t)$ look similar. At short contact times (<1 min), that is, momentarily after the first contact between adhesive and substrate, the adhesive interaction P_i appears. Its value is illustrated in Figure 6.13 as "initial P."

Despite the fact that P_i depends on the conditions of patch-to-skin bonding (contact time and pressure value, number of pressings, etc.), the average value of P_i is $\cong 0.2 - 0.3P_{max}$ for commercial patches, whereas it is as high as $0.4P_{max}$ for Duro-Tak 87–900A. The initial adhesion may be slightly enhanced by the increase of bonding pressure (~15%). However, in the latter case a visible deformation (stretching) of the skin occurs. After unloading, shrinking stress appears and leads to the formation of the wavy surface of the adhesive joint, which changes deformation conditions within the failure zone under peeling of the adhesive film. Thus, the scattering in P_i values increases. With the increasing observation time (at $t > t_i$), the initial strength of the skin adhesive joints gradually increases in the absence of load and asymptotically approaches the limit value $P_i \rightarrow P_{max}$. For most systems under study, P_{max} was approached over 18–20 hours. However, in some cases (Figure 6.14) it was not possible to reach stable P_{max} values because within that

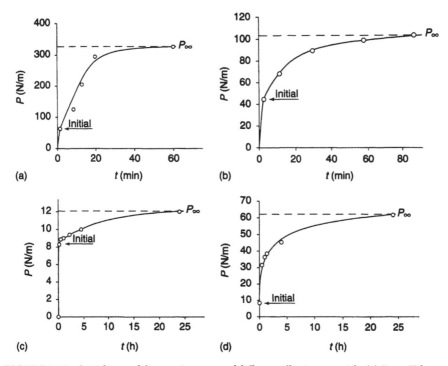

FIGURE 6.13 Initial part of the wearing curve of different adhesive materials: (a) Duro-Tak 87–900A; (b) Sample 7; (c) Sample 1; (d) Sample 2 (see Table 6.1).

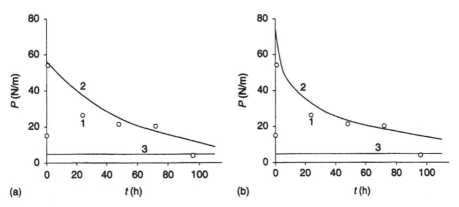

FIGURE 6.14 Experimental wear data (1); Curve (2) – math approximation based on the final part of wearing data 1 using Equation 6.10 (a) and Equation 6.12 (b); and (3) the limit of the adhesive strength $\approx$ 5 N/m.

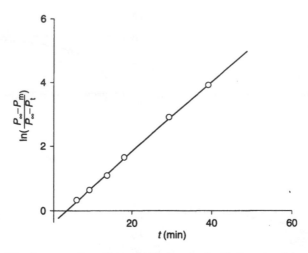

FIGURE 6.15   Wearing curve from Figure 6.13a, described with Equation 6.5.

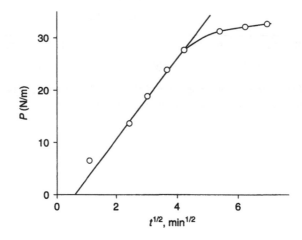

FIGURE 6.16   Wearing curve from Figure 6.13a described with Equation 6.6.

time frame the process of the adhesive joint failure began. Thus, in contrast to the schema given in Figure 6.12, the first stage is immediately followed by stage III.

The analysis of the kinetics of peel strength change during patch wearing demonstrates that the most reasonable fit of experimental data with theoretical curves may be achieved using Equation 6.5 (Figure 6.15). Equation 6.6 provides good agreement with experimental data only at $(P - P_i)/(P_\infty - P_i) < 0.8$ (Figure 6.16). For this reason Equation 6.5 was further used for the calculations of P_∞ and τ in all studied systems. With this purpose, Equation 6.5 was rearranged in the following form:

$$P = P_\infty - (P_\infty - P_i)\exp\left(-\frac{t}{\tau}\right) \qquad (6.8)$$

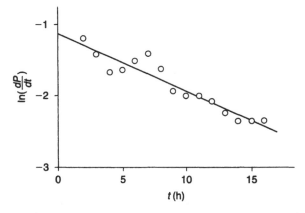

FIGURE 6.17 Time dependence of the rate of peel strength growth presented in semilogarithmic coordinates for the initial part of the wearing curve shown in Figure 6.13c.

TABLE 6.2 Characteristic Time of the Adhesive Contact Formation for Different Adhesives

Adhesive (see Table 6.1)	Slope (tangent)	Characteristic Time (h)
1	0.007	145
2	0.16	6.3
7	1.2	0.8
Duro-Tak 87–900A	6.67	0.15

After differentiating and taking the logarithm, the linear Equation 6.9 was derived.

$$\ln\left(\frac{\partial P}{\partial t}\right) = \ln\left(\frac{P_\infty - P_i}{\tau}\right) - \frac{t}{\tau} \tag{6.9}$$

Experimental data presented in $\ln(\partial P/\partial t) - t$ coordinate system (Figure 6.17) are well treated by log regression, allowing us to calculate the relaxation times and the limit of peel strength from the slope and intercept of corresponding linear relationships for patch–skin adhesive joints. The results of the calculations are given in Table 6.2.

Using Equation 6.5 and the calculated relaxation time and P_∞ values, we have reconstructed the kinetic curves for the change in peel strength of examined acrylic adhesive joints with skin (dashed lines in Figure 6.18). Experimental maximum values of the peel strength in a certain time range are somewhat lower the hypothetical values, which depend on the chemical structure of the adhesive and substrate. Interestingly, the time to achieve the limit of adhesion is also somewhat smaller than t_∞. The closest values for t_{max} and t_∞ were observed for DT 87–900A. This difference in relaxation time is related to the relaxation time necessary to reach the equilibrium adhesion strength, which in turn relates to the adhesive viscosity. Thus, by changing the adhesive viscosity it is possible in principle to reach such values of parameters at which the equilibrium state may be attained in a different necessary time.

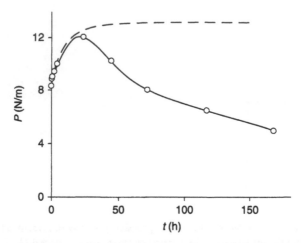

FIGURE 6.18 Kinetic curves of the change of peel strength at the stages of adhesive joint formation and wearing. The dotted curve is calculated from the initial section of the kinetic curve and parameters of Equation 6.9.

6.6   Aging Kinetics of Adhesive Joints under Wearing

Summarized experimental wear curves are presented in Figure 6.19. These data illustrate the change in the strength of patch–skin adhesive joints during a long-term wear study, virtually to the moment of spontaneous detaching of the patch,* that is, without applying any external force. For comparison, in Figure 6.19 hypothetical kinetic curves of achieving limiting P_∞ values are also presented, which were calculated from initial portions of kinetic curves as described previously.

Let us analyze now the obtained results from purely formal point of view, that is, without invoking any information concerning adhesive joint failure mechanism, moisture migration, and morphology of debonded surface. First, a decrease in the adhesive joint strength begins prior to achieving the P_∞ value in the system. The actual decrease in adhesive strength is observed at contact times somewhat longer than 10 h. When designing the adhesive formulation, it would be reasonable to target the interface interaction close to the P_∞ value in the skin–adhesive system. This will enable researchers to realize better the adhesive properties of material, which are mainly governed by the chemical structure of macromolecular chains of the adhesive.

Second, the intensity of the decrease in adhesive joint strength slows down with time. The most pronounced change in adhesion performance, that is, the highest rate of adhesive strength decrease, is observed within the first 2 days of wearing. For all adhesives, this performance period is accompanied by a 1.5- to 2-fold decrease in adhesive strength, independent of P_{max}, which was obtained at the stage of contact formation. A drop of adhesive strength slows down and amounts to just 50–60% within the following 2 days.

---

* Debonding is caused by the friction force between patch surface and clothes.

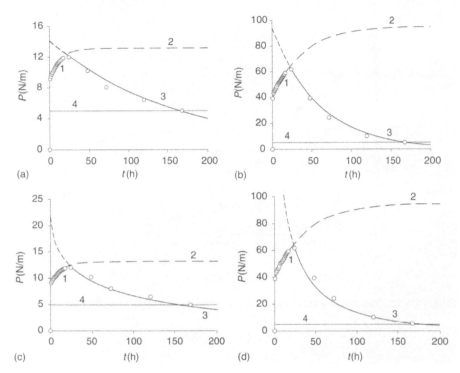

FIGURE 6.19 Peel strength–wearing time relationships for various volunteers: (a) female volunteer, 23 years old; (b) female volunteer, 30 years old; (c) female volunteer, 45 years old; (d) male volunteer, 65 years old. (1) Experimental points; (2) hypothetical curves of limiting P_∞ values established based on the initial part of the wearing curve; (3) calculated curve based on the final part of wearing curve; (4) the limit of the adhesive strength ≈ 5 N/m.

In general, it is conceivable that kinetic curves representing the decrease in adhesive strength during long-term wearing observations asymptotically tend to a certain minimal value associated with the aforementioned spontaneous debonding of a patch. This effect was most often observed at $P \cong 5$ N/m.

Third, information presented in Figures 6.20 and 6.21 is of fundamental importance for describing the decreased kinetics of peel strength in patch–skin joints. The obtained data provide well-defined evidence of a common trend in adhesion performance during wearing. All results presented in a relative coordinate system lay within a sufficiently narrow area of P/P_∞ confined within two parallel curves (Figure 6.21). These boundary curves are shifted to each other by a value equal to the difference between the maximum adhesion strengths achieved at the stage of adhesive contact formation.

Thus, we can assume that the failure mechanism of the patch–skin adhesive joint depends only slightly on the nature of the adhesive and is determined rather by processes that take place within volunteer epidermis.

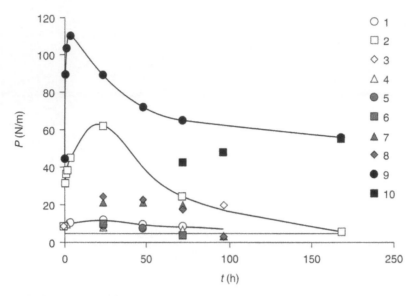

FIGURE 6.20   Summarized experimental curves of wear study on 10 different volunteers.

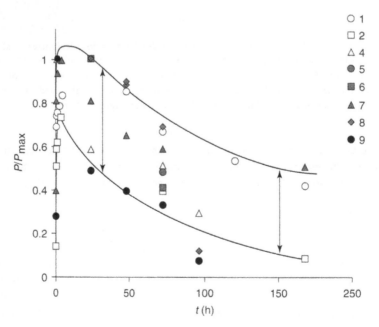

FIGURE 6.21   Deduced experimental curves from Figure 6.20 in relation to P_{max} value.

6.7 Theoretical Description of Adhesive Joint Aging during Wearing

Theoretical analysis of the above-presented experimental curves indicates that they could be mathematically described by two types of analytical equations, either by simple exponential function:

$$P = P_{max} \exp(-t/\tau) \tag{6.10}$$

or by the following expression:

$$P = P_{\infty} \exp(-kt^n) \tag{6.11}$$

Here k, n, and τ are empirical constants. Treatment of experimental data using Equation 6.6 illustrated that this expression adequately describes a trend of the change with time of patch–skin adhesive joint strength during wearing (Figure 6.14a, curve 2). It is worth pointing out that extrapolation of the kinetic curves to zero time yields $P_{\infty}(t \rightarrow 0)$ value, which is very close to P_{∞} calculated from the initial part of the wearing curve (the stage of adhesive bond formation). However, the correlation coefficient is relatively low, $R^2 \cong 0.85$ (Figure 6.14a), implying unsatisfactory quantitative descriptions of experimental data.

Equation 6.11, rearranged in the form

$$P = P_{\infty} \exp(-kt^{1/2}) \tag{6.12}$$

describes experimental data to a greater precision ($R^2 \cong 0.93$; Figure 6.14b).

Numeric values of constants in Equation 6.11 were obtained by the treatment of experimental wear results in a reduced co-ordinate system.

$$\ln(P_{\infty}/P) = \ln k + n \ln t \tag{6.13}$$

According to Equation 6.13, the segment intersected by the lines on the ordinate axis is equal to constant k, whereas their angle of slope is related to exponent n (Figure 6.22). Parameter k is presented in Table 6.3. As a first approximation, parameter k is independent of time and equal to 0.16 ± 0.03. Probably, after obtaining a larger body of experimental wear data, it would be possible to introduce and construct a coordinate system with certain reduced time t/τ plotted along the x axis, where $\tau = 1/k$ is a parameter accounting for true time, temperature, and interaction of adhesive material with human skin.

Using critical value P_{cr}, at which spontaneous debonding of a patch occurs during wearing, and Equation 6.11, it is possible to derive a simple expression for calculating and predicting long-term wearing time

$$t_{\kappa p} \cong \frac{1}{k} \ln \left(\frac{P_{\infty}}{P_{\kappa p}} \right) \tag{6.14}$$

and for definite k and P_{cr} values we finally get:

$$t_{\kappa p} = 7.6 \ln P_{\infty} - 12.3 \tag{6.15}$$

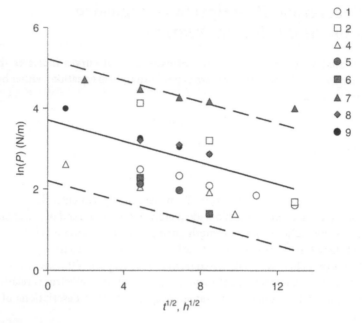

FIGURE 6.22 Experimental wear curves (from Figure 6.20) after rearrangement using the Equation 6.12.

TABLE 6.3 Characteristic Time and Parameter k for Different Sites of Wearing

Adhesive (see Table 6.1)	Parameter (k)	Characteristic Time (h)
Wearing site: arm		
1	0.11	9.1
2	0.33	3.3
4	0.12	8.3
5	0.19	5.1
9	0.15	6.7
7	0.06	15.9
6	0.24	4.0
Duro-Tak 87–900A	0.10	10.0
Average value	0.16	7.6 ± 1.5
Wearing site: loin		
Proprietary patch	0.37	2.7
Proprietary patch	0.28	3.5
Proprietary patch	0.23	4.3
Average value	0.29	3.5 ± 1.0
Wearing site: foot		
J&J (blister)	0.30	3.3
Dr. Scholl's (blister)	0.42	2.4
Average value	0.36	2.9 ± 0.5

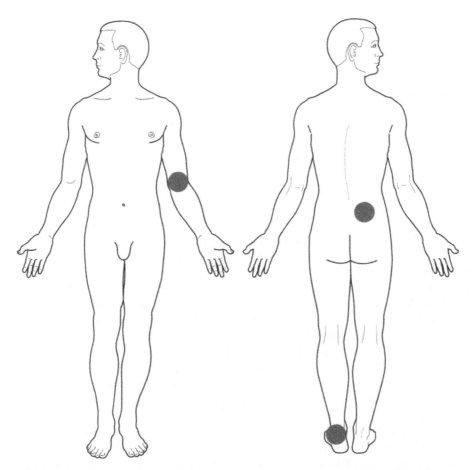

FIGURE 6.23 Illustration of the patch application sites during wear study (for Table 6.3).

The main parameter that determines wear time duration of patches based on polymeric adhesives is the highest value of peel resistance achieved at the patch–skin interface during the first 10 hours of contact. Figure 6.21 presents the correlation between P_{max} and time of wear until debonding.

Information on the adhesion strength growth kinetics during the first 10–12 hours of contact is of vital importance for selecting adhesives. This is illustrated in Figure 6.21, where various kinetic curves of adhesion growth and decrease are presented. Adhesives characterized by high adhesion performance (P_∞) but low rate of adhesive strength growth at initial contact with skin eventually fail to ensure long-term wearing. At the same time, adhesives with significantly lower adhesion performance (lower interfacial adhesion) but with a higher rate of establishing adhesion equilibrium can meet the requirements of long-term wear.

We expect that summarizing the wear data obtained for application to various body sites (Figure 6.23), including the feet, will allow us to formulate more general requirements for long-term skin contact adhesives.

References

1. Bach M., B.C. Lippold. Percutaneous penetration enhancement and its quantification. *Eur. J. Pharmaceut. Biopharmaceut.*, 46, 1–13, 1998.
2. Degim T. New tools and approaches for predicting skin permeability. *Drug Discov. Today*, 11(N11/12), 517–523, 2006.
3. Kinloch A.J. *Adhesion and Adhesives Science and Technology.* Chapman & Hall, London, NY, 1987, p. 441.
4. Furuzawa K. in *Advances in Pressure Sensitive Adhesive Technology-3*, Satas D., ed. Satas & Associates. Warwich, RI, 1998, pp. 12–41.
5. Venkatraman S., R. Gale. Skin adhesives and skin adhesion. *Biomaterials*, 9, 119–1136, 1998.
6. Madison K. Barrier function of the skin. *J. Invest. Dermatol.*, 121, 231–241, 2003.
7. Norlen L., Al-Almoudi A. Stratum corneum keratin structure, function and formation the cubic rod-packing and membrane templating model. *J. Invest. Dermatol.*, 123, 715–732, 2004.
8. Kurz G. Kosmetische Emulsionen und Cremes. Formulierung. Herstellung. Prufung. Verlag fur Chemischen Industrie, GmbH. Angsburg, 2001, p. 272.
9. Wu S. Interphase energy, surface structure and adhesion between polymers. in *Polymer Blends*, Paul D.R., S. Newman, eds. Vol. 1 Academic Press, New York, London, 1978, p. 324.
10. Bright W.M. in *Adhesion and Adhesives*, Rutzler J.E., J. Savaga R.L., eds. Wiley, New York, 1954. pp. 130–138.
11. Gul V.E., V.N. Kuleznev. *Structure and Mechanical Properties of Polymers* [in Russian]. Labirint, Moscow, 1994, pp. 300–329.
12. Gul V.E., L.L. Kudryashova. in *Adhesion of Polymers* [in Russian]. Nauka, Moscow, 1963, p. 134.

7

Competitors for Pressure-Sensitive Adhesives and Products

István Benedek
*Pressure-Sensitive
Consulting*

As a result of the continuous development in the technology of pressure-sensitive adhesives (PSAs), adhesive-coated, monoweb constructions fulfill the most sophisticated end-use requirements (see also Chapter 4). Such products can be used for almost all pressure-sensitive product (PSP) classes. Mainly economic considerations forced the development of uncoated self-adhesive PSPs. Some decades ago, PSPs were only manufactured as PSA-coated products (see also Chapter 1). Coating was the sole manufacturing method used for PSPs (see also *Technology of Pressure-Sensitive Adhesives and Products,* Chapter 10). In this time period, mostly permanent adhesives and products were used. Changes in end use forced the development of removable adhesives and products. Progress in application technology (e.g., for labels) and the discovery of the temporary protective function of such products (tapes and protective films; see also Chapter 4) required the precise control and reduction of their adhesivity to values that are similar to the autoadhesion of plastic films and that allowed the manufacture of adhesive-free self-adhesive products. In this range, there are products (e.g., tapes, protective films, or forms) with a peel strength of 0.5–1.5 N/25 mm. Such low peel resistance values (comparable to the unwinding resistance of common tapes, see Chapter 8) can be achieved for

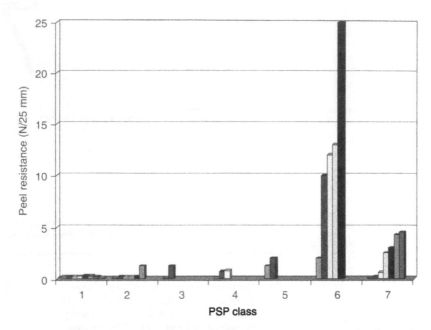

FIGURE 7.1   Peel resistance of various self-adhesive and PSP classes with and without adhesive. (1) Hot-laminating protective films for coil coating; (2) self-adhesive protective films for PMMA plates; (3) self-adhesive protective films for PC plates; (4) EVAc-based self-adhesive protective films for PC plates; (5) adhesive-coated protective films for PC plates; (6) adhesive-coated films for labels; and (7) adhesive-coated films for tapes.

adhesive-coated products by formulation (plasticizing or cross-linking) of the adhesive and by changes in the coating geometry. Figure 7.1 illustrates that with the exception of labels, various PSPs have adhesion values that can be achieved using adhesive as well as adhesiveless constructions.

As a result of such development, PSPs possessing very low adhesion were produced that were comparable to the self-adhesion of certain plastic films and therefore must be applied under pressure and at elevated temperatures, like self-adhesive plastic films. From the point of view of their application, such adhesive-coated PSPs demonstrate more similarities to soft plastic films than to common PSPs. Their manufacture and application, together with adhesiveless self-adhesive competitors, imposed the mutual development of carrier films and PSAs. In the first period of development, PSAs were coated on a relatively hard carrier material to manufacture a PSP. Later, plastic films were used as the carrier material and their deformability (compliance) improved the adhesion of the PSA, allowing the use of a lower coating weight. In the past decade soft plastic films were developed, which adhere without a PSA.

Both paper and common plastic films are nonadhesive carrier materials [1]. Such behavior is due to their partially ordered structure. Crystalline, ordered structures generally do not allow rubber-like high elastic deformations. Their flow occurs in the amorphous domain only. Therefore, the deformational behavior of plastics is different from that of rubber; although ordered and cross-linked, natural rubber exhibits slight self-adhesion.

The "conformational gas" of ideal rubber, with long elastic bridges between the cross-link points, allows a high degree of deformation. In addition, its self-adhesivity can be improved by tackifying, that is, by transforming the cross-linked, elastic structure in a mobile construction that exhibits elastic deformation and viscous flow (due to enhanced chain entanglement and molecular motion; see also *Technology of Pressure-Sensitive Adhesives and Products*, Chapter 8). Rubber can act as an adhesive; a plastomer cannot (at room temperature). Therefore, for classic PSP constructions the plastomer is the nonadhesive carrier and rubber is the adhesive component.

In some cases, for special applications it is possible to produce a virtually adhesive-free (noncoated) carrier film based on a common plastomer, which displays self-adhesion. Various industrial methods can be used to manufacture such bulky self-adhesive products. Polymer synthesis, formulation, and application conditions together ensure self-adhesivity. In practice, the self-adhesivity of such materials is achieved by special chemical composition, by using viscoelastic raw materials, or by physical treatment [1].

In some cases the finished product is a carrier with "true" built-in "pressure-sensitivity." As discussed in Ref. [2], plastic films may undergo cold flow and, therefore, plastic-film-based carrier materials with a special chemical composition may work as a one-component PSP, where pressure sensitivity is the bonding ability at high temperatures under increased pressure. In this case, manufacture of the finished product by extrusion is actually processing plastics. The most common plastic films are manufactured via extrusion [1]. In a similar manner, PSPs with built-in adhesivity are produced by extrusion. Their manufacturing technology is extrusion technology, characterized by the formulation of the raw materials and the equipment used for their manufacture. This is a one-step procedure, which may include on-line physical treatment (see also *Technology of Pressure-Sensitive Adhesives and Products*, Chapter 10). Principally, plastomer-based, tackified plastomer-based, and viscoelastomer-based self-adhesive films can be manufactured. Physically treated, extruded plastomer films are used as hot-laminating films, that is, their adhesion is improved by pressure and increased temperature. By tackification of the plastomer with noncompatible tackifiers, the application temperature can be reduced. Room temperature laminatable films, which compete with adhesive-coated PSPs, can be manufactured using viscoelastic polymers, such as ethylene–vinyl acetate (EVAc) and ethylene–butyl acrylate (EBAc) copolymers.

Tackification of natural rubber with resins, plasticization of cellulose derivatives, and plasticizing of polyvinyl chloride (PVC) demonstrated the possibility of transforming an elastomer or a plastomer into a viscoelastomer (see also *Technology of Pressure-Sensitive Adhesives and Products*, Chapter 8). Advances in macromolecular chemistry demonstrated that internal plasticizing can compete with external plasticizing (see also *Technology of Pressure-Sensitive Adhesives and Products*, Chapter 1). The synthesis of vinyl chloride copolymers, EVAc, propylene copolymers, and branched polyethylenes (PEs) led to macromolecular compounds with greater chain mobility. Such compounds exhibit viscoelastic flow and are self-adhesive.

The first application domain of adhesive-free, self-adhesive films was the lamination of packaging films. This development continued in other areas and led to PSPs without a coated adhesive layer. Hot-laminable, corona-treated, common plastic films and special plastic films have been introduced. These films do not contain special viscous components (tackifiers) or viscoelastic components [thermoplastic elastomers (TPEs) or low-molecular-weight elastomers] and are applied at a temperature above or near the softening range of the products. Adhesive-free laminating films are not new; they have been used for many years. They have been applied on different substrates. For instance, a thermoplastic heat-activated adhesive film has been designed for laminating calendered soft PVC foils to PE foam, on aluminum, fiberboard, and woven fabric. Bonding has been achieved with conventional flame bonding machines at speeds of 10–20 m/min. Coating paper with film-forming (tack-free) EVAc has been used to manufacture packaging materials with barrier and sealing properties.

As discussed in *Technology of Pressure-Sensitive Adhesives and Products*, Chapter 10, PSPs can be manufactured using coating, extrusion coating, or coextrusion. Each of these methods exhibits advantages and disadvantages. Therefore, their applicability for a given product must be rigorously examined. The manufacturing possibilities of PSPs by coating, extrusion coating, and coextrusion were examined comparatively in Ref. [1], with the free choice of carrier material and the running speed as the main evaluation criteria. Coating allows free choice of the product components (carrier material and adhesive), allows on-line manufacture, and ensures a high running speed of the production line. Extrusion coating does not permit the free choice of product components and only ensures a lower productivity. In a similar manner, coextrusion does not allow free choice of the product components and also leads to lower productivity than coating. However, extrusion coating, as well as coextrusion, allows the manufacture of self-adhesive products through plastics processing, which can compete with adhesive-coated products in special applications where the main technical criterion is removability. From the theoretical point of view, taking into account the "trivial" definition of PSAs as products with high instantaneous tack or the recent scientific definition of pressure sensitivity as a process, conditioned by internal cohesion and free volume on a macromolecular scale [3,4] (see also *Fundamentals of Pressure Sensitivity*, Chapter 7), such adhesiveless films are self-adhesive (like a PSA), but are not pressure sensitive. On the other hand, taking into account the application practice of protective webs, which constitute the main class of adhesive-coated, removable PSPs but do not display instantaneous high tack (they must be laminated under pressure), self-adhesive plastomer films and self-adhesive (adhesive-coated) pressure-sensitive films are equivalent in end use.

There is no doubt that common (adhesive-coated) protection films are PSPs; they have an adhesive that displays aggressive tack if used with a high coating weight (due to its special stress frequency-dependent viscoelasticity). However, in some industrial practices this adhesive is coated as a thin (1–5 μm) layer, and its tack is very low (see Chapter 4 in this book and *Technology of Pressure-Sensitive Adhesives and Products*, Chapter 10). Therefore, it can be applied under pressure only, that is, its application tack is the result of intrinsic tack and external pressure [5,6]. In a quite different manner from common PSPs, which work due to their instantaneous tack and peel resistance, bonding of such products can be characterized by peel resistance only, and spontaneous peel build-up (described commonly by dwell time) is replaced in this case by forced wetting through pressure. Thus, although it is incorrect from the theoretical point of view, in practice one can accept the fact that adhesive-coated protection films, as well as adhesiveless films, both function as pressure-sensitive self-adhesive products. Supposing that lamination pressure, as a technological aid for common or adhesive-less protection films, improves adhesive flow and bonding, a similar effect can be admitted for high temperatures used for hot laminating.

The manufacturing technology and application technology of self-adhesive films is an organic part of plastics processing; thus, with the exception of tests of such products, specialists in classic PSPs must be familiar with terms and techniques that belong to the plastics industry.

7.1   Classification of Self-Adhesive Films

Generally, self-adhesive films are used as protective, removable webs. As illustrated in Table 7.1, protective webs include adhesive and adhesiveless constructions.

The adhesiveless, self-adhesive films differ according to their physical treatment and chemical composition. They can be classified according to their build-up, application conditions, and end use. The formulation of one-component (carrier only) PSPs is a problem in plastics processing. It displays some aspects in common with the formulation of carrier films [1]. For self-adhesive products the flat-die (chill roll) procedure is the preferred manufacturing technology. The raw materials and recipes suggested for such a formulation concern the manufacture of a solid-state carrier material with well-designed mechanical and adhesive characteristics to allow the end use (application and deapplication). In some cases, the adhesive characteristics of such product are given by the thermoplasts used in the manufacture of the carrier film (together with a physical surface activation); in other cases, adhesive components must be built in. In certain cases, the product possesses a monolayer construction and in other cases a multilayer sandwich is designed [7] (see also Chapter 1). In such cases (sometimes) the reciprocal adhesion of the layers should be improved using primers (see also *Technology of Pressure-Sensitive Adhesives and Products*, Chapter 10) or barrier layers must be built in to avoid both-side migration of the adhesive component. For many applications, no release layer is required. But some products have a built-in release layer. A short discussion of the build-up of self-adhesive films follows.

TABLE 7.1   Build-Up and End-Use of Protective Films

| | Construction | | | |
| | With Adhesive | | Without Adhesive | |
Application Domain	Adhesive in Carrier	Adhesive on Carrier	Untreated Carrier Film	Treated Carrier Film
Plastic surface protection	•	•	•	•
Metal surface protection	•	•	—	•
Protection of other materials	•	•	—	—

TABLE 7.2 Major Raw Materials and Manufacturing Procedures for Adhesiveless PSPs

	Manufacturing Procedure			PSP			
Raw Materials	Extrusion	Coextrusion	Casting	Protective Film	Business Form	Decoration Film	Tape
Polyethylene	•	—	—	•	—	—	—
Ethylene–vinyl acetate copolymers	•	•	—	•	•	—	—
Ethylene–olefin copolymers	•	—	—	•	—	•	—
Ethylene–acrylate copolymers	—	•	—	•	—	—	•
Butene–isobutene copolymers	—	•	—	•	—	—	•
Vinyl acetate copolymers	•	—	—	—	•	—	—
Vinyl chloride copolymers	•	—	•	—	—	—	•

7.1.1 Classification According to Build-Up of Self-Adhesive Films

According to their build-up, self-adhesive films (SAFs) are classified as adhesive-less and adhesive-formulated products. Adhesiveless SAFs are special, unformulated (uncompounded) plastomers, that is, their recipe includes the components required for the manufacture of a common self-supporting plastic film. Adhesive-formulated self-adhesive films (the term *formulated* is used instead of *coated*, which characterizes common adhesive PSPs) are films tackified in bulk with viscous components or visco-elastomers. Generally, both types of SAF can have a monoweb construction. Table 7.2 presents the major raw materials and manufacture procedures for adhesiveless PSPs.

7.1.1.1 Unformulated Self-Adhesive Products

Unformulated self-adhesive products do not have a coated or embedded adhesive; their self-adhesion is given by the plastomer used for processing the film. According to Djordjevic [8], "films are planar forms of plastics, thick enough to be self-supporting but thin enough to be flexed, folded, or creased without cracking." The upper dimensional limit for a film is diffi-cult to define and is situated between 70 and 150 μm, depending on the polymer used as raw material. The first application domain of adhesive-free, self-adhesive films was the lamina-tion of packaging films. These films have been applied on different substrates. For instance, a thermoplastic heat-activated adhesive film was designed for laminating calendered soft PVC foils to PE foam, aluminum, fiberboard, and woven fabric. Coating of paper with film-forming (tack-free) EVAc has been carried for the manufacture of packaging materials with barrier and sealing properties. This development continued in other areas and led to PSPs without a coated adhesive layer. Hot-laminable, corona-treated, common plastic films and special plastic films have been introduced. These films do not contain special viscous com-ponents (tackifiers) or viscoelastic components (TPEs or low-molecular-weight elastomers) and are applied at a temperature above or near the softening range of the products.

Later, plastic films with viscoelastic properties (e.g., very-low-density polyolefins, copolymers of ethylene with polar vinyl, or acrylic monomers) or common films containing viscoelastic components [e.g., polybutene (PB) copolymers and TPEs] were developed for use (lamination) at lower temperatures or room temperature. Classic plastic films without an adhesive layer cannot possess the tack of PSAs. Plastics can flow, but such flow is a time (stress)- or temperature-related performance [9] (see also Section 7.1.2.2). High application pressure, temperature, or special surface treatment alone are not sufficient to allow hot laminating of plastic films. A special chemical or macromolecular composition that ensures an improved flow is also required. Thus, true adhesive-free hot-laminating protective films are products with a special carrier film composition.

Another need for such an "adhesive" carrier film originates from the field of adhesive-coated PSPs, where the requirement for improved adhesive anchorage became evident as a result of the development of polyolefin carrier materials. In this case, a minimum adhesive level is required that ensures anchorage of the adhesive on the carrier. As discussed in Chapter 1, the most common film carrier materials used for PSPs are based on polyolefins. They are generally nonpolar, chemically inert, partially crystalline plastomers. Crystallinity ensures the required mechanical properties, but reduces surface adhesivity. To improve surface adhesivity, "external" or internal tackification can be used. For instance, monolayer and multilayer polypropylene carrier materials with embedded special resins have been developed for this purpose. This leads to formulated adhesive films (see Section 7.1.2). On the other hand, to improve their surface wetting properties and elasticity without postcompounding, copolymers of olefins with polar comonomers (e.g., vinyl acetate, acrylics, and maleic derivatives) have been developed. Copolymerization of ethylene with polar vinyl and acrylic derivatives gives rise to polar copolymers, the properties of which greatly differ from those of low-density PE (LDPE) (see also Section 7.1.1.1.2). The presence of functional groups in ethylene copolymers makes them chemically more reactive than LDPE.

According to the nature of the self-adhesive polymer, the main types of SAFs are based on polyolefins, EVAc, EBAc, and polyisobutene (PIB) copolymers. Principally, unformulated self-adhesive films include nonpolar and polar films. The test of self-adhesion of such films is discussed in Chapter 8 (see also Figure 8.5).

7.1.1.1.1 Nonpolar Self-Adhesive Films

Nonpolar olefin polymers were used as corona-treated, self-adhesive, hot laminating films. If coextruded, they have an adhesive layer designed to adhere to the surface to be protected, while having a surface layer for printing, slip characteristics, and protection. Later, functionalized polar olefins and other polymers were developed as self-adhesive films. The first unmodified, common films that could be used as self-adhesive products were based on special polyolefins. Warm-laminated films based on very-low-density PE (VLDPE) are the main representatives.

Stretch films were introduced for wraparound packaging 2 decades ago. One-side adhesive-oriented (400%) thin films have been used for this scope. The low level of crystallinity of VLDPE provides an intrinsic cling to extruded films that are used in coextruded structures for industrial wrapping stretch film. The stretch film is usually a multilayer structure in which at least one of the external layers is composed of very VLDPE or a blend of ultra-low-density PE (ULDPE) and linear low-density PE (LLDPE). For a standard VLDPE, the difference between melting and softening range is about 60°C; for ULDPE it is about 80–90°C. This means that such films are very conformable at 20–40°C. The raw materials, as well as the manufacturing procedure of the film, determine its self-adhesivity. Although the tackiness of the stretch wrapping film manufactured by the chill-roll casting process can also be achieved through the use of special raw materials in conjunction with the very fast rate of cooling on the chill roll, in the case of blown film extrusion it is necessary either to use compounded raw materials that already incorporate the necessary additives or to feed the additives into the system during extrusion.

Cast (chill roll) films based on VLDPE and common LDPE are applied as self-adhesive protective films (without PSA coating). Such films have a thickness of 50–70 μm. These products must adhere to extruded and cast plastic [poly(methyl methacrylate) (PMMA), polycarbonate (PC), etc.] plates (see also Section 7.1.3). Generally, they are corona treated. For cast plastic plates the application is carried out at room temperature and for extruded plates the application occurs at 60°C. In some cases, the film must support thermal forming where the protective film-coated plastic plate is heated (ca. 5 min at 180°C). Polyolefin films containing EVAc, EBAc, or PB have also been manufactured and tested. For blown films the incorporation of liquid PIB offers additional technical superiority over EVAc films, with PIB slowly migrating to the film surface (see also Section 7.1.1.1.2).

Warm-Laminating Films Based on VLDPE The use of polyolefins as self-adhesive PSPs (without coated or embedded adhesive) is based on the flow properties of low-modulus polymers. As discussed in Ref. [3], such behavior is the result of special chain build-up and amorphous structure. LLDPE and VLDPE exhibit self-adhesion. This characteristic is utilized in their application as shrink film. LLDPE has been in the market in the United States since 1977. Copolymers of ethylene with butene (gas phase polymerization), hexane, or octene (liquid phase polymerization) have been manufactured. Copolymers with hexene yield LLDPE with a low melt flow index (MFI) and better tear resistance. Such films may also be used as self-adhesive materials. VLDPE is defined as an LLDPE with a density lower than 0.915 g/cm³. Common VLDPE possesses a density of 0.905–0.915 g/cm³. VLDPE (0.905 g/cm³) has a low modulus of 87.5 N/mm² in comparison with a value of 170 N/mm² for common LDPE (ASTM-D-882), which provides low seal initiation temperature. Such very-low-density, low-modulus films can be used as warm-laminated protecting films as well. VLDPE and ULDPE, with densities below 0.900 g/cm³, are used when super high flexibility and autoadhesion are required. VLDPE, with a density of 0.885 g/cm³, is self-adhesive and it is difficult to process for blown films with hot-seal adhesivity.

Warm-Laminating Films Based on LLDPE Butene-based LLDPE may be used as a cling (adhesion) modifying component in slot cast stretch films for pallet wrapping, wrapping of paper reels, and bundling. It should be applied in blends at a level of 2–30% LLDPE for 20- to 25-μm monofilms or as a 3- to 5-μm layer for coextruded films with the same gauge. The film manufacture technology (cast or blown film) determines the adhesivity. The formulation with special raw materials, in conjunction with the very fast rate of cooling on the chill roll (amorphous structure), provides enough tackiness for stretch-wrapping film.

In the case of blown film extrusion of butene-based LLDPE, it is necessary either to use compounded tacky raw materials that already incorporate the necessary additives or to feed the additives into the system during extrusion.

7.1.1.1.2 Polar Self-Adhesive Films

Increasing the polarity of polyolefins enhances their adhesion. For instance, ethylene–propylene block copolymers were grafted with maleic anhydride [10]. Such films, when hot-pressed for 10 min at 200°C, yield a peel adhesion of 8.5 kg on stainless steel and 3.8 kg on aluminum. Copolymers of ethylene with VA and VA and acid–ethylene copolymers were also manufactured to achieve self-adhesion (i.e., low-temperature sealability). Generally, such comonomers provide polarity in the chain and decrease the crystallinity of the material. Both favor the wetting ability of the substrate. At the same time, the melting point of the film decreases. For instance, BA decreases the crystallinity of PE, increases the tack, and provides good mechanical properties. Maleic anhydride increases the adhesion on polar surfaces and allows the initiation of covalent chemical bonding with some polymers. Terpolymers of ethylene, acrylic ester, and maleic anhydride have also been synthesized. Ethylene–acrylic acid (EAA), ethylene–maleic anhydride (EMAA), EBA, and ethylene–methyl acrylate copolymers have been produced. Because of their lower crystallinity, these are very-low-modulus polymers (the value of the flexural modulus of such polymers is situated at 8, respectively 18 MPa). Consequently, they also have a greater degree of tack. EVAc have a processing temperature limitation of approximately 230°C. At higher temperatures, acetic acid can split off from the functional group. EAA copolymers can be processed at 330°C.

Self-Adhesive Protective Films Based on EVAc Copolymers   EVAc display PE-like, or adhesive-like, properties as a function of their VA content (see also *Technology of Pressure-Sensitive Adhesives and Products*, Chapter 1). In practice, their processing for SAF is the processing of a plastomer, which itself can build up a film-like carrier material. Unlike PIB-based self-adhesive films, where the pressure-sensitive properties are given by a low-molecular-weight tackifier component (see Section 7.1.1.2; incompatible with the carrier polymer), EVAc copolymers possess pressure-sensitive properties and are compatible with the carrier material.

Ethylene copolymers with polar monomers are used as self-adhesive carrier materials. EVAc were the first of these materials from this product range. The sealing temperature of an EVAc copolymer film with 12% VAc is 130–150°C; an increase in the VAc content lowers the temperature. Such raw materials are recommended for cold sealing as well. Copolymers with 15–18% VAc were suggested for packaging films to modify their seal properties. Commercial EVAc grades contain 10–40% VAc. Their properties are strongly influenced by the molecular weight and side chains of the polymer. Only high-molecular-weight EVAc polymers exhibit adequate adhesion to plastics. A higher content of 32% VAc in EVAc copolymers leads to a partially crystalline polymer; at the 40% VA level, a completely amorphous polymer is achieved. Polymers with 15% VAc display PE-like properties. Polymers with 15–30% VAc possess PVC-like performance, whereas polymers with greater than 30% VAc are elastomer-like products.

For self-adhesive protective films, both the mechanical and the blocking properties of the film play an important role. The elongation of VAc copolymers increases with VAc concentration. The main increase is given by up to 15% VAc content. The optimum tensile strength is achieved for a content of 20–30% VAc. Blocking is reduced by slip and antiblocking agents (see also *Technology of Pressure-Sensitive Adhesives and Products*, Chapter 8) and by cooling during manufacturing. In the manufacturing of such (blown) films, the suggested blow-up ratio (BUR) is 4/1. For this ratio the MD/CD shrinkage values are about 50%.

The main application of EVAc-based polar SAF films is the protection of plastic plates and films (PC, PMMA) laminates, profiles, furnitures, sanitary items, etc. Such films possess the advantage of a deposit-free removability, resistance to environmental stress cracking, and cuttability. Their main disadvantages are that it is difficult to regulate their adhesion (peel resistance); their adhesion depends on the processing (application) conditions, and peel may build up.

Such films replaced paper-based protective webs for PC plates (with a thickness of 750 µm to 13.3 mm). EVAc-based protection films have a thickness of 50–100 µm and are applied at a temperature of 60 to 135°C. They must exhibit a peel resistance of 140–290 g/25 mm. Films with a thickness of 50 µm are recommended for the protection of plastic films (plates) with a thickness of 175 µm to 1.5 mm. Such films are coextruded, with an adhesive layer of 12.5 µm with 7.0% EVAc. The adhesion build-up for such films is a complex phenomenon, and the formulation alone (without exact processing) does not provide a controllable product quality.

The main advantage of EVAc copolymers for the manufacturer of such films is the potential for a one-step production procedure. On the other hand, the manufacturer of such films strongly depends on the raw material suppliers; that is, the manufacturing know-how of such films is provided by the suppliers of raw materials. Special extrusion-winding and confectioning conditions are required. To avoid blocking, high-speed machines with fine regulated winding characteristics (see also *Technology of Pressure-Sensitive Adhesives and Products,* Chapter 10) are necessary. Such EVAc-based film is a mass and end product; no supplemental working steps (i.e., added value) are possible.

Generally, protective films based on EVAc are manufactured using polymers with a content of 3–28% VAc. An MFI of 1.5–3 is recommended. For protection of PMMA plates, EVAc with less than 7% VAc have been suggested. The VAc content of the polymer used depends on the application temperature of the substrate to be protected, and it decreases with increased application temperature, that is, the adhesion temperature, the macromolecular "dilution" of the ordered structure, and the increased polarity compete to achieve better flow and bonding. The BUR used for blown film manufacturing [1] is a function of the application temperature and decreases with increased application temperature. To achieve a homogeneous mixture, the output of the extruder should be limited. Polypropylene (PP) can be used as a carrier material for EVAc-based SAF. The self-adhesive layer is built up by coextrusion of an EVAc layer (EVAc with 18% Vac and a layer thickness of about 10 µm).

Bilayer heat-shrinkable insulating tapes for corrosion protection of petroleum and gas pipelines have been manufactured from photochemically cured LDPE with an EVAc as adhesive sublayer [11]. The linear dimensions of the two-layer insulating tape decreased

10–50% depending on the curing degree (application temperature 180°C). The adhesive strength of such tapes decreased 10–45% after 1 year.

7.1.1.2 Formulated Self-Adhesive Products

Like the manufacture of PSAs, in which polymer synthesis and the use of polymer blends, that is, formulation, serve as a common method to design pressure-sensitive recipes, the manufacture of self-adhesive films consists of the synthesis of special, low-modulus polymers without a tendency to crystallize (e.g., VLDPE, EVAc, etc.; see above) and the formulation (tackification) of such films. Such tackification requires adequate tackifiers, and in a similar or quite different manner or adhesive tackification (see also *Technology of Pressure-Sensitive Adhesives and Products*, Chapter 8), their compatibility or incompatibility with the carrier-forming main polymer must allow migration and build-up of a tackifier-rich surface layer.

Incompatibility between the carrier and an embedded additive, as related to molecular weight and structure, can be used to build up "working" constructions where the incompatible additive is released from the carrier material. The best known examples are SAFs with embedded tackifying components, but in some cases for common PSPs the release agent is incorporated as an incompatible compound in the release binder. The film is cured to a flexible solid layer by irradiation, and the abhesive substance migrates to its surface. In a similar manner, combining fluoropolymers with polyamic acid leads to antiadhesive coatings with improved performance due to the spontaneous separation into two layers: the bottom layer is polyamic acid and the top layer is a fluoropolymer.

Isobutene derivatives are used as incompatible tackifiers for SAFs. In the case of tackification of a carrier for self-adhesive films, the mechanism can be quite different from that of the tackification of PSAs. As mentioned earlier, in the range of SAFs some products are tackified with compatible tackifiers, but in some products the tackifier incompatibility is used to achieve the final performance. VAc and EVAc are compatible with PE, low-molecular-weight polyolefins, and some tackifier resins. Such compounds are used to prepare adhesive-free protective films. On the other hand, PB is not compatible with certain polyolefins. Its incompatibility with and migration to the film surface is used to manufacture cling films.

Butene or halogenated butene derivatives were suggested as (compatible) tackifiers for elastomers [12]. For instance, tack and green strength of blends of bromobutyl and ethylene–propylene–diene multipolymer have been improved by increasing the level of bromobutyl. In a similar manner, isobutene derivatives work as tacky additives and lead to adhesive films if embedded in a plastomer carrier material. For instance, flexible adhesive tapes for the temporary protection of fragile surfaces have been manufactured by coextrusion (50 μm) of a polyethylene/synthetic resin adhesive mixture and a mixture of butyl rubber (BR) and synthetic resin adhesive (15 μm). Butene and isobutene derivatives and their use in the formulation of PSPs (as base elastomer and tackifier) are described in detail in Refs. [13,14]. Screening formulations for isobutene-based tapes are given in Ref. [15] (see also *Technology of Pressure-Sensitive Adhesives and Products*, Chapter 8). The performance properties of PIB-based adhesives are discussed in detail by Willenbacher and Lebedeva in *Technology of Pressure-Sensitive Adhesives and Products*, Chapter 4. In the range of elastomer-based PSPs, PIB-based PSAs constitute a special class with an old and large application field. Due to their classic synthesis, allowing the manufacture of various, well-characterized products, PIBs (as homo- or copolymers, e.g., BR), were used for pressure-sensitive tapes in different end-use domains [16]. On the other hand, due to their systematic investigation (including the means of macromolecular chemistry and physics and mechanics), their application performance characteristics can be easily related to their fundamental aspects, that is, they serve as model compounds. In *Technology of Pressure-Sensitive Adhesives and Products*, Chapter 4, the roles of high- and low-molecular-weight PIB fractions in adhesive performance are described, along with the impact of molecular weight distribution, chain entanglements, M_e (cross-linking). Advantages and drawbacks of PIB adhesives compared with others PSAs are discussed.

7.1.1.2.1   Self-Adhesive Protective Films Based on PB and PIB

PB as a tackifier of plastomers has been used for the manufacture of silage wrap films as well as for stretch and cling films. The pressure sensitivity of such products is achieved using PB as a viscous, tackifying component. Because of its incompatibility with the main polymer, the viscous component migrates to the surface of the film and imparts the required adhesion. PB (3–6%) is used together with LDPE or LLDPE. For LLDPE a level of 3–5% tackifier is generally used; its concentration depends on the adhesion required, the grade of PB, and the processing conditions.

PBs are not plastomers; they are viscoelastic compounds. Their processing is the mixing of a plastomer with a viscoelastic or viscous compound. As discussed in Ref. [1], normal feed-in of such components is not possible. PB (mainly an isobutene–butene copolymer) can be added to the recipe as a masterbatch (10–20%), as tackified pellets (4–6%), or directly (2–8%). The use of a masterbatch is less expensive than compounding with pellets and has the advantage of being blendable with the polymers of choice. Unfortunately, high masterbatch concentration is required; the masterbatch is sticky and its dosage is difficult. When using tackified pellets, there is no converter responsibility. On the other hand, like the masterbatch the granules are tacky and difficult to handle. The extrusion of such raw materials is difficult, and during storage block building may occur. The upper limit of the PIB level is limited. Masterbatch and pellets are expensive special products. Taking into account these disadvantages, many variances of direct feed-in were developed. The direct feed-in occurs as a direct addition of a solid-state component, in the molten state with the aid of a cavity mixer, or via other injection possibilities. Direct injection into a hopper is less expensive. Unfortunately, screw slip limits injection. Injection at the hopper throat uses a mixture of PB and solid PE. In this case, reduced throughput screw is required.

Direct injection into the polymer melt avoids problems with screw slip, but high pressure injection, involving high capital costs, is required. For addition in the molten state, extruder injection is proposed. Such equipment should allow the heating up and dosage of PB. A heated storage tank, precision speed-controlled pump, connecting hose, and lance are required. A mixture of PB and molten PE is injected into the extruder body. Direct injection into polymer via the center of the screw is also possible by means of a lance that injects the additives through the screw shank into the screw channel. A special technical modality has been developed using a cavity mixer that injects the molten materials between the extruder barrel and the screen changer die. A low-cost process for injecting PIB uses a specially designed screw and grown feed sections.

Stretch films with one-side cling effect have been manufactured for packaging that uses such formulations and technologies. In their manufacture, PIB is blended with LLDPE. The adhesive is on the inside and the outer side possesses slip properties. Such a product can be manufactured as a one-layer film, but a three-layer coextrudate is better. In this case, the LLDPE layer provides the stretch properties and another layer ensures the adhesive characteristics. Generally, the addition of PB to PE film depends on the film manufacturing process, the type of PE, the build-up of the film, and the application of the film. The main parameters that affect the quality of the product include the type of PB, the type of PE, the processing conditions, the concentration of the components, and storage of the finished film.

Peel cling [5] increases with the increased molecular weight of PB and increased addition level. Lap cling [5] decreases as the molecular weight of the PB and the addition level increase (see also Chapter 8). The density of the PE influences the migration of PB; therefore, the choice and dosage of PE is very important. The suggested PB level depends on the film manufacture process as well. Generally, for blown film a higher PB level of 2–8% and for cast film a level of only 2–4% PB are suggested. According to Ref. [17], for blown film 3–5% tackifier and for cast film 1–3% tackifier are recommended. Die-gap, frost line height, melt temperature, and winding tension influence the adhesivity of the film. Of these parameters, winding tension and frost line height are the most important. The storage time and temperature also influence the migration of the PB and the self-adhesive properties. The finished product should be stored under well-defined conditions (in a heated room at least 5 days after manufacture). It should be taken into account that (because of the more amorphous nature of the film) migration of PB in cast film occurs more rapidly.

Self-adhesive protective films based on PB are applied for the protection of plastic plates, in coil coating, and in the automotive industry (see later). For the customer, SAFs based on PB possess the advantage of better adhesion in comparison to EVAc-based films. Therefore, laminating conditions are less important for PB-based SAFs. On the other hand, due to their higher adhesion, their application and handling are more difficult and blocking may appear. For the manufacturer this is a relatively new product class that is more expensive than PE-based hot-laminating films. The manufacturer of such films requires special machines and know-how. Therefore, the production of such SAFs is not recommended for small firms or newcomers to the industry.

In some cases, a coated PSP laminate can have an enclosed adhesive layer as well. For certain products an adhesive modification of the bulky carrier material is required to improve the anchorage of the adhesive without a supplemental primer coating. In such cases, the improved self-adhesivity of the top layer requires the use of an adhesive repellent (release) layer. For instance, biaxially oriented multilayer PP films for adhesive coating have been modified to improve their adhesion to the PSA by mixing the PP with particular resins. For such films the preferred tackifiers (15–25% w/w) are nonhydrogenated styrene polymer, methyl styrene copolymer, pentadiene polymers, α-pinene or β-pinene polymers, rosin or rosin derivatives, terpene resins, and α-methylstyrene–vinyl–toluene copolymers (see *Technology of Pressure-Sensitive Adhesives and Products*, Chapter 8).

To improve their adhesion even further, corona discharge treatment has been suggested (see also *Technology of Pressure-Sensitive Adhesives and Products*, Chapter 10).

7.1.1.2.2   Tackified Self-Adhesive Films Based on Other Compounds

Various polymers were tackified and processed as self-adhesive films. PVC was the first tackified self-adhesive polymer. Its tackification was really plasticization. The autoadhesion of soft PVC films depends on the polymer nature, plasticizer, slip agent, and filler. Self-adhesive PVC coatings were developed first as so-called plastisols. Such products replaced common plastisols with the disadvantage of requiring a primer coating for anchorage. Self-adhesive laminating of PVC is practiced for manufacture of films as well. For instance, a 70-μm PVC film can be laminated with a 10-μm LDPE film.

Self-adhesive, sealable polypropylene films with a level of up to 20% terpene resins have been tested [18]. Chlorinated pE, acrylonitrile–butadiene rubber, liquid chlorinated paraffin, and fire retardant fillers (e.g., hydrated alumina, $CACO_3$, Zn borate, Sb_2O_3, etc.) were compounded and molded into oil-resistant tapes (0.7 mm thick) for electric wires and cables [19].

Peelable protective films (30 μm) comprising methacrylate (e.g., EMAA) copolymers, an organic filler, and slip have been blow molded. Such films, which are useful for the protection of rubber articles, demonstrate a peel strength of 165 g/25mm in comparison with 500 g/25 mm for EEA copolymers [20]. Production of a protective film from PMMA solution is described in Ref. [21].

Tackified self-adhesive films based on EVAc copolymers were suggested for roofing insulation and laminating of dissimilar materials [22]. Such formulations contain EVAc–polyolefin blends, rosins, waxes, and antioxidants. They are processed as hot melts and cast as 1- to 4-mm films. Such films can be considered tackified constructions. PVA films coextruded with LDPE have been recommended for three-layered SAFs with a core layer of PE mixed with PB. For instance, such a film may have the following build-up: LDPE (9 μm)/LDPE–PB (37 μm)/EVAc (4 μm). Partially hydrolyzed PVAc (ester number 50–80) has been proposed as a peelable protective film for rough surfaces [23]. EVAc with chlorinated PP has been suggested as a removable protective film [24]. EEA copolymers have been proposed for removable clear protective films on metals [25].

A cold, stretchable, self-adhesive film is based on an ethylene–α-olefin copolymer (88–97% w/w) with PIB, atactic PP, *cis*-polybutadiene, and bromobutyl rubber [26]. This polymer has a density of less than 0.940 g/cm^3 and exhibits an adhesive force of at least 65 g (ASTM 3354-74).

According to the nature (chemical composition and macromolecular characteristics) of the tackifying additives formulated, tackified self-adhesive films can be pressure sensitive in the classic sense of pressure sensitivity if they contain a viscoelastomer as tackifier. Such SAFs can be considered highly filled PSAs that are formulated for removability (see also *Technology of Pressure-Sensitive Adhesives and Products*, Chapter 8).

7.1.1.3   Physically Treated Films

Physically treated, extruded plastomer films are used as hot-laminating films, that is, they are applied at high temperatures under pressure. Nontackified PE (LDPE), PE–EVAc blends, and PB-tackified PE (coextrudate) were proposed as warm- or hot-laminating

SAFs for plastic plates. Nonpolar, common PE films are corona treated; blends with polar monomers can be corona treated as well (see also *Technology of Pressure-Sensitive Adhesives and Products*, Chapter 10). Generally, such corona-treated films are laminated under pressure at elevated temperatures (see also Section 7.1.2.2). The adhesion of the corona-treated PE film to the plastic plate substrate increases with increased application temperature.

Such methods (i.e., high pressure and elevated temperature) to improve adhesion are used for other technologies as well. For instance, a polyolefin laminate can be manufactured by laminating together two corona-treated surfaces at a temperature that is lower than the softening temperature of the films. In this case, high temperature and high pressure ensure better conformability and contact of the films. For perfect adhesion, full-surface contact and full-surface treatment are absolutely necessary. Therefore, corona treatment of very thin films of PE and PP requires the use of a pressure cylinder to eliminate air bubbles (see also *Technology of Pressure-Sensitive Adhesives and Products*, Chapter 10). The type of substrate surface to be protected determines the choice of pretreatment method. For instance, with anodized aluminum the kind of surface treatment and its age influence adhesion.

LDPE-based hot-laminating films (see also Section 7.2.2) used for coil coating, protection of unvarnished and varnished metallic webs, and plastic plates are manufactured from common PE. Their self-adhesivity is due to a special physical treatment. Because of their one-component formulation and their manufacture as a monolayer blown film, such films are the most inexpensive protective webs. However, the mechanism of their bonding and the technology of their manufacture are not clear. Their production is based mainly on empirical thesaurus.

The use of electrical charges to ensure instantaneous adhesion is a general technical procedure. It is suggested for sheet handling and temporary adhesion as well. A high-tension (100,000 V) electrical field is applied to charge such films. Generally, the adhesion to a metal surface depends on the energetic status of the metal (wetting out, adsorbtion), the electrical potential (electrical double layer), the morphology and geometry of the metal surface (roughness), and the chemical structure. The same is valid for the plastic film surface to be applied, where supplemental viscoelastic parameters interfere. Physical treatment modifies some of these surface characteristics (see also *Technology of Pressure-Sensitive Adhesives and Products*, Chapter 10).

The adhesion of corona-treated surfaces is well known; plastic films with a surface treatment higher than 41 mN/m demonstrate autoadhesion. The treatment quality of such films is usually tested as surface tension. In the production praxis, films with a given surface tension are used to obtain a given debonding resistance (peel). Therefore, for practical use a correlation is admitted between the peel resistance (P) and the treatment quality (Q_s) of the film surface:

$$P = f(Q_s) \tag{7.1}$$

where the surface quality is evaluated as surface tension (ζ):

$$P = f(\zeta) \tag{7.2}$$

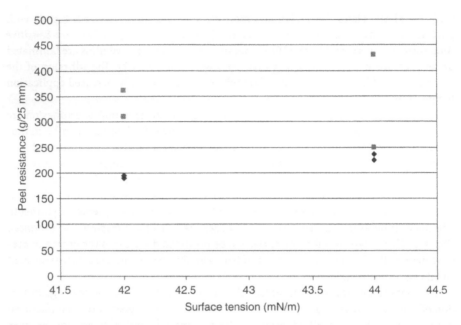

FIGURE 7.2   Peel resistance and peel build-up over time as a function of the film surface tension.
(♦) Instantaneous peel resistance; (▪) peel resistance after 1 month of dwell time.

As illustrated in Figure 7.2, for a given polymer (plastic film) and adherent surface there
is a correlation between the peel resistance and surface tension of a corona-treated PE
film, meaning that the adhesion of the hot-laminated film increases with the surface
tension (treatment degree) of the film. Unfortunately, Figure 7.2 demonstrates, peel
build-up over time does not develop in a similar manner.

In production practice the surface tension of a polymer film depends on its chemical
nature, formulation, manufacture, storage, and treatment. The treatment regulation has
its limits; therefore, the peel resistance level is controlled by regulating both the surface
characteristics of the film (i.e., surface tension) and the intrinsic properties of the web
to be coated. The resultant peel resistance for a hot-laminated, corona-treated polyolefin
film can be written as a function of the surface tension of the film and surface quality of
the adherend (Q_a):

$$P = f(\zeta), (Q_a)$$ (7.3)

This means that practically, in the case of coil coating, the desired peel resistance is
achieved by regulating the surface tension of the protective film and by modification of
the lacquer recipe for the coated coil. Unfortunately, the surface tension alone is not sufficient to characterize the adhesion properties of the treated plastic film. Table 7.2 indicates
quite different peel values are obtained for the same initial surface tension of samples of
polyolefin protective film on the same adherend after a period of time. Therefore, one
should acknowledge (for a given adherend quality) the existence of another supplemental

parameter (β) that takes into account those surface characteristics that are not characterized by surface tension alone.

$$P = f(\zeta, \beta), (Q_a) \tag{7.4}$$

The bonding depends on the laminating conditions (C_l) as well,

$$P = f(\zeta, \beta), (Q_a), (C_l) \tag{7.5}$$

where C_l takes into account the influence of the laminating temperature, pressure, and
time. For a given laminating pressure and temperature the adhesion of a given film is a
function of the treatment degree and of the nature of the surface to be protected. A similar adhesion behavior exists in the packaging industry where self-adhesive heat-sealable
films are used. Peel of heat-sealable polar ethylene copolymers depends on the corona
treatment, coating weight (i.e., layer thickness), and aging.

Corona treatment also depends on the crystallinity of the surface and the relative
humidity of the air. As a result of the treatment, different functional groups, such as
hydroxy, ketone, carboxy, epoxy, ether, and esther, are formed. Each of the new polar
groups ensures a different surface energy level, that is, adhesion. Theoretically, the same
"apparent" surface tension can be achieved with different induced functional groups
(depending on their concentration). Their tailored synthesis is not possible. The oxygen
concentration attains its maximum at 20 J/cm² (depending on the experimental conditions). For the same surface energy level per square centimeters and different experimental conditions, the same surface tension but quite different oxygen level can be obtained.

It is possible to corona treat a substrate and have no indication of a change in surface tension, although adhesion is improved. The treatment effects depend on the nature and age
of the film. Slip agents decrease the effect of surface treatment. Surface tension possesses a
polar and a nonpolar component. Their increase, depending on the treatment energy, can
be quite different. As stated by Yao et al. [27], the polar component increases nonlinearly
with exposure time; there is a rapid change for short exposure times, followed by saturation.
The measured lap shear strength maximum corresponds to the maximum surface energy.

Another problem is the dependence of the treatment degree, measured as the surface
tension on the level of the treating energy. According to Prinz [28,29], the treatment of
PE is a surface phenomenon up to a treatment energy of 10^4 J/m². The autoadhesion of
PE attains a maximum at 500 J/m². The surface tension for this energy level is situated at
60×10^{-3} J/m², which means for plastic films there is a limit to the treating energy situated at about 500 J/m². The required corona energy level (D) depends on the generator
power (G_p), web width (W_w), and web speed (s_w), as follows.

$$D = G_p/(W_w \cdot s_w) \tag{7.6}$$

Figure 7.3 illustrates a dependence of the surface tension on the treating energy level
(generator output). The dependence is almost linear up to a saturation level. Unfortunately,
the adhesion between the adherent and treated film increases over time (several months).
Such a delay in adhesion build-up increases with treatment level. The saturation of the
surface tension depends on the polymer nature. Such saturation is achieved for PP with

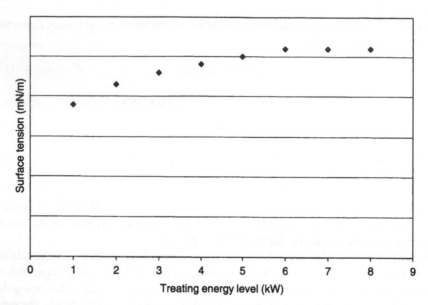

FIGURE 7.3 Dependence of the surface tension (mN/m) on the treating energy level (kW).

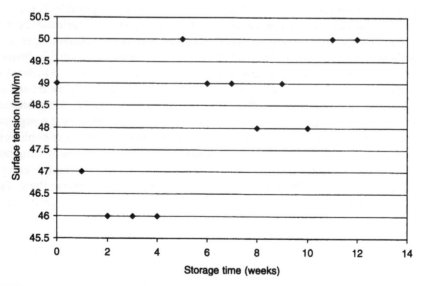

FIGURE 7.4 Discrepancy between the surface tension values as a function of the storage time of the SAF.

an energy level of 2 J/cm². It is astonishing that the surface tension attains its maximum at a relatively low treatment energy level.

Treatment degree depends on the constructional characteristics of the machine and treatment conditions. The effect of corona treatment is temporary and occurs after storage discrepancies appear between the initial and "aged" values (see Figure 7.4).

The manufacturer of corona-treated hot-laminating films must account for the fact that the exact nature of the dependence of adhesion on treatment degree is not yet fully understood. For final bond strength, the laminating conditions (temperature and pressure) and the substrate surface are more important than the chemical nature of the film. Therefore, corona-treated hot-laminating films require perfect cooperation of the lacquer manufacturer, film manufacturer, and laminator (coil coater). For the coil coater, the parameters influencing the adhesion of the protective film are the electrostatic treatment (polarization) of the film, storage time of the film (stability of the treatment), type of lacquer, color and gloss of lacquer, lamination temperature, period of time between the lacquering oven and lamination (running speed dependent), period of time between laminating and cooling (running speed and equipment build-up dependent), temperature of the coil after cooling, and the evolution of adhesion during storage. As stated by Bhowmik et al. [30], the lap shear strength of the PE/steel composites depends mostly on the exposure time, keeping all other parameters constant.

Taking into account the above parameters, hot laminating is a complex technology. Special apparatus and test conditions are required for quality assurance of such products. The versatility of a given corona-treated hot-laminating film (tested on a given surface) to be used on slight differently adherend is minimal. Application requires lengthy preliminary testing of adhesion build-up because adhesion increases over time.

7.1.2 Classification According to Application Conditions

Generally, the end-use performance of protective films manufactured using different technologies, for example, plastomer-extrusion/corona treatment, adhesive-filled plastomer extrusion, modified plastomer-extrusion/corona treatment, or coating of non-adhesive carrier material, is different. The application conditions to achieve adhesion, the value of the peel resistance by debonding, and its build-up in time differ. For adhesiveless PSPs, the regulation of peel build-up is more difficult than for common PSAs, because it strongly depends on the manufacture and application parameters.

The application conditions depend on the self-adhesive character of the film. Its influence on the end use is complex. Bonding is a function of deformability or plasticity (liquid-like behavior) of the product; on the other hand, laminating and de-laminating require high mechanical resistance (solid state-like behavior). Protective films are laminated on the product to be protected. Unlike the manufacture of classic composite structures, the lamination does not occur simultaneously with the manufacturing of the adhesive film (i.e., converting), but after it, like labeling or the application of a pressure-sensitive tape. Protective films are manufactured as preformed materials and undergo laminating after an indefinite storage time. Therefore, they must possess permanent adhesivity given by the cold flow of macromolecular compounds above the T_g.

In principle, the self-adhesive character of a film may be achieved for the whole material (bulk) or for a built-in, or built-on, layer of it; that is, the carrier itself can possess pressure-sensitive properties (see Section 7.1.1.2) or it can be coated with a layer of pressure-sensitive material (see also *Technology of Pressure-Sensitive Adhesives and Products*, Chapter 10). Independent of the construction of the self-adhesive layer, for protective use the material should demonstrate removability.

Some years ago, the face stock materials used by converters in the PSA label and tape industry were common materials developed for the packaging industry (e.g., PVC, cellulose derivatives, and, later, polyolefins [1,31]). The main requirements for these packaging materials include mechanical strength, flexibility, printability, and the lack of self-adhesivity (blocking). Only later did the development of plastic films for a new packaging technology (shrink and cling films), in parallel with the development of tack-ifiers for hot-melt formulations, allowed the manufacture of plastic films with a self-adhesive surface. Unlike classic PSAs, which are fluid (cold flow) at normal temperature, the adhesivity, that is, the laminating ability of the new SAFs, can be used under well-defined temperature and pressure conditions; that is, these materials are adhesive only under exactly defined laminating conditions. Moreover, their removability depends on the laminating conditions as well. Another disadvantage of such uncoated protective films is their substrate-dependent adhesivity, that is, they do not have (actually) a gen-eral usability for different application fields. The manufacture of such films requires the skill of specialists in the plastic film production (extrusion) field.

Self-adhesive films can be imparted in room temperature laminated films and hot-laminated films.

7.1.2.1 Room Temperature Laminated Films

As discussed previously, some tackified self-adhesive films do not need high tempera-tures to improve their bonding; in this case, high-pressure laminating allows bonding. High-pressure application is used for other self-adhesive products as well. For instance, transfer printing materials like Letraset™ can be considered carrierless and temporary laminates with a monoweb character (after application). In such cases, the printing ink includes a pressure-sensitive component and (sometimes) a release component. It may also contain a vinyl polymer with a high tensile strength. This type of product is applied under pressure.

7.1.2.2 Hot-Laminated Films

In common thermal laminating, two thermoplastic substrates (or substrates with ther-moplastic coating) are combined with the application of heat and pressure. The protec-tive film is thermoplastic and may or may not have an adhesive coating (or embedded PSA). Generally, the adhesion of the corona-treated PE film to the plastic plate substrate increases with application temperature.

Laminating equipment determines the laminating temperature and pressure, which are the main laminating parameters. Figure 7.5 summarizes the main parameters of hot laminating.

The application temperature of the protective film depends on the manufacturing tech-nology used for the protected item. Extruded plastic plates have an elevated temperature after extrusion. This is used to improve the adhesion of the postmanufacture laminated protective film. Therefore, such plastic plates can be laminated at high temperatures with adhesiveless protective films. As an example, extruded double-layered PMMA and PC plates are laminated with a (40-μm, EVAc-based) SAF at 60–70°C. Cast items or plates are laminated with an SAF at room temperature, but they must support the processing temperature (170°C, 30 min for PMMA; 160°C, 10 min for PC). In this case, a higher film

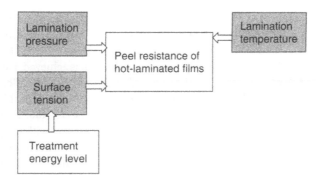

FIGURE 7.5 The main parameters that affect the peel resistance by hot lamination.

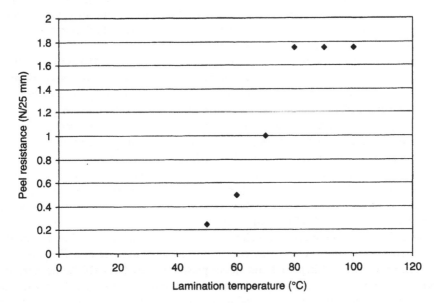

FIGURE 7.6 Dependence of the peel resistance of hot-laminated film on the lamination temperature.

thickness (80 μm) is suggested. Such high-temperature extruded plates may be laminated using high pressure with adhesive-coated protective films. In this case, the adhesion of the film does not require an elevated laminating temperature. The temperature is imposed by the manufacturing procedure of the substrate to be protected. The laminating speed may vary with the equipment and the products to be protected. This influences the high-temperature contact time of the film. For instance, the same coil coating production line may run at a speed of 20–120 m/min. Figure 7.6 illustrates the dependence of the peel resistance of a hot-laminated joint on the lamination temperature.

Cast (chill roll) polyolefin films are used as self-adhesive protective films (without PSA). They have a thickness of 40–80 μm. They must adhere to extruded and cast plastic plates.

After cooling, the film must be removed from PC and AC substrates deposit-free and with low peel resistance. For cast plates the application is carried out at room temperature; for extruded plates it is carried out at 60–70°C. LDPE may be sealed at 110°C. If the temperature is increased to 120°C, a seal strength of 1.7–3.4 N/25 mm is achieved. Hot-laminating ability and temperature resistance may be required by the end-use conditions of the protected item. In many cases such films must survive thermal forming. In this process the protective film-coated plastic plate is heated (for 5–15 min at 170–180°C). The adhesion of LDPE on steel (measured as the coefficient of friction) increases with temperature and attains a maximum at 220–250°F. The weldline strength of LDPE measured at N/15 mm (time 0.5 s, pressure of 0.5 N/mm^2) possesses a value of about 10 at 110°C, but this value increases to 13 at 130°C.

The roughness of the surface (determined by manufacturing and treatment procedures) influences the type of protective film used. For coil coating the gloss of the lacquer is a determining parameter for the choice of protective film. For instance, for hot-laminating films, 30–40% gloss is recommended. The best known adhesive-free protective films in this field of coil coating are hot-laminating films.

Common hot-laminating films used since the early 1970s are LDPE based. Such films are thermoplastic and do not display autoadhesion at room temperature and low pressure. Their application is possible due to the improved conformability and self-diffusion of a thin, polarized, molten surface layer during their application under high temperature and pressure. Such polymer deformation (conformability) is a viscous flow. The bonding-related deformation of elastomers or viscoelastomers is a viscoelastic flow, which means that viscoelastically bonded surfaces ensure instantaneous debonding resistance given by the elastic characteristics and viscosity of the material. Plastic-bonded macromolecular compounds (e.g., hot-laminated polyolefin films) exhibit an instantaneous debonding resistance that is a function of the (temperature-dependent) viscosity of the material. Mutual interpenetration of the contact surfaces, regulated by the diffusion of the macromolecular compound, remains for true PSAs, a mostly hydrodynamically and rheologically controlled phenomenon (see also *Fundamentals of Pressure Sensitivity*, Chapters 3 and 4). For plastomers used as (self) "adhesives," the chemical nature of the contacting surfaces is more important in comparison with true PSAs. The surface treatment must produce an attraction between the contacting surfaces of hot-laminated films (based on the double layer of electrical charges) and to ensure a chemical affinity between the polymer and the contacted surface due to the induced polar functional groups.

7.1.3 Classification According to Application Field

The simplest way (at least theoretically) to build up a protective film is to manufacture an uncoated plastic film that possesses the mechanical strength of a carrier film and has a built-in adhesivity. The monoweb, as a challenge for label producers and a reality for tape manufacturers, is undergoing further development for protective films. To understand the direction of this development, we will look at the requirements for protective films. Conformability, mechanical strength, balanced adhesion to the protected surface, and removability are the main application criteria for protective films.

A number of protective materials fulfill these requirements at least partially. As discussed previously, many types of materials for surface protection exist. Some, like packaging materials, are self-supporting, mechanically resistant carrier materials that do not need to be in intimate contact with the product to be supported; thus, their removability does not affect the product. Other products, like varnishes or lacquers, develop intimate contact with the surface to be protected. Generally, these are permanent coatings that do not need a self-supporting carrier. In some cases, even permanent coatings may display a sheet-like character (laminates). Some packaging materials (e.g., cover films) autoadhere to the protected product, although they need no lamination to perform their function.

Self-adhesive protective films can be classified according to their main application fields (and conditions) into SAFs for hot laminating and SAFs for warm or room temperature laminating. The self-adhesive films for hot laminating can be used for metal, plastics, and paper surfaces. Hot-laminating self-adhesive films can be divided into films for uncoated and films for coated adherent surfaces. Films for coated surfaces can be applied to coated lacquered or coated laminated adherents.

According to the nature of the protected surface, self-adhesive films can be classified as SAFs for plastic surfaces, paper, mold release casting, back protection of glass or special surfaces (fabric), and coil coating.

7.1.3.1 SAF for Hot Laminating of Plastic Surfaces

Nontackified polyethylene (LDPE), PE–EVAc blends, and PB-tackified PE (coextrudate) were proposed as SAFs for plastic plates. Nonpolar, common PE films are corona treated; blends with polar monomers can be corona treated as well. For instance, a corona-treated LDPE (density of 0.923 g/cm³) film with a thickness of 50–60 μm is manufactured using a blow procedure, with a BUR of 1:1.8 to 1:2.4 depending on the dimensions. The LDPE is pretreated for a surface tension of 46–48 dyn/cm. The adhesion of such films attains 350–750 g/50 mm.

For extruded PC and PMMA plates, a 70/40 μm SAF with 18% EVAc is suggested. Such films are laminated at 60–70°C. For instance, a 40-μm coextrudate possesses an 8-μm EVAc adhesive layer (based on a copolymer with 7–9% VAc). Extruded PC or cast PMMA is processed as a sanitary material. The carrier film for such applications must be deep drawable to support the processing conditions (160–170°C, 10–30 min) of the protected item. For such applications, self-adhesive or adhesive-coated films with a minimum thickness of 70–80 μm can be used.

Generally, thin (less than 6–10 mm) plastic plates are protected with (70 μm) SAF, based on LDPE/EVAc (80/20), with an EVAc with 9–12% VAc. For thicker plates, adhesive-coated protective films with a higher thickness (80–100 μm) have been proposed. Protective films with a thickness of 25–100 μm (9–12% VAc) have been tested for deep-drawn PC plates.

Passive protective films for PC plates have a thickness of 50 μm; processible protective films have a thickness of 70 μm. They should be applied at both 50 and 100°C. PC plates are hygroscopic and are generally dried before use. Drying is carried out at 130°C for 0.5–48 h, depending on the plate thickness (0.75 to 12 mm). Therefore, short-term adhesion build-up is tested after storage at 130°C. Shrinkage at 130°C is very important.

TABLE 7.3 Application and Processing Characteristics of Protective Films Used for Plastic Plates

Protective Film Grade	Carrier Thickness (μm)	Protected Plastic Plate Material	Protected Plastic Plate Manufacture	Laminating Conditions Temperature Plate	Laminating Conditions Temperature Press Roll	Processing Temperature/ Time (°C/min)	Peel Resistance (N/25 mm)
Adhesive coated	50	PMMA	Extrusion	70	70	150/15	1.25–2.0
Adhesiveless	40	PMMA PC	Casting Extrusion	70	70	150/15 170/15 180/5	0.12–1.25
Adhesiveless	80	PMMA	Casting	70	30	170/30	0.02–0.08

Long-term adhesion build-up is tested after storage at room temperature and elevated temperature. The protective films for such plastic items are generally applied by in-plant lamination. A hot-roll laminator nip, or a hot-roll press, is used, and heat and pressure cause the film to conform and adhere to the surface. Table 7.3 demonstrates how the grade of the protective film used for a plastic plate depends on the plate material, manufacture process, and geometry.

7.1.3.2 SAF for Coil Coating

In many applications, the coil (0.4–1.5 mm) to be protected by a film is chromated, primed (5 μm), and lacquered (20 μm) continuously at a running speed of 50–80 m/min. The lamination temperature of the self-adhesive polyolefin protective film on this web is 200–250°C. After laminating, the film is cooled in a water bath. The debonding force (90° peel resistance) of such films is about 150–210 g/25 cm (with tolerances of 50 g). Generally for coil coating, relatively thick (120,150, or 170 μm) protective films are used. Such films are corona pretreated at 44–50 mN/m.

The protection of aluminum, especially anodized aluminum surfaces, is a complex domain of coil coating. Standard tests of adhesion are generally made on stainless steel. Although the correlation between peel resistance on stainless steel and aluminum can be evaluated, the increase in adhesion with the dwell time on stainless steel and aluminum may be quite different (Figure 7.7).

All lacquered surfaces (unpolished, polished, and bright polished) should be pretested before coating. The ability to bond through oil is essential in many processing applications.

For certain acrylic coil coating formulations, 1–5% of special PE dispersions are added to the recipe to impart better deformability and to prevent blocking of the lacquer surface. Therefore, the application of a PE-based hot-laminating film for such coils may cause PE/PE adhesion, that is, a higher adhesion level. Such limitations exist for adhesive-coated protective films as well. Rubber–resin adhesive-coated protective films cannot be used for copper or brass (bright or mill finish). Sulfur-containing copper reacts with rubber.

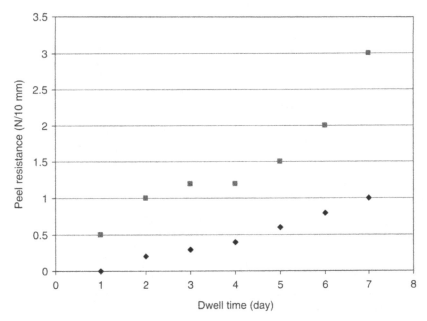

FIGURE 7.7 Build-up of the peel resistance of a hot-laminated film as a function of the dwell time and substrate surface; (♦) stainless steel; (■) aluminum.

7.2 Comparison of Coated/Uncoated Monoweb Constructions

In principle, a PSP can be manufactured in a one-step procedure if the plastic film demonstrates built-in pressure sensitivity. Such a procedure offers primarily economic advantages.

On the other hand, a fluid PSA layer (that is coated) with a thickness of a minimum of 1.0–1.5 μm ensures better conformability and contact with a solid substrate surface than a self-supporting plastomer (or plastomer/viscoelastomer) with a minimum thickness of 7–25 μm. The conformable carrierless (transfer) tapes are also manufactured using coating technology.

Coating allows the combination of adhesive and nonadhesive areas on the same surface; this is not possible for extruded PSPs.

In general, noncoated monowebs allow lower bond strength than coated products. On the other hand, because of the bulk monolayer nature of the "adhesive" and its limited flow in such constructions, the nature of the adhesive break can be easily controlled.

A big difference exists between the adhesive PSAs and carrier plastomers. The main characteristic of PSAs is their flow; their mechanical strength is generally not sufficient for them to be self-supporting. In some cases, their cohesion is not sufficient to ensure acceptable shear resistance. The main characteristic of plastomers is their excellent mechanical resistance; their flow is weaker and allows self-adhesion only through the use of high-temperature or high-pressure lamination.

At a time when end user demand for low-cost, removable products is high, SAFs are finding their way into the world market as competitors for adhesive-based PSPs.

7.2.1 Other Competitors

Other competitors include competitors to labels, tapes, and protective films. Such products may have a web-like construction or are used as lacquer-like products.

7.2.1.1 Competitors to Labels

Competitors for pressure-sensitive labels are described in Chapter 4. Such products include wet adhesive labels, heat-adhesion labels, in-mold labels, shrink labels, and wraparound labels.

7.2.1.2 Competitors to Tapes

Competitors for tapes are also described in Chapter 4. Such products include wet tapes and heat-seal tapes.

7.2.1.3 Competitors to Protective Films

Competitors for protective films are described in Chapter 4. Such products include protection coatings and non-pressure-sensitive protection films (see Table 7.4).

TABLE 7.4 Main Removable Products for Surface Protection

Removable Protective Product			Application			De-Application		
Product	Chemical Basis	H (μm)	Spray	Dipping	Lamination	Peel Off	Delamination	Washing
Varnish	Wax Low-molecular-weight EVAc copolymer Low-molecular-weight AC copolymer Polyvinylbutyral	5–10	•	•	—	•	—	Water alkalia solvent
Low Strength Film	PVC plastisols Polybutadiene cross-linked Polybutene/resin Polypropylene/resin VC–AC copolymer	10–70	•	•	—	•	—	•
Paper	Paper	30–100	—	—	•	—	—	•
Film	PE, PP, PET, PVC	12–120	—	—	•	—	—	•
Film*	PE, PP, PVC	25–0.50	—	—	•	—	—	•

* *Separation Film.*

References

1. Benedek I., Manufacture, in Developments in *Pressure-Sensitive Products*, Benedek I., Ed., Taylor & Francis, Boca Raton, 2006, Chapt. 8.
2. Benedek I., Physical Basis of PSPs, in *Developments in Pressure-Sensitive Products*, Benedek I., Ed., Taylor & Francis, Boca Raton, 2006, Chapt. 3.
3. Feldstein M.M., Molecular Fundamentals of Pressure-Sensitive Adhesion, in *Developments in Pressure-Sensitive Products*, Benedek I., Ed., Taylor & Francis, Boca Raton, 2006, Chapt. 4.
4. Feldstein M.M., Pressure-Sensitive Adhesion as a Material Property and as a Process, in *Pressure-Sensitive Design, Theoretical Aspects*, Benedek I., Ed., VSP, Utrecht, 2006, Chapt. 2.
5. Benedek I., Adhesive Properties, in *Developments in Pressure-Sensitive Products*, Benedek I., Ed., Taylor & Francis, Boca Raton, 2006, Chapt. 7.
6. Benedek I., Adhesive Properties, in *Pressure-Sensitive Adhesives and Applications*, Marcel Dekker, New York, 2004, Chapt. 6.
7. Benedek I., Build-up and Classification, in *Developments in Pressure-Sensitive Products*, Benedek I., Ed., Taylor & Francis, Boca Raton, 2006, Chapt. 2.
8. Djordjevic D., Tailoring Films by the Coextrusion Casting and Coating Process, in *Proc. Speciality Plastics Conference '87, Polyethylene and Copolymer Resin and Packaging Markets*, Dec. 1, 1987, Maack Business Service, Zürich, Switzerland.
9. Benedek I., Comparison of PSAs, in *Pressure-Sensitive Adhesives and Applications*, Marcel Dekker, New York, 2004, Chapt. 4.
10. Dow, *XZ87131.32, Experimental Data Sheet*. Dow Chemicals, Horgen, Switzerland.
11. Neste Chemicals, Technical Information, Polyethylene, FO133, 1991.
12. Benedek I., The Role of the Design of Pressure-Sensitive Products, in *Pressure-Sensitive Design, Theoretical Aspects*, Benedek I., Ed., VSP, Utrecht, 2006, Chapt. 1.
13. Benedek I., Design and Formulation Basis, in *Pressure-Sensitive Design and Formulation, Application*, Benedek I., Ed., VSP, Utrecht, 2006, Chapt. 1.
14. Benedek I., Chemical Composition, in *Pressure-Sensitive Adhesives and Applications*, Marcel Dekker, New York, 2004, Chapt. 5.
15. Benedek I., Design and Formulation Basis, in *Pressure-Sensitive Design and Formulation, Application*, Benedek I., Ed., VSP, Utrecht, 2006, Chapt. 6.
16. Benedek I., End-Use of PSPs, in *Developments in Pressure-Sensitive Products*, Benedek I., Ed., Taylor & Francis, Boca Raton, 2006, Chapt. 11.
17. Polyethylene 89, Maack Business Service, Zürich, Switzerland, p. 53.
18. *Neue Verpackung*, (2) 61 (1991).
19. Toshio M. (Furukawa Electric Co. Ltd.), Japan Pat., in *CAS, Adhesives*, 24, 6 1988.
20. Yamada K., Miyazaki K., Owatari Y., Egami Y., and Honma T. (Sumitomo Chem. Co., Ltd.), EP 257803/02.03.1988, in *CAS, Coatings Inks & Related Compounds*, **17**, 6 (1988).
21. Khoklovkin A.E., Vladimirovkaya N.V., and Bushkova E.S., *Sovrem. Lakokrasokh. Mater. i Tekhnol.ikh Primeneniya, Mater. Semin. N.*, in *CAS, Coatings Inks & Related Products*, **18**, 2 (1988).

22. Mitsui K., PCT, WO8802767, Mitsui Petrochemical Ind. Ltd., in *CAS, Adhesives*, **22**, 4 (1988).
23. Eastman Kodak Co., US Pat., 3718728, in *Coating*, (6) 154 (1974).
24. Nitto Electr. Ind. Co. Ltd, Japan Pat., 24229/70, in *Coating*, (2) 38 (1972).
25. Litz R.J., *Adhes. Age*, (8) 38 (1973).
26. Haas A. (Société Chimique des Charbonnage-CdF Chimie, France), US Pat., 4624991/25.11. 1986, in *Adhes. Age*, (5) 26 (1987).
27. Yao Y., Liu X., and Zhu Y., *J. Adhes. Sci. Technol*, **12** (11) 1181 (1998);**12** (11) 1181 (1998), (7) 63 (1993).
28. Prinz E., *Coating*, (10) 360 (1979).
29. Softal Electronic, Report, Nr.102, Softal Electronic GmbH, Hamburg, Germany, 2002.
30. Bhowmik S., Ghosh P.K., Ray S., and Barthwal S.K., Surface modification of HDPE and PP by DC glow discharge and adhesive bonding to steel, *J. Adhes. Sci. Technol*, **12** (11) 1181(1998).
31. Benedek I., Chemical Basis, in *Developments in Pressure-Sensitive Products*, Benedek I., Ed., Taylor & Francis, Boca Raton, 2006, Chapt. 5.

8

Test of Pressure-Sensitive Adhesives and Products

István Benedek
*Pressure-Sensitive
Consulting*

As in the case of many other finished products, a wide range of test methods were developed for pressure-sensitive adhesive (PSA)-based labels, tapes, and coatings. A number of organizations such as FINAT, AFERA, PSTC, and TLMI have established (standard) test methods that are widely used in the industry, although there remain significant differences between the various methods used. These methods provide a good basis for the evaluation of adhesives, but some modifications or additional tests are required when testing materials for specific applications. Therefore, the principal PSA manufacturers and converters have developed their own methodology. It is not the aim of this chapter to discuss the standardized PSA testing methods in detail. It is of greater interest to describe here the special methods worked out for special end uses or specific PSA applications. For a pressure-sensitive product (PSP) manufacturer, the properties of the liquid adhesive, its coating behavior, and the PSP performance characteristics are more important. From the point of view of methodology, these areas differ significantly.

The test methods for PSAs and PSPs serve to control the manufacturing process, the materials used in the process, the technological discipline, and the finished products. Some of them are equipment-related measurements that belong to process control and others are laboratory tests.

Test methods for PSAs and PSPs were discussed in detail by Benedek in Refs. [1,2]. The fundamental aspects PSAs testing are described in *Fundamentals of Pressure Sensitivity*, Chapters 6–8. In this chapter, the manufacture technology-related tests will be discussed first.

8.1   Manufacture-Technology-Related Tests

Quality assurance is an organic part of the manufacturing technology of PSAs and PSPs. This technology uses various formulation components for the adhesive (see also *Technology of Pressure-Sensitive Adhesives and Products*, Chapter 8) and various carrier materials and adhesives (see also *Technology of Pressure-Sensitive Adhesives and Products*, Chapter 10, and Chapter 1 in this book) to build up the PSP. As discussed previously, various raw materials are used for the formulation of the PSA, which must be tested. They include the main adhesive components, as well as the chemical and technological additives used in the formulation (see also *Technology of Pressure-Sensitive Adhesives and Products*, Chapter 8).

8.1.1   Test of Formulation Components

The test of product components includes the raw materials used for the formulation of the PSA (e.g., base elastomer, tackifier, etc.), the test of the main PSA components (e.g., viscoelastomers), and the additives. The main raw materials used for PSAs are described in *Technology of Pressure-Sensitive Adhesives and Products*: rubber-based adhesives are discussed in Chapters 2–4; silicones are the subject of Chapter 6; viscoelastomers are described in Chapters 5–8; and hydrophylic adhesives are discussed in Chapter 7.

8.1.1.1   Test of Raw Materials for PSA

The test of raw materials for PSA evaluates the various non-pressure-sensitive components of the recipe. Principally, the test includes the main components of the recipe, that is, the base elastomers and the tackifier resins. Both are synthesized off-line and supplied ready to formulate. Therefore, for the base elastomer, supplemental quality control is generally not required, except in some special cases where the elastomer must be masticated (see also *Technology of Pressure-Sensitive Adhesives and Products*, Chapter 10).

8.1.1.1.1   Test of the Solid-State Resin

Resins must be tested concerning their color and softening point. Resin color was discussed by Benedek in Ref. [3]. The color stability after 72 h aging at 180°C is tested also (see also Section 8.2.2.4). Various methods used for color characterization are described by Martinez in *Technology of Pressure-Sensitive Adhesives and Products*, Chapter 2. According to O'Brien et al. [4], the tackifying resins are characterized using test methods typical for the industry of PSAs: (1) methylcyclohexanone/aniline (MMAP) and diacetone alcohol/xylene (DACP) cloud point measurements, (2) softening point, (3) dynamic mechanical analysis (DMA), (4) differential scanning calorimetry, and (5) and molecular weight distribution, determined by size exclusion chromatography. O'Brien et al. also caution the reader that when comparing formulations with different resins, it is important to first characterize the resin (aromaticity, softening point, and molecular weight) to determine the source of any observed differences.

According to Hu and Paul (see *Technology of Pressure-Sensitive Adhesives and Products*, Chapter 3), a simple way to test compatibility is to dissolve an equal amount of tackifier and polymer in a solvent such as toluene and then cast a film to check the clarity of the blend. An incompatible polymer–tackifier blend typically loses clarity and its tackiness as well. The compatibility of tackifier and polymer can be further confirmed by rheologic tests (see also Section 8.2.1.7.1). The importance of tackifier compatibility is discussed in *Technology of Pressure-Sensitive Adhesives and Products*, Chapter 8. Cloud point measurements are made to determine the relative aromaticity, polarity, and compatibility of the tackifying resin [5–7] (see also *Technology of Pressure-Sensitive Adhesives and Products*, Chapter 2, where details concerning this method are given). In these tests, the tackifying resin is dissolved in a 25-mm-diameter test tube at 140°C in either a solution of MMAP or DACP. The solution is then allowed to cool to ambient temperature, and the temperature at which the solution becomes cloudy and the resin becomes insoluble is recorded. To determine the relative aromaticity, 5 g of resin is dissolved in 15 mL of MMAP, a 1:2 mixture (by volume) of methylcyclohexane and aniline. To determine the relative polarity, 5 g of resin is mixed with 10 g of DACP, a 1:1 (bw) mixture of xylene and diacetone alcohol (4-hydroxyl-4-methyl-2-pentanone). As the DACP cloud point decreases, the polarity increases. The relative aromatic and aliphatic content of tackifying resins are expressed by plotting the MMAP cloud point as a function of DACP cloud point. MMAP and DACP cloud points are linearly related; the lower the value of the MMAP cloud point, the lower the value of the DACP cloud point [7]. The more aromatic and less aliphatic the resin is, the lower the cloud points. The standard deviation in MMAP cloud point is typically 2°C, but can be as much as 6°C [4]. The standard deviation of the DACP cloud point is much more than the MMAP and can be as much as 12°C. Nuclear magnetic resonance (NMR) is also suggested to determine the aromaticity of the resin. NMR avoids the differences introduced by the operator and the age and moisture content of the DACP, which affect the cloud point measurement. However, NMR cannot be used alone to predict resin–polymer compatibility because NMR does not account for molecular weight effects between the resin and the polymer. Although these results are not published, the relative aromaticity measured by NMR correlates very well to both MMAP and DACP cloud points [4].

The ring and ball softening point (RBSP, see the description of the method in *Technology of Pressure-Sensitive Adhesives and Products*, Chapter 2) of the resin is actually a viscosity measurement and does not correspond to a change in physical state of the test material, unlike a T_g or melting point. It is widely accepted in the adhesives industry that due ease of measurement, low equipment cost, and a low level of technology are required [4]. Widely used instruments for measuring softening points are the Herzog ring and ball tester and the Mettler drop point tester. For the Herzog method, the RBSP is defined as the temperature at which a disk of the sample 19.7×4.4 mm (0.775×0.172 in.) imbedded in a ring is forced downward under the weight of a 9.5-mm (0.375 in.)-diameter steel ball. The specimen is heated in silicone oil at a rate of 5 ± 0.5°C/min. The Mettler drop point uses a sample ring 1.0 mm in diameter and 0.8 mm deep and a ball 0.8 mm in diameter. The test is run in air, heated at a rate of 5 ± 0.5°C/min, and the end point occurs when a drop of the resin falls from the ring. Various methods and devices used to determine the resin softening point are described by Martin-Martinez in *Technology of Pressure-Sensitive Adhesives and Products*, Chapter 2.

A glass transition measurement obtained by a differential scanning calorimetry (DSC) or a dynamic mechanical analysis (DMA) is more meaningful, but is generally not used as a control parameter because few tackifier users have that equipment. Furthermore, DSC results of resins can be complicated by physical aging, which will manifest as an endotherm prior to passing through the T_g. According to Ref. [4], RBSP generally correlates well to the glass transition temperature and, in fact, the crossover at high temperature (tan $\delta = 1$ max) measured by rheology is often near the RBSP of the resin (see also Section 8.2.1.8.2). The glass transition for hydrocarbon resins measured by DSC is typically 50°C below the RBSP. Therefore, it is recommended to use all three measurements, DSC, rheology, and the Herzog RBSP, to characterize the thermal transitions of the tackifying resin.

According to O'Brien et al. [4], the thermal transition between the glassy and the rubbery state is not a well-defined measurement and the ranking order of RBSP among the different tackifiers cannot be clearly established. In other words, in many cases there is not enough difference in the T_g's, and consequently the RBSPs, of the tackifiers tested to resolve their differences within the precision limits of the four methods used to characterize the thermal transitions. Furthermore, for styrene–isoprene–styrene (SIS) copolymers the relative ranking of the T_g of the adhesive formulations could not be repeated due to batch-to-batch variations (± 1°C) in the glass transition of the isoprene-rich phase. These results demonstrate that the differences in RBSP of the resins are not significant enough to observe a correlation with the Fox equation (see also *Technology of Pressure-Sensitive Adhesives and Products*, Chapter 3).

8.1.1.1.2 Test of Resin Dispersions

Aqueous dispersions are shear-sensitive systems in which mechanical influences during storage, handling, and coating operations may cause the formation of grit or foam (see Section 8.1.3.1.2); therefore, examination of the mechanical stability remains very important. Resin dispersions are secondary dispersions with increased mechanical sensitivity, which can undergo sedimentation. Solid-state sediments (grit) in resin dispersions are examined as sieve residue, which is a measure of how much deposit the user can expect to find in the filters. Normally, the specification allows a maximum of 0.05%; a more realistic figure, however, is a maximum of 0.015%. Sieve residue can be tested using a metallic filter element (0.18 DIN 4188), according to German Norm DIN 53786 [8].

Density is a measure of how much air is entrained in the dispersion. A low density (i.e., high air entrainment) causes problems through dehydration of the emulsion at the surface. Other routine tests are carried out during mixing and dissolving of the formulation components.

8.1.1.2 Test of Main Pressure-Sensitive Formulation Components

As discussed in detail by Benedek in Ref. [3] (see also *Technology of Pressure-Sensitive Adhesives and Products*, Chapters 1 and 3–8), PSAs can also be formulated using viscoelastomers as raw materials. The components of the formulation are pressure sensitive, and they must be tested concerning their pressure-sensitive performance characteristics (i.e., tack, peel resistance, and shear resistance) for the design of a screening recipe. The fundamentals of such measurements are described in *Fundamentals of Pressure Sensitivity*, Chapters 6–8. Their practical aspects will be discussed in Section 8.2.1.

8.1.1.3 Test of Additives

As discussed in *Technology of Pressure-Sensitive Adhesives and Products*, Chapter 8, formulation includes additives as well. Such technological or chemical additives are supplied per se, and they are tested together with the formulated adhesive (see also Section 8.2.2).

8.1.2   Test of Carrier Material

Various web-like materials are used as carriers for PSPs. Some are manufactured by the producer of PSPs; other materials are modified only. The manufacturing of a non-paper or nonmetallic carrier generally involves processing of plastic webs; modification includes physical or chemical treatment and coating methods (see also *Technology of Pressure-Sensitive Adhesives and Products*, Chapter 10). The manufacture of the carrier material was described by Benedek in Ref. [9]. Its choice for various PSPs is discussed in *Technology of Pressure-Sensitive Adhesives and Products*, Chapter 10, and in Chapters 1 and 4 in this book. The mechanical characteristics of the carrier material were discussed in Ref. [10]. Here, only those test methods belonging to the manufacturing process of PSP and control of the application performance characteristics of the PSP will be described. The main part of the tests carried out for carrier materials (e.g., tests of printability, dimensional stability, etc.) are correlated to the PSA (due to the interaction adhesive carrier) and are used for the end control of the pressure-sensitive laminate (see also Section 8.2). In the case of in-line manufacture of carrier materials (films) and PSPs, quality assurance must include the film polymers, adhesive, additives, printing inks, and related troubleshooting [11]. Problems encountered in the processing of films include gels, poor opticals, plate out and blocking, poor slip antiblocking, sealing, and printing. Converters are looking for lower gel, better gauge uniformity, consistent layers by coextrudate, and better roll geometry (see also *Technology of Pressure-Sensitive Adhesives and Products*, Chapter 10).

8.1.2.1   Test of Coatability of the Carrier

As discussed in *Technology of Pressure-Sensitive Adhesives and Products*, Chapters 8 and 10, most PSPs are manufactured by coating, in which various carrier materials (see also Chapters 1 and 4) are coated with a PSA or release material and other liquid components (e.g., primer, printing ink, lacquer, etc.) or pretreated. Such pretreatments or pre-coatings (see also *Technology of Pressure-Sensitive Adhesives and Products*, Chapter 10) are necessary to allow wetting-out and anchorage of the next coating on the web. Therefore, the coatability test of the carrier includes wettability, as well as printability. As mentioned previously, coatability depends on the carrier and coating method. Formulation for coatability is discussed in *Technology of Pressure-Sensitive Adhesives and Products*, Chapter 8. Test methods for coatability will be discussed in detail in Section 8.1.3.1.

8.1.2.1.1   Test of Wettability of the Carrier

Formulation for wetting-out was discussed in *Technology of Pressure-Sensitive Adhesives and Products*, Chapter 8. The test of wettability is carried out using general methods based on the measurement of the surface tension of the material or by special methods based on the wetting ability of the specific materials to be coated on the carrier (e.g., adhesive, lacquer, etc.). Direct (on-line) measurement of the surface tension during corona treatment has been developed. The method is based on the continuous evaluation of the friction between the treated web and a special traductor surface. As stated by Hosseinpur et al. [12], contact angle measurement based on the velocity of a moving droplet is in the range of 5% difference with that measured by static contact angle analysis.

8.1.2.1.2 Test for Printability of the Carrier

Printability of the laminate was discussed by Benedek in Ref. [13], where the ability to lay flat and dimensional stability were described. General and special printing considerations, together with performance characteristics of the main printing methods used for PSPs and special printing considerations (e.g., printing of plastics and nonpolar carrier materials), were discussed. Special printing parameters should be taken into account depending on what carrier material is used, the product class, and the type of ink. Printing-related performance characteristics of the carrier material (e.g., shrinkage, ability to lay flat, smoothness, stiffness, elongation in printing, and wrinkle build-up) were discussed in detail by Benedek in Ref. [14]. Flagging or winging up of labels on curved surfaces was described in Ref. [6].

The printability of the carrier material and the printability of pressure-sensitive laminates are discussed in *Technology of Pressure-Sensitive Adhesives and Products*, Chapter 10. The test for printability covers the measurement of various printing-related performance characteristics of the carrier material. Some special performance characteristics of the carrier material allow the evaluation of its printability. The characteristics are related to the dimensional stability of the material, which includes the stability of the geometric dimensions and shape (form) of a PSP. The form and shape are characterized by various parameters, e.g., shrinkage, the ability to lay flat, etc. Smoothness and stiffness affect printability as well. Direct printability tests are also necessary to evaluate printability itself. For instance, label printability can be evaluated through examination of the print quality (visible appearance), color tone of material, and curling of the printed material. Unfortunately, whereas the printability of polyethylene (PE) tapes can be evaluated by testing the surface tension (wettability), the printability of polyvinyl chloride (PVC) tapes cannot. Print-quality evaluation is carried out automatically at productivity rates as high as 100,000 labels/h. Some tapes must be writable as well [15].

8.1.2.2 Test for Dimensional Stability

The influence of the formulation on converting properties (e.g., shrinkage, ability to lay flat, curl, migration, etc.) and confectioning properties (slitting, cutting, die-cutting, winding, etc.) was discussed in detail by Benedek in Ref. [13].

The converting properties of PSAs (e.g., the ability to lay flat, dimensional stability, surface quality, cuttability, etc.) are described in Ref. [11]. Benedek [9,14] discussed convertibility as the sum of the convertibility of the PSA and that of the pressure-sensitive laminate, that is, dimensional stability as a result of adhesive–carrier interaction. The convertibility of the PSA has been described as its coatability and is a function of adhesive properties, the solid-state components of the laminate, the coating technology, and the end-use properties [14]. Dimensional stability as an index of convertibility includes various parameters, such as shrinkage, migration, plasticizer resistance, the ability to lay flat, and curl.

The various conversion-related parameters of PSPs depend principally on the flow of the PSA (see Table 8.1).

TABLE 8.1 Effects of Adhesive Flow

Flow in Laminate Cross-Direction (Vertically)		Flow in Laminating Direction (Horizontally)	
Performance	First Effect	Performance	First Effect
Adhesive properties		Adhesive properties	
Peel resistance	PSA anchorage Dwell-time	Shear resistance	Creep
Removability	Residue	—	—
Shear resistance	Creep	—	—
Converting properties		Converting properties	
Printability	Migration	Printability	Shrinkage (curl, ability to lay flat)
	Shrinkage	Cuttability	Bleeding
Cuttability	Stiffness	—	—
	Telescoping	—	—
Origin of cross-directional PSA flow		Origin of machine-directional PSA flow	

Carrier	PSA	Processing Technology	Carrier	PSA	Processing Technology
Porosity	Rheology (G'/G'')	Laminating pressure	Smoothness	Rheology	Laminating pressure
Chemical composition	Composite structure (additives, humidity)	Unwinding tension Drying conditions	Processing (built-in tensions)	Composite structure	Unwinding tension

On the other hand, PSA flow is a function of the PSA, the solid- state components of the laminate, and the manufacturing parameters. PSA flow affects the adhesive performance as well. Both adhesive flow in cross-sections and in the machine direction influence the various adhesive and conversion-related product characteristics directly (e.g., effect on shear resistance) or indirectly through migration of the adhesive in the carrier material (e.g., change of stiffness) or along the carrier material, which leads to a change in dimension of the PSA layer and causes edge oozing, smearing, telescoping, and inadequate cuttability. Quality assurance must test all mentioned product characteristics. Tests for dimensional stability were discussed by Benedek in Ref. [1]. The test for dimensional stability was standardized according to FINAT FTM 14.

8.1.2.2.1 Test of Shrinkage

Shrinkage is a phenomenon that changes the original dimensions of the PSP with or without changing its original shape. Such dimensional changes can be tested on the solid-state components of the pressure-sensitive laminate (e.g., carrier, liner, and adhesive), but they are the result of interactions between the laminate components, including the formulation additives for the carrier, adhesive, and web pre- and postprocessing (see also *Technology of Pressure-Sensitive Adhesives and Products*, Chapters 8 and 10). The various liquid components of the pressure-sensitive laminate can migrate as a function of their chemistry, environmental conditions (time/temperature), and processing conditions (e.g., laminating pressure, web tensioning, etc.). Such migration produces dimensional and formal changes in the PSP (e.g., shrinkage, elongation, and curl) and aesthetic changes (e.g., staining, haze, discoloration, etc.). Such phenomena must be discussed in reciprocal correlation. Shrinkage is a general phenomenon and is tested for other adhesives as well [16].

As noted in *Technology of Pressure-Sensitive Adhesives and Products*, Chapter 8, formulation affects the converting properties related to dimensional stability (characterized by shrinkage, the ability to lay flat, and migration) and the confectioning properties.

The chemical and macromolecular characteristics of the adhesive and its cross-linking strongly affect shrinkage. Formulation for converting properties was described in Ref. [17], accounting for the influence of the formulation on the parameters of dimensional stability (i.e., shrinkage, the ability to lay flat, and migration; see also *Technology of Pressure-Sensitive Adhesives and Products*, Chapter 8). Shrinkage and the ability to lay flat were discussed in detail in Refs. [13,14] as printing-related performance characteristics of the carrier material. Filler effects on shrinkage are discussed in *Technology of Pressure-Sensitive Adhesives and Products*, Chapter 2.

The nature of the material, its manufacture, and its processing influence its dimensional stability. Thus, the converting operations affect the ability to lay flat and shrinkage. For instance, different printing procedures require different paper qualities. Paper grades supplied for roll-offset printing possess a lower humidity content than papers for sheet-offset and gravure printing. In a roll-offset machine the humidity content of the paper decreases up to 10% or less, that is, paper shrinkage appears.

The main components of the plastic carrier shrinkage (i.e., without PSA) include the manufacturing-induced component, the environmental component, and the coating-related component. This is the result of built-in and "processed-in" tensions (Figure 8.1).

Polycarbonate films are laminated with a protective film using a laminating station (with rubber-coated cylinders) at 60°C and a lamination pressure of about 50 kPa/cm².

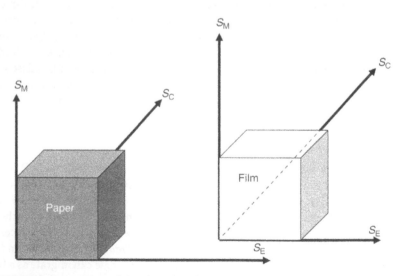

FIGURE 8.1 Components of shrinkage for paper-based and film-based PSPs. S_M, carrier manufacture-induced shrinkage; S_E, environment-induced shrinkage; S_C, coating-induced shrinkage.

The laminating pressure, cylinder hardness, cylinder cooling, temperature, running speed, and web tension influence the final adhesion level and film shrinkage. In the manufacture of a PSA laminate (e.g., label) and during PSA coating or laminating, the plastic film is stretched and strained; the strained film will be laminated with the dimensionally more stable paper-based release liners. During storage, relaxation occurs and the strained material returns to its original length (i.e., it shrinks). This is the manufacture-induced component (S_M) of the shrinkage. Such built-in tensions may produce shrinkage of the taped or protected substrate. For instance, regarding the use of protective films for plastic plates it should be taken into account that when the poly(methyl methacrylate) (PMMA) or polycarbonate (PC) items (plates) are first heated, manufacturing technology-dependent shrinkage occurs (see also Chapter 4). Such material suffers de-tensioning (relaxation) as a function of time and temperature. For instance, the shrinkage of extruded PVC is partially the result of tensions arising during extrusion of the film. Cast or calendered films are more dimensionally stable than extruded films. On the other hand (independent of the manufacturing procedure), each material is influenced by the environmental factors: the temperature and substances that migrate from the atmosphere into the carrier or from the carrier in the environment. Generally, the balance of the manufacture-, environment-, and coating-induced shrinkage components differs for film-based and paper-based PSPs. Principally, films are plastic, that is, deformable materials, which suffer easier elongation during manufacture, coating, and conversion. Their coating causes increased chemically and thermally induced deformation as well (see Figure 8.1). Therefore, shrinkage of soft PVC face stock, for instance, requires the use of shrinkage-resistant adhesive systems, such as cross-linked acrylics. On the other hand, paper dimensions are environment dependent.

In the case of paper, there is an equilibrium with the humidity of the air. In the case of plastics, solvents (used in printing and plasticizers) can interact with the film, and in special cases (e.g., cellulose derivatives, polyamide, polycarbonate, etc.) water interacts with the film also. This is the environmental component (S_E) of the shrinkage. Other influences result from coating the carrier with adhesives or printing inks (S_C). For instance, the film can be printed by means of screen printing, which uses solvents (see also *Technology of Pressure-Sensitive Adhesives and Products*, Chapter 10). The solvents can migrate into the adhesive and dilute it. Certain solvents can penetrate into the film as well. Therefore, screen printing of PVC increases its shrinkage. In a similar manner, the special elastic modulus of paper (E_M^*) is related to the humidity elongation coefficient (β) in the machine (M) and cross (C) directions according to the correlation

$$E_M * \beta_M = E_c * \beta_C = \text{constant} \qquad (8.1)$$

where the humid elongation coefficient is a function of the relative humidity of the air, its variation, and the variation of the length of the sample.

Shrinkage, that is, dimensional stability, is also a function of the temperature resistance of the carrier. Generally, the suggested processing temperature for a polymer-based carrier material is specified by the film supplier.

The importance of shrinkage differs for different PSP classes. Acceptable shrinkage values vary depending on the postconversion steps or end use of the product. Because of the high printing quality of labels and the mutual interaction between the label and printing, the control of shrinkage plays a important role in their quality. The liner for name plates must exhibit trouble-free release, dimensional stability, and good die-cutting, perforating, prepunching, and fanfolding (see also Chapter 4). For self-adhesive, EVAc-based protective films (see Chapter 7), shrinkage is used as a quality criterion.

In a quite different manner from labels, tapes have applications in which shrinkage is required. For instance, for heat-shrinkable insulating tapes (see Chapter 7) shrinkage of the carrier is used during application, when the dimensions of the tape decrease by up to 10–50%. As an example, bilayered heat-shrinkable insulating tapes for anticorrosion protection of oil and gas pipelines have been manufactured from photochemically cured low-density polyethylene (LDPE) with an ethylene–vinyl acetate copolymer (EVAc) as the adhesive sublayer. The linear dimensions of the two-layered insulating tape decrease as a function of the curing degree (application temperature 180°C). The application conditions and physical–mechanical properties of the coating were determined for heat-shrinkable tapes with 5% shrinkage at a curing degree of 30% and shrinkage test force of 0.07 MPa [18].

Performance characteristics of different PSAs as a function of the face stock material (paper/film) were examined by Benedek in Ref. [19]. Shrinkage is also a function of the adhesive. Several EVAc-based adhesives demonstrate high shrinkage. Thus, the most suitable adhesive materials for film coating are acrylic PSAs. Acrylics and carboxylated butadiene rubber (CSBR) display better shrinkage resistance than EVAc copolymers. The shrinkage of solvent-based acrylics cross-linked chemically or with ultraviolet (UV) light was evaluated in Ref. [20]. Ionic UV systems result in less shrinkage and less toxicity. The type of cross-linking agent influences shrinkage as well. Shrinkage values for different face stock materials are summarized in Table 8.2.

Table 8.2 illustrates how shrinkage values of various films depend on the pretensioning of the material, that is, values in the machine and cross directions differ. Shrinkage tests can also be considered aging tests in which only the dimensional changes are examined after high-temperature storage (see also Section 8.2.2.4). For instance, the shrinkage of a closure tape with 98-μm oriented polypropylene (OPP) carrier is tested

TABLE 8.2 Shrinkage Values for Different Carrier Materials

Carrier Characteristics			Test Conditions			Shrinkage (%)	
Material			Thickness (μm)	Temperature (°C)	Time (min)	MD	TD
Paper	—	—	—	0.8–1.0	—		
Film							
	PE	LDPE	30–200	70	60	—	1–3.0
		HDPE	—	80	15	1.0	2.0
	PP	CPP	100	90	5	1.0	—
		BOPP	50–75	130	5	<8	<4
			—	135	7	5.0	5.0
	PVC	—	40–200	70	60	—	0–2.0
	PET	—	30–350	190	5	3.0	2.0
		—	—	—	—	3.3	3.5

after 10 min of storage at 125°C. Test conditions and shrinkage of PSPs are discussed in detail by Benedek in Refs. [21,22]. The tests for shrinkage characterize the plasticizer resistance as well; therefore, they will be described in Section 8.1.2.2.2.

8.1.2.2.2 Test for Plasticizer Resistance

Most plastic films (e.g., PVC) are compounded materials and contain liquid additives (e.g., plasticizers). Possible interactions exist between such micromolecular components of the film and the adhesive. Plasticizers and emulsifiers from the film can migrate into the adhesive. On the other hand, monomers, oligomers, and surface agents from the adhesive can migrate into the carrier, causing stiffening or shrinkage of the face stock and the loss of adhesive properties.

Different practical tests are described in the literature that deal with the plasticizer resistance of common adhesives [23–25]. Most use a PVC film (soft PVC with a well-defined plasticizer content) and store the PVC between the adhesive surfaces to be tested. The plasticizer migration is estimated as a weight difference.

Tests for shrinkage are used as an indirect method for plasticizer resistance. For instance, the shrinkage of a plasticized vinyl film with a transfer-coated adhesive film, conditioned for 24 h at standard humidity and temperature, is measured as the change in length and width; then it is heat aged (7 days at 158°F) on the liner, reconditioned, and measured again. The plasticizer migration resistance of the polymer is acceptable if the shrinkage when heat aged on a release liner is in the range of 0.3–0.9% [26]. For transfer tapes (see also Chapter 7) used for structural bonding, plasticizer resistance is tested at elevated temperatures (70°C, 30% plasticizer).

Principally, unmounted and mounted shrinkage tests are carried out. In a mounted shrinkage test a PSA-coated PVC face stock is bonded to a test surface, often stainless steel, aluminum, or release liner. The laminate then is exposed to elevated-temperature accelerated aging, typically 70°C for 1 week, and the percentage shrink back of the PVC film from its original dimensions is measured. Both machine direction (MD; subject to the greatest stress-induced elongation) and cross-machine direction (CD) measurements are carried out. In another test, a 4 × 4 in. coated sample is adhered to a Formica panel and aged for 7 days at 110°F [26]. Evaluation criteria include shrinkage in the MD of less than 0.5% and shrinkage in the CD also below 0.5%. According to Ref. [27], plasticizer migration is tested indirectly (as a decrease in peel resistance) through a test for 180°F peel resistance after 24-h dwell time, 24-h aged peel resistance, and 1 week of aging on a liner at 158°F. The peel retention after 7 days at room temperature and 158°F, the shrinkage on the liner (6–8%), and mounted shrinkage (11–17%) are measured (24 h at 158°F) as well. For transfer tapes used for structural bonding, plasticizer resistance is tested as shrinkage at elevated temperatures (70°C) [28].

8.1.2.2.3 Test of the Ability to Lay Flat and Curl

The ability to lay flat is the shape stability of a PSP. For laminated products it is the result of the geometric stabilities of each component. Inadequate ability to lay flat is generally manifested as curl of the solid-state components. Like shrinkage and most other dimensional changes during processing and application, the lack of ability to lay flat, that is, curl, is the result of the tensions or their relaxation in a carrier material. Such tensions may appear as the result of the composite structure of the laminate or laminate components (see also *Technology of Pressure-Sensitive Adhesives and Products*, Chapter 1). Some films have an anisotropic build-up. Paper itself is a composite built up from fibers. Its humidity balance produces changes in the fiber diameter (across the original MD). These changes may be different in the middle and in the external paper layers. Therefore, paper curl can be caused by humidity changes as well (see also Equation 8.1). The stress, σ, in the paper layer is the result of deformation by mechanical forces, ε, and humidity (e),

$$\sigma = E(\varepsilon - e) \tag{8.2}$$

where E is the modulus. The ability to lay flat for paper requires control of the humidity with a precision of 0.3–0.6%. To achieve a better humidity balance, carbamide, glycerine, or carbamide–calcium nitrate can be added to paper (see also *Technology of Pressure-Sensitive Adhesives and Products*, Chapters 8 and 10).

The edge curl (corner drag, edge lifting) of printed films depends on the film thickness, thickness of the printing ink layer, differences in the elasticity of the printed and carrier layers, and diffusion of low-molecular-weight substances in the carrier film. The ability to lay flat of printed PVC film (coated with PSAs) depends on the thickness of the film, the thickness of the ink layer, the bonding forces between label and release, and the elasticity of the film and ink. In particular, in the case of a printed PVC carrier, the printing ink film is hardened more at a high drying speed and is less extensible at room temperature than the PVC film. This can lead to curl. "Frozen" processing tensions in the film may lead to curl as well (see also shrinkage). When molten plastic is forced through a slit in a die, subjected to pressure by finish rolls, and cooled rapidly, it develops strains [9]. When the sheet is later subjected to heat or solvents, it relaxes and warps, buckles, bows, or "dishes." The larger the piece and the heavier the gauge, the greater the potential for problems. The shear rate during the processing of raw materials is quite different for calendering (e.g., $10–10^2$ s^{-1}) and extrusion ($10^2–10^3$ s^{-1}); therefore, the tensions in the materials are different also. Tensions can be built into the PSP during coating and laminating of the web. Therefore, nip pressure and web tension should have low values to avoid curl (see also *Technology of Pressure-Sensitive Adhesives and Products*, Chapter 10).

Web preprocessing (see also *Technology of Pressure-Sensitive Adhesives and Products*, Chapter 10) and postprocessing (see also *Technology of Pressure-Sensitive Adhesives and Products*, Chapter 10) can cause tensions as well. Printing methods differ according to their side effects on the PSP. Principally, nonimpact printing methods introduce less tensioning in the carrier material. Impact procedures differ as well. For instance, curl of film edges may be due to or amplified by screen printing (see *Technology of Pressure-Sensitive Adhesives and Products*, Chapter 10). The drying of labels printed by screen printing may produce tensions in the material depending on the printing ink used. These tensions cause curl. Ideally, printing inks must allow rapid drying and optimum screen opening. Very elastic acrylic inks with alcohol as a solvent may be processed with low curl. If a printing ink based on PVC (copolymer) is used for PVC, low curl occurs. High-gloss printing inks based on an acrylic yield low curl as well. Unfortunately, mixed compositions that allow high printing speed and open screen produce high curl for PVC. Curl depends on coating weight, machining conditions, and the type of adhesive. For instance, because of the lack of sensitivity to atmospheric humidity, poly(vinyl methyl ether) has been added to formulations for paper tapes to avoid curl. Adhesives with good cohesive strength prevent adherends from recovering from bending.

Curl is a function of the laminate build-up as well. The use of a stiff release liner (e.g., 80-lb kraft liner) for labels helps to avoid curl. The liner plays an important role in the functionality and cost of most PSPs. Tear strength, dimensional stability, ability to lay flat, good gauge-control hardness (lack of compressibility), and surface characteristics are the most important features of the release liner. Tear strength is important for label converting and dispensing. Dimensional stability (the ability to maintain the original dimensions when exposed to high temperature and stresses) is important for print-to-print and print-to-die registration. If the liner stretches under heat and tension, the labels may be distorted. Liner stretch can affect label dispensing as well (see also Chapter 4). The ability of the liner to lay flat is very important for sheet labels or pin-perforated and folded labels. Labels that are butt cut, laser printed, and fan folded must lie flat to allow them to be properly fed and stacked. The ability of a liner to lay flat is affected by resistance to humidity, dimensional stability at different temperatures, and the stiffness (thickness) of the liner. The use of radiation-cured or low-temperature thermally cured silicones allows a better humidity balance of carrier paper and increased ability to lay flat.

When unwound from the roll, tapes should have no tendency to curl. Curl of tapes may be due to inadequate carrier design (e.g., use of a two-layered coextrudate film with different mechanical properties of the layers), processing, and transport tensions or printing. Shrinkage and the ability to lay flat are important quality characteristics for heat-laminated protective films. The ability to lay flat influences the control of delamination, the separation of the protective film from the protected item.

Films made with a MD orientation display outstanding stiffness in the MD and flexibility in the CD (see also Chapter 1). Because an isotropic material is necessary for labels, the carrier films for labels ideally should be biaxially oriented. In certain applications for labels conformability is required. Therefore, paper labels should be laid out in the long grain direction to provide maximum conformability. Unfortunately, the ability to lay flat and conformability of labels may be contradictory (conformability is especially required in medical applications). The ability to lay flat in printing must be complemented by on-pack wrinkle-free squeezing ability. The influence of the adhesive formulation on the ability to lay flat was discussed by Benedek in detail in Ref. [17]. Special products require improved ability to lay flat. For instance, for an excellent ability to lay flat the carrier for application tape (see also Chapter 4) contains high-density PE (HDPE) and is manufactured by coextrusion. To avoid curl if unwound by application, adhesion on the back side should be minimal. Curl must be avoided for security labels (see also Chapter 4) and for name plates (see also Chapter 4).

8.1.2.2.4 Test for Flagging

Flagging or winging-up of the labels on curved surfaces due to insufficient adhesion of the PSAs to the substrate is different from flagging after coating (i.e., the loss of the ability to lay flat). As discussed previously, this phenomenon is due to the chemical influence of the adhesive or printing inks. Wing-up depends on the adhesive/printing ink composition, the thickness of the film, the adhesive properties of the PSAs, the thickness of the adhesive/ink film, the differences in the modulus of coating/coated material, and the coating width (full or limited).

8.1.2.2.5 *Test for Elongation*

As discussed above, during manufacture of the solid-state components of PSP and web finishing, tensions may build up in the materials and cause dimensional changes. They depend on the forces applied in the machine and on the temperature in the machine and are also a function of the material characteristics. As discussed previously, a correlation generally exists between machining conditions and dimensional changes in different films. A correlation exists among tensile strength during printing, tensile strength during print drying, and elongation during printing for various films. For instance, OPP exhibits a higher elongation at 80°C than polyamide at the same temperature. There is a correlation between the gravure-roll inserting-tensile strength and elongation during printing. The tensile strength of the material influences its elongation during lamination and the lamination temperature affects elongation.

8.1.2.2.6 *Test for Wrinkle Build-Up*

Wrinkle build-up during printing is due mainly to overdrying the paper carrier. The humidity content of paper may decrease to 1–2%, changing its dimensions. This phenomenon depends on the stiffness of the paper. For papers of lower weight (40–80 g/m²), the build-up of wrinkles is more accentuated. Such papers wrinkle with shorter wavelengths than papers of 100 g/m² weight. Wrinkle build-up depends on machine construction as well. Tension fluctuations on the machine can cause wrinkles. In this case, winding equipment plays a decisive role. Rewinders with independent indexing arms allow reel changeovers at full speed without any tension flutters (see also *Technology of Pressure-Sensitive Adhesives and Products*, Chapter 10).

8.1.2.3 Test for Other Properties

Other performance characteristics (e.g., smoothness and stiffness) that influence the processibility of the carrier and the PSP must also be tested.

8.1.2.3.1 *Test for Smoothness*

Liner and face stock smoothness affect several performance features. The adhesive surface is a replica of the face stock or release surface. If it is rough, the level of initial adhesion will be reduced. If a rough label is laminated with a clear face stock material, its clarity will be negatively influenced. Optical image of the transparent adhesive layer is decisive for "no-label-look" labels (see Chapter 4). If the liner is too smooth, air entrapment during dispensing can become a problem. The roughness of the label back side can affect web tracking on the printing press. Smooth liners can weave, causing cross-web print registration. In laser printing the roughness of the liner plays a critical role in sheet-feed ability and tracking through the printer. Smoothness may influence telescoping as well (see also Section 8.2.2.5.1). Mounting tapes used in the fashion and textile industry are protected against telescoping during storage with a woven or nonwoven supplementary layer.

8.1.2.3.2 Test for Stiffness

Stiffness influences the conformability of the PSP in the printing press and wrinkle build-up. It affects die-cutting (see *Technology of Pressure-Sensitive Adhesives and Products*, Chapter 10) and label dispensing (see Chapter 4) as well. Cuttability and its dependence on the adhesion/cohesion balance of the adhesive, the chemistry and rheology of the PSA, the adhesive properties (tack, peel resistance, and hot-shear resistance), coating weight, anchorage of the adhesive, laminate thickness and stiffness, and product build-up were described and evaluated quantitatively in Refs. [13,16,29] (see also *Technology of Pressure-Sensitive Adhesives and Products*, Chapter 8). For instance, glassine liners are harder and more densified; therefore, they are used for high-speed dispensing. Tests for stiffness, according to DIN 53121, DIN 53123 and DIN 53350, were described in Ref. [30]. PSTC-37 describes a test method for stiffness (Taber stiffness test) based on the bending moment necessary to deflect the free end of the specimen of a tape.

8.1.2.3.3 Test for Migration/Bleeding

Bleeding (staining) means that the adhesive penetrates (migrates) through the carrier material. (The release coating can migrate as well; see *Technology of Pressure-Sensitive Adhesives and Products*, Chapter 11.) The main parameters of the bleeding are the carrier material and the adhesive. Bleeding is correlated with plasticizer resistance as well. The surface quality of the laminate can be influenced through adhesive migration in the face or migration of the face components in the adhesive layer. For paper labels, adhesive penetration changes the visual appearance of the face. Porous, primerless papers as face stock material need nonmigrating adhesive formulations. Generally, migration of the low-molecular-weight components from the adhesive (e.g., plasticizer, tackifier, surface active agent, water, etc.) into the paper or from the film (plasticizer) or paper (water, special chemicals like components of thermal papers) into the PSAs may alter the chemical composition of the coated adhesive. For instance, when the amount of oil in hot-melt PSAs (HMPSAs) exceeds the capacity limit that mid-block polymers can hold, oil bleeding or plasticizer migration occurs (see also *Technology of Pressure-Sensitive Adhesives and Products*, Chapter 3). The oil bleed can cause staining, especially on paper face stocks, or can even form a weak adhesion layer that destroys the adhesive bond. When PSAs were coated onto porous substrates such as paper, an assessment of the tendency of the adhesive to migrate or bleed through the paper should be made. Carrier materials can also act as protective surfaces. Stain-resistant, nonpaper face materials help with the staining problem.

Migration must be tested by the adhesive manufacturer (formulator), because it is a technological characteristic of the PSA (see also Section 8.1.3). The migration of the adhesive can discolor the face surface and reduce the effective adhesive film thickness, thus changing various adhesive properties. On the other hand, the migration may dilute the adhesive and decrease the adhesive properties. The influence of the adhesive design and formulation on the migration was discussed by Benedek in Ref. [17]. Migration is a function of the base polymers used for PSA, their chemistry, and their macromolecular characteristics, but tackifying components and micromolecular additives affect migration as well. The processing conditions and technological characteristics of the adhesive also influence migration. For instance, to avoid penetration (i.e., bleed-through) of HMPSAs, the application viscosity should amount to 17,000–20,000 mPa/s at 180°C. Batch-to-batch consistency and staining are two problem areas or disadvantages of rubber–resin PSAs. For easy-to-wet, directly coated face stock materials, low-viscosity dispersions may be used only if bleed-through is not an important issue. For difficult surfaces (release liners) or porous materials, a higher application viscosity appears to be required. As noted in Ref. [13], stress distribution in the adhesive layer (i.e., anchorage of the adhesive and stiffness of the solid-state components) and geometry of the PSA layer (coating weight) are more important for cold flow than the viscosity of the adhesive. In a similar manner, plasticizers and temperature promote wall slip, whereas fillers decrease it. The same interdependence may be supposed for the cold flow (bleeding) of the adhesive (see also Section 8.2.2.4.1). Carrier-related formulation is discussed in *Technology of Pressure-Sensitive Adhesives and Products*, Chapter 8.

A test method for the resistance against bleeding is described by Czech in Ref. [31]. A test for bleeding of HMPSAs is described in *Technology of Pressure-Sensitive Adhesives and Products*, Chapter 3. Migration must be tested for the finished PSP also as a time/temperature (and manufacture quality)-related performance characteristic (see Section 8.2.2.4).

8.1.2.3.4   *Test of Oozing/Edge Bleeding*

Temperature-sensitive cold flow of the adhesive (edge bleeding, oozing, or smearing) influences printability (especially for heat-generated laser printing or Xerox printing). Edge bleeding affects cuttability and die-cuttability (see also *Technology of Pressure-Sensitive Adhesives and Products*, Chapter 10). Edge bleeding is a cold flow phenomenon related to the adhesive, coating weight, laminate components, and laminate construction and processing technology. Low coating weights provide "clean" converting conditions (no oozing). Edge bleeding was discussed in detail in Ref. [13] (see also Section 8.2.2.4.1).

8.1.3   Test for PSA

Coaters mainly work with PSAs formulated according to their in-house requirements. Such adhesives must be tested concerning their technological and end-use properties.

8.1.3.1   Technological Test of the Adhesive

As discussed in detail by Benedek in Ref. [17], the adhesive is formulated for coating technology as well (see also *Technology of Pressure-Sensitive Adhesives and Products*, Chapter 8). First, its coatability must be tested.

8.1.3.1.1   *Test of Coatability of PSA*

Wettability, control of viscosity (thinning, diluting response), mechanical stability, the desired coating weight under required conditions (running speed, temperature), and optical appearance are factors of coatability behavior. Viscosity and surface tension influence wettability; they should be tested *a priori*.

The coatability test includes the control and adjustment of the flow parameters that affect processibility of the adhesive, principally the viscosity and wetting-out. The properties of the liquid adhesive to be tested can include the solids content and the free monomer content. The flow properties of solvent-based PSAs (stable systems), water-based PSAs (shear-sensitive systems), and hot-melt PSAs (highly viscous, temperature-sensitive systems) differ.

Test for Viscosity   Viscosity control is carried out using a common apparatus. The importance of viscosity in PSA formulation and processing was discussed by Benedek in Ref. [32].

The most important processing parameter of HMPSAs is their viscosity; thus, this parameter should be tested according to common methods. Not only the viscosity, but also its time/temperature dependence (aging) is very important. According to Hu and Paul (see *Technology of Pressure-Sensitive Adhesives and Products*, Chapter 3), the viscosity of the adhesive ideally should be controlled at below 25,000 mPA at 160°C to ensure good coatablity and maintain normal processing temperatures. Melt viscosities are often determined in the temperature interval range of 140–200°C using a Brookfield viscometer (spindle LVF 4). In Ref. [33], a test with a Brookfield viscosimeter RVT at 180°C, spindle 7, and 20 rpm is suggested. The viscosity of styrene block copolymers (SBCs) is very temperature and shear sensitive. Therefore, as the shear rate and temperature increase, the differences in the viscosity of different formulations decrease. Thus, low-shear viscosity measurements do not constitute a correct basis for real flow behavior of HMPSA under practical coating conditions. Recent developments in the measuring techniques/devices, using cone and plate geometries, allow a more rapid determination of the viscosity curves at different shear rates and temperatures requiring a minimal amount of adhesive.

Generally, the processibility of solvent-based PSAs depends on their viscosity and solids content. Coating devices are designed for a given viscosity, whereas the coating weight and the drying speed depend on the solids content. In general, the solids content and viscosity should be measured for formulated compositions. Because of the nonideal flow behavior of the adhesive solutions (dispersions), viscosity measurements at different flow rates should be made. However, the most common industrial method is the cup-flow method, in which a one-point viscosity measurement (flow time versus cup orifice) is carried out; Ford cup or DIN cup 3 or 4 are commonly used.

Generally, the properties of aqueous dispersions to be tested include the viscosity and its stability in time, the thinning (diluting) characteristics, the solids content, and the temperature stability. The particle size and particle size distribution influence the viscosity, surface tension, and mechanical stability of the dispersion [19]. The emulsions typically contain about 40–70% polymer when manufactured, whereas preferred latexes typically have a content of about 40–60 wt% polymer solids. The correlation between solids content and viscosity includes the particle size and particle size distribution (PSD). A bimodal or multimodal PSD increases the solids content. Most often, the latexes have particles with a diameter in the range of about 120–1000 nm. By decreasing the particle size (to very low particle size), the viscosity increases exponentially. Introducing polydispersity in an acrylic dispersion might yield a lower viscosity at the same volume fraction as monodisperse spheres. Dispersions with spherical particles display lower viscosities than other particle forms.

Mechanical stability influences the viscosity; viscosity and surface tension affect coatability. For PSA conversion, coatability remains the main property of the liquid adhesive because wettability also leads to the choice of adequate coating geometry of the face stock and release liner. For water-based PSAs viscosity (Brookfield RVT, spindle 4, 20 rpm) at room temperature should be tested immediately after compounding; in some cases, the thixotropy index is also measured. The thixotropy index is the ratio of the Brookfield viscosity at 0.5 rpm to the Brookfield viscosity at 50 rpm. This index demonstrates an increase in low-speed viscosity of 5% for conventional polyacrylic PSAs used as a thickener compared to 33% for some special products.

The most commonly used test methods for viscosity measurements of water-based dispersions are listed by Benedek in Ref. [1].

The thixotropic behavior of silicone PSAs can be easily measured on a rheometer using a frequency sweep routine with controlled strain, followed by calculating a ratio of viscosity at two different frequencies (e.g., 0.5 rad/s versus 100 rad/s) to give a useful thixotropic index. This delivery system is limited to products that can be applied by screen- or stencil-print operations [34,35].

Viscosity measurement can serve as control method for raw materials. For instance, copolymers of acrylic esters (40–85%) with vinyl esters (60–15%) have been suggested for medical tapes (see Chapter 4). Such polymers are used as PSA solutions in ketones, alcohols, esters, and aliphatic and cycloaliphatic solvents. The molecular weight of such a composition is controlled by the viscosity in isopropyl acetate (50% solids, 3000–11,000 mPa/s).

Test for Wettability   Wetting-out is tested by coating the ready-to-use adhesive on the specific carrier material (face stock or release liner) to be used. This test includes control of the coating image in the laboratory and on the coating machine. Wetting-out can be characterized by the contact angle, surface tension, and dynamic surface tension. In practice, wetting out can also be characterized by crater number and wetting defects [36].

The special features of wetting-out for aqueous dispersions were discussed in Ref. [32] concerning the rheology of water-based dispersions. The wetting characteristics of a dispersion are examined either directly or indirectly. The direct evaluation of the wetting characteristics implies the subjective examination of the wet-out of the original or slightly diluted dispersion on a release liner or on a face stock material to be coated. Depending on the nature of the coating (direct/transfer), either the face stock or the release liner should be used as a test surface. The test of wet-out on a face stock material is mostly used for film carriers. Through coating of a release liner, both solvent-based and solventless silicone liners can be included in the evaluation. After drying, a comparison of the coatings should be made based on the visual inspection of the presence (or absence) of defects (cratering, pinholes, "fish eyes," etc.). The wetting-out can be characterized by crater number and wetting defects (fish eyes).

The indirect characterization of wetting properties refers to the measurement of the surface tension and surface energy of the dispersion or the wetting angle. Evaluation of the wetting-out and the existing norms have been described by Benedek in Refs. [37,38]. The coating performance and wetting-out depend on the surface tension, but the surface tension measurement alone is not sufficient to characterize the actual behavior of an adhesive on a coating machine. In the coating of a latex at high speeds, the dynamic aspect of wetting also appears to be critical (see also *Technology of Pressure-Sensitive Adhesives and Products*, Chapters 8 and 10).

The measurement of the contact angle of a latex on a face stock material or the surface tension of the latex with different surfactant systems determines the choice of surfactant. Surface tension tests were standardized according to FINAT methods. Similar test methods were published by DIN and ASTM (e.g., D2578–84) and TLMI. Surface tension measurements and (wet) film-weighing machine measurements allow the evaluation of adequate surfactants, their diffusion velocity, the reduction of the surface tension, and evaluation of the required surfactant concentration [25].

Pitzler [39] proposed dynamic contact angle measurements with advancing and receding angle measurements as a more adequate method (than the surface tension test) and indicated that, for wetting-out, the advancing contact angle is decisive. This depends on the coating speed, which means that both static wet-out and dynamic tests under shear are necessary. Evaluation of surface tension, with the aid of a wettability/ wetting angle measuring device, has been described in Ref. [40]. A simple method to test the surface tension uses a stalagmometer (capillary tube). The simple pendant drop (or stalagmometer) method for the measurement of the surface tension is described by Benedek in Ref. [1].

Measuring of the wetting angle may be used to test the face stock, the release liner, or the PSAs. The same method applies to determine the nature (solvent-based, water-based, or hot-melt) of an unknown PSAs (or label). In this case, it is assumed that the water-based adhesive always leaves traces of surface agents on the release liner. Therefore, a delaminated release liner that was originally coated with a water-based adhesive allows a better wet-out for pure water. To determine the water-based or solvent-based nature of an unknown PSA sample, the wetting angle of water on the delaminated release liner should be measured (deionized water should be used). For example, typical wetting angle data for a solvent-based rubber–resin and water-based tackified acrylic PSA-coated release liner are 101–103° and 91–93°, respectively [1]. Water extraction of the water-soluble surface-active agents from a PSA-coated laminate (adhesive nature unknown) and comparison of the wetting angle of the extract with that of a known hot-melt or solvent-based PSA makes it possible to identify the nature of the adhesive.

In the evaluation of the wettability of an adhesive on a release liner, care should be taken because of the change in wettability of the siliconized paper over time. Following a small increase after a few days, the wettability decreases with storage time (i.e., the wettability increases with increasing age of the siliconized paper).

Surface tension may be controlled during manufacture by using special test liquids or crayons between 30 and 72 mN/m. The test liquids are standard, according to DIN 53364. The normal range of crayons is 40, 42, 44, 46, 48, and 50 dyn. A broad range of plastic films like PE, PP, BOPP, LDPE, linear low-density PE, HDPE, polyamide (PA), Surlyn, and polyethylene terephthalate (PET) may be tested.

8.1.3.1.2 *Test for Processibility of the Adhesive*

The test for processibility of the PSA, that is, test for runnability, is an industrial test, carried out on the coating machine and with the coating device to be used, to control foaming, coating weight, long-time shear stability (coagulum build-up), coating image, and drying ability of the adhesive (see also *Technology of Pressure-Sensitive Adhesives and Products*, Chapter 10).

Some properties allow screening of water-based PSAs from the point of view of the coating properties. Parameters such as solids content, particle size, and density yield indirect information about the machine properties of the adhesive. With regard to the processing properties of the dispersion, there are several requirements: absence of foaming, no drying on the rolls, wet-out, mechanical stability when being pumped, and lack of odor. Aqueous dispersions are shear-sensitive systems in which mechanical influences during storage, handling, and coating operations may cause the formation of grit or foam; therefore, the examination of mechanical stability remains very important. For handling of such materials, special pumps with minimum agitation turbulence or pulsation are recommended. On the other hand, water-based adhesives are difficult to coat because of their low viscosity and high surface tension and because of the low surface tension of the release liner or face stock web (see also *Technology of Pressure-Sensitive Adhesives and Products*, Chapters 8 and 10). Density and particle size of the dispersion may influence the stability, coating properties, and drying ability. Foam properties determine the processing speed and coating image.

Test for Coating Weight   The coating weight influences the adhesive, converting, and end-use properties of PSAs. The dependence of the adhesive properties on the coating weight was extensively discussed by Benedek in Ref. [41]. The influence of the coating weight on the converting and end-use properties was discussed by Benedek in Ref. [13]. Many different test methods were established to characterize the adhesive, converting, and special properties of the PSAs. Because these depend on the coating weight, the measurement of this parameter is discussed next. The common test of the coating weight is an industrial test carried out continuously on the coating machine. Control is achieved by means of special equipment and by using common laboratory tests (see also *Technology of Pressure-Sensitive Adhesives and Products*, Chapter 10). The test of the coating weight possesses a decisive importance for the finished product (pressure-sensitive laminate).

Industrial in-line measurement and control of the coating weight, as well as spot checks of the coating weight, is carried out. In-line thickness measurement is carried out with sensors working contactless (see also *Technology of Pressure-Sensitive Adhesives and Products*, Chapter 10). Optoelectronic (laser triangulation, laser scanner, light transmittance), capacitive (condensator), inductive roll sensor, contact, pneumatic (pressure-die), radiometric (β-radiation), or acoustic (ultrasons) infrared (IR) principles are used. The most used sensors for measuring film thickness usually fall into one of the three categories: caliper, nuclear, or infrared measurements. Caliper gauges make a direct, physical measurement of the total thickness of the web. Such gauges can contact the product on both sides or may have a relative contact layer of air on both sides. Such devices are not capable of measuring the individual layers of the laminate. Nuclear sensors pass β- or γ-radiation through the material. They cannot measure the thickness of individual layers. Infrared sensors use the 1.30- to 2.70-μm portion of the near infrared spectrum and work like a spectrometer (see also *Technology of Pressure-Sensitive Adhesives and Products*, Chapter 10). The limited number of filters in the sensor restricts the number of polymer layers in a laminate that can be measured simultaneously. In-line gravimetric measurements of the coating weight can be carried out by inserting siliconized paper or aluminum foils in the laminating station.

For laboratory purposes the gravimetric method is generally used. To avoid the discrepancies caused by the sensitivity of the face stock material toward the liquid adhesive, aluminum foil may be used. A simple gravimetric method can help to keep the coating weight specifications within ± 0.5 g/m^2, resulting in considerable savings. The coating weight is tested by dissolving the (coated) adhesive layer in a solvent and determining the weight difference between the coated and uncoated face material. The apparatus and test method are described in detail by Benedek in Ref. [1].

Test of Foaming   Foaming affects the surface quality of PSPs (and the adhesive properties); thus, methods to avoid foaming are very important. Foaming depends on the formulation of the PSA (see also *Technology of Pressure-Sensitive Adhesives and Products*, Chapter 8) and on the coating device used (see also *Technology of Pressure-Sensitive Adhesives and Products*, Chapter 10). The most important tests for the characterization of the foaming ability of a dispersion include stirring and measuring the foam density, high-speed stirring and measuring the foam height or weight of the foam column, foam generation by air bubbling, and examination of foam dynamics as a function of time [1]. Foaming tests are used for water-based dispersions or for the choice of defoamers.

According to the procedure for testing foam height, foaming is characterized by the volume of the foamed liquid. Such a test is described in Ref. [1]. The surfactant solutions (or water-based formulations) should be subjected to high-speed agitation at 25–60°C for 3 min in a blender, and the foam volume should be measured.

According to another procedure for testing foaming, a volume of 50 mL of the diluted dispersion is stirred for 1 min with a high-speed mixer, foamed, and weighed. The weight of the liquid (dispersion/air mixture) is an index for the foaming characteristics.

In another test discussed in Ref. [1], foam volumes should be recorded at the 15-s mark and then each minute thereafter. With the aid of the weight/volume correlation, density values can be calculated. With use of this method the sample contains both entrained air and froth foam, and the lowest density value corresponds to the case of the highest amount of foam. Errors related to the transfer of the foam into another vessel are eliminated (see also *Technology of Pressure-Sensitive Adhesives and Products*, Chapter 8).

Test of Drying Speed   Test of drying speed is carried out on the coating machine to control the maximum running speed. Such tests are used in the laboratory to compare various screening recipes. While screening different water-based PSAs with similar adhesive but different processing conditions, it appears very important to estimate their drying ability. The drying ability is the drying speed (loss of water with time) under real conditions.

Particle size and distribution, as well as solids content, strongly influence the drying and mechanical stability of water-based dispersions. A method for particle-size testing is given in Ref. [1] and measurement of the solids contents is also discussed.

Unfortunately, there is no standard method or apparatus for drying test of PSAs. Preliminary screening tests may be carried out by the weight loss determination of the coating over time. ASTM developed methods of determining "drying, curing, or film forming of an organic coating at room temperatures" (TE5-1740-65T).

Test of Shear Stability   The term mechanical stability refers to the ability of the emulsion polymer to resist coagulation under the influence of shearing stresses as encountered during agitation or pumping. Shear under machine conditions depends on the running speed and viscosity and the shear characteristics of the coating device. In the case of secondary dispersions (e.g., resin dispersions), stability includes resistance to phase separation and mechanical stress. Phase separation stability is important for storage and mechanical stability is important for coating. Density, particle size and distribution, viscosity of the continuous phase, particle charge, and compounding ingredients influence phase separation. Particle charge, surfactant type and level, ionic strength and pH, particle deformation, and compounding ingredients affect mechanical stability. Phase-stable emulsion may be mechanically unstable and mechanically stable emulsion may be phase separate.

Test of shear stability (mechanical stability) is carried out on the coating machine and on the industrial handling equipment of the adhesive, that is, pipelines and pumps. Such a test is performed in the laboratory using long stirring times. Industrial tests are tests in which the adhesive is evaluated under real conditions. For companies producing emulsion PSAs it is recommended to perform coating tests with a reverse gravure coater (see also *Technology of Pressure-Sensitive Adhesives and Products*, Chapter 10) before choosing equipment. The leading machine manufacturer and raw material suppliers are able to offer such tests.

In the most frequently employed laboratory test the polymer is stirred at high speed in a blender for a specific time (typically 10 min). The amount of coagulum (instability) resulting from this treatment is determined by filtering the latex through an 80-mesh screen; a 100-mesh screen is sometimes used [1]. There are standard test methods for determining grit, lumps, or undissolved matter in water-borne adhesives. For instance, the International Organization for Standardization (ISO) standard ISO 705 (third edition, 1985-08-15) describes the determination of coagulum content (sieve residue) for rubber latex.

A rapid measure of mechanical stability is obtained by filtering a sample through a 100-mesh screen and then stirring it at high speed in a blender for 5–10 min. The emulsion is again passed through the screen and washed with water until only solid residue, if any, remains on the wire. If there is no solid residue the emulsion is considered to possess very satisfactory mechanical stability. If so desired, the amount of solid can be weighed after the screen is dried in a circulating draft of a vacuum oven at 80–100°C. The Hamilton Beach blender test is another way of testing mechanical stability. According to this method, the percentage residue created by subjecting the emulsion to high-speed agitation for a set period of time is a measure of the stability of the emulsion (or the lack thereof).

According to Ref. [41], the grit represents the coagulum that did not pass through a 200-mesh screen. The test can be visual, either by coating a thin (2–5 g/m^2) adhesive layer on a transparent face stock material or by using a screen (gravimetrical).

According to Ref. [42], the blender test is not useful in assessing high shear due to the lack of confined geometry to produce high shear field. Earlier rheometers normally lack the ability to remove viscous heating during high shearing, which could produce shear-induced flocculation instead of coagulation. Better results are obtained using a high shear rheometer (e.g., Hake Rheostress 300) to test the viscosity stability for 600 s at 70 K s^{-1}. Such laboratory methodology for high-shear stability of acrylic emulsions was confirmed by pumping with a positive displacement progressive cavity pump. The IR spectrum of the coagulum demonstrated that it is three times higher in the tackifier than in the formulation, that is, tackifier dispersions possess low-shear stability.

Grit-like gel particles may disturb hot-melt coating. SIS polymer has a predominantly chain scission degradation mechanism that minimizes cross-linking and consequent gelation. Gel particles must be avoided to provide smooth, uniform, and streak-free PSA coatings.

8.1.3.1.3   Test of Solid-State Adhesive

Other properties of PSA raw materials to be examined consist of the softening point, glass transition temperature, cold flow, and tensile strength of the adhesive.

Test of Softening Point and Glass Transition Temperature   The softening point provides information about the nature of the adhesive (hot-melt/solvent-based, competitive samples) and the formulation (tackifier amount and nature). Tackified dispersions differ with regard to their softening (melting) point. The tack/shear balance of an adhesive is related to the softening point of the tackifier resin (see also *Technology of Pressure-Sensitive Adhesives and Products*, Chapter 8).

Various methods for testing the softening point exist (range); see also Section 8.1.1.1. The softening range can be determined according to ASTM E-2867. Softening points as measured by the Mettler device are 5–10°C higher than ring-and-ball softening points. According to Ref. [43], the softening temperatures of the adhesives were determined from indentation profiles obtained by static thermomechanical analysis. Other subjective methods for the examination of the softening behavior of a coating have been developed; an example (using the Koffler heating stand) is given in Ref. [1].

The glass transition temperature, T_g, of amorphous blends provides indications about the extent of miscibility between the components, at least at a molecular level. Hence, thermal analysis, and especially DSC, is well suited for the study of the phase behavior of a polymer and a tackifying resin (see also Section 8.1.1.1). According to Ref. [4], a good rule of thumb for determining whether two adhesive formulations are different is if the difference between the glass transition temperatures is 3°C or greater.

Additional data were obtained using tapping mode atomic force microscopy (AFM), which is an established technique for the characterization of the phase morphology of block copolymers [4,43–47].

Test of Cold Flow   The role of cold flow in converting PSPs was discussed in Ref. [13] (see also Section 8.2.2.3.4). In this test the dimensional change of an adhesive sample after storage under load over a long time (1 week) is measured. The test is described in Ref. [1]. According to this test the cold-flow resistance is equal to the diameter of the squeezed disc of the experimental adhesive (loaded with 100 g weight or using 3 psi at 120°F), divided by the diameter of the squeezed disc of the standard adhesive.

Creep is an index for cold flow. According to Yarusso et al. [48], the apparatus used for the shear creep measurements was similar to that described by van Holde and Williams. The applied shear stress was 3.5×10^4 dyn/cm. The creep displacement was measured with a capacitance probe transducer for a period of 100 min following the application of the load.

Test of Tensile Strength   Tensile strength and its correlation with the adhesive properties were described by Benedek in detail in Ref. [10] and can be used for the evaluation of cohesive strength of the adhesive layer according to ASTM Test Method D-3759. Old supplier specifications for classic PSAs (e.g., Acronal 500 D, BASF) included the value of tensile strength. A good correlation existed between tensile strength and peel values for SBC and HMPSAs based on such copolymers [49]. The test of the tensile strength at break yields unrealistic values (too high) for cohesion; the tensile strength at 300% elongation provides data that are in good agreement with measured shear values. Tensile tests were used to study strain hardening observed during fibrillation (see also *Fundamentals of Pressure Sensitivity*, Chapter 4). The strain hardening observed in the fibrillation part of the probe test curves can be correlated to tensile tests [50]. Using analysis of PSA fracture mechanics in the course of peeling, Feldstein correlated peel force to tensile strain and stress σ [51]. Generally, axis-symmetric adhesion tests are based on tensile measurement.

The (ASTM D-638) method used for the evaluation of adhesive films by tensile tests is described by Benedek in Ref. [1]. For this purpose, 3.2 ± 0.4 mm thick and 19 ± 0.5 mm wide dumbbell-shaped specimens are prepared and strained in a tensile tester until break occurs. The tensile force at break is registered and divided by the original cross-section area of the narrow section of the specimen. Tensile strength can also be measured according to ISO-R 527:1966 (plastics-determination of tensile properties) or ASTM D882-91.

Test of Food and Drug Administration (FDA) and European Norm (EN) Compliance   Generally, PSAs should comply with the following FDA and European regulations:

FDA 21 CFR*	175 105 (adhesives)
	176 170 (aqueous food)
	176 180 (dry food)
	178 3400 (emulsifier)
BGA	Proposal XIV for polymer dispersions, 01.08.85 170
	Mitteilung Bundesgesundheitsblatt 28, 305 (1985)
	Prüfung auf Physiologische Unbedenklichkeit

*Code of Federal Regulations.

Current European regulation 1935/2004 is valid for materials intended to come into contact with foodstuffs [52]. The European Framework regulation 1935/2004 states general guidelines that also apply to adhesives, printing inks, and lacquers. For the first time, a manufacturer in Europe could be prosecuted for the influence of packaging on foodstuffs. In Germany the Foodstuffs and Consumer Goods Code (Lebensmittel und Futtermittel Gesetzbuch) of 1 September 2005. For instance, the pure monomer aromatic tackifier resins comply with FDA direct food contact regulations (see also *Technology of Pressure-Sensitive Adhesives and Products*, Chapter 3).

8.1.3.2   Test of End-Use Properties of the Adhesive

The end-use properties of the PSA include its adhesive properties and special performance characteristics related to its use as a PSP (e.g., label, tape, etc.). Although the adhesive properties of a PSA are currently tested for the adhesive per se, that is, on standard carrier materials, their decisive control is carried out for the finished product (see Section 8.2). The test of other end-use properties will be described in Section 8.3.

8.1.4   Test of Release Liner

Test of the release liner is discussed by Jones and Schmidt in *Technology of Pressure-Sensitive Adhesives and Products*, Chapter 11. The release force from the liner can be measured according to FINAT FTM 10. This is a 180° peel test with a test tape that is stored for 20 h at 70°C under 200 N/cm² pressure [53] (see also Section 8.2). Special release tests that imitate the practice-related label die-cutting conditions are described in *Technology of Pressure-Sensitive Adhesives and Products*, Chapter 10.

8.2   Test of the Finished Product

Most properties of PSAs may be examined when coated (i.e., in the PSA laminate). All PSA samples should be examined for their main characteristics; PSAs manufactured for special purposes should also be examined for their special application-related properties (see also Chapter 7). The test of the finished product includes the test of the coated and laminated web and the test of the confectioned PSP. Thus, such tests must include the general performance characteristics, the product class (e.g., label, tape, etc.)-related characteristics, and application-related characteristics (e.g., roll label, sheet label, permanent or removable label, etc.). As mentioned previously, tests of the finished products cover the adhesive properties and the other end-use-related performance characteristics.

First, the test of adhesive performance characteristics will be discussed. Independent of their end use, all PSA laminates possess an inherent degree of tack, peel resistance, and shear resistance, that is, an adhesion/cohesion balance (see also *Technology of Pressure-Sensitive Adhesives and Products*, Chapter 8), as well as certain converting properties (see also *Technology of Pressure-Sensitive Adhesives and Products*, Chapter 10). The coating weight influences all these properties [54]. On the other hand, the adhesive and converting properties depend on the aging response of the adhesive (see also *Technology of Pressure-Sensitive Adhesives and Products*, Chapter 8).

8.2.1   Test of Adhesive Performance

Generally, the following properties of the dried PSA coating should be tested: the tack, peel resistance, cohesion (shear resistance), Williams plasticity, dimensional stability, storage properties, temperature resistance, peel from the release liner (release force), and aging on the substrate. As described in detail in Refs. [2,17,54,55], the main adhesive properties of PSAs are characterized by tack, peel resistance, and shear resistance. The fundamentals of tack are described by Creton in *Fundamentals of Pressure Sensitivity*, Chapter 6; the theory of peel resistance is discussed by Kim et al. in *Fundamentals of Pressure Sensitivity*, Chapter 7; and shear resistance is evaluated in *Fundamentals of Pressure Sensitivity*, Chapter 8, by Antonov and Kulichikhin. This chapter discusses the practice-related aspects of such characteristics.

8.2.1.1 Test Conditions

The first step in the evaluation process of adhesive (or other) properties involves the preparation of the specimen. Generally, the PSA label specimen to be tested is manufactured in the laboratory when tests concern the PSA, the face stock material, or the release liner (i.e., when screening the laminate components).

Quality control tests are carried out on specimens taken from industrial production. Generally, laboratory samples are prepared by laboratory (direct/indirect) coating, drying, and conditioning of the coated sample. However, the laminate components, the equipment, and the drying and conditioning parameters used by different PSA suppliers or converters can vary quite a bit.

Test methods have been established in the United States (ASTM) or in Germany (DIN). EN also have been defined. Several adhesive manufacturing associations have developed test procedures that are internationally accepted and are used in the trade as reference methods. The test procedures for PSAs generally examine three key aspects: tack, peel adhesion at 90° and 180°, and shear resistance (cohesion). Procedures for measuring adhesion of self-adhesive materials to a test surface normally rely on a peel test (see also Chapter 7). Such test procedures were published by a number of organizations such as FINAT, AFERA, and PSTC. They also have been incorporated in buying specifications of the German military procurement office (BWB).

Standard test methods for adhesives were described by MacDonald [56] and Symietz [57]. Test methods for PSAs are derived from those used originally for gummed papers [54]. Shear resistance, tensile strength, and peeling tests were conducted on gummed papers to obtain performance characteristics with respect to the three classic stress modes.

In some cases, reference materials, like tapes, were used for the evaluation of adhesive properties. Thus, as a reference ordinary office-grade Scotch tape was selected; it has a rolling ball tack (RBT) of about 180 mm. On the other hand, it is difficult to compare the shear strength of an ordinary adhesive tape with that of a standard tape, because the backing of this commercial product will not support the test weight.

8.2.1.1.1 *Preparation of the Specimen*

Preparation of the specimen includes coating the adhesive on a solid-state material, drying the coated material, and in some cases laminating it on the release liner/face stock material. Preparation of the specimen also includes, in most cases, conditioning the samples. Different kinds of laboratory equipment are used to coat the adhesive. The coating procedure and the hand-operated coating device are described by Benedek in Ref. [1].

8.2.1.1.2 *Face Stock and Substrate Used*

A series of performance tests can be carried out on experimental adhesives using essentially end-use substrates. For reasons of comparison, most tests are made on standard plates. The largest single factor in the measurement of bond strength is the test plate. However, normalized test procedures differ in their definition of the required surfaces. FINAT uses float process plate glass; AFERA suggests brushed steel, whereas PSTC and BWB propose polished steel. FINAT specifies glass as the most suitable material; the surface smoothness is well defined, and the chemical nature of the material is stable and constant. Other parameters influence the results (e.g., drying of the test plate after cleaning). Different procedures and surface agents will leave different deposits on the test plate, which may influence adhesion of the self-adhesive material to the plate.

8.2.1.1.3   Drying Conditions

Drying conditions influence the content of liquid components in the adhesive layer, the composite structure (coalescence), and the roughness of the adhesive layer. They can also affect the mechanical properties of the solid-state laminate components. In most cases, different coating weights, different test surfaces, and different drying conditions are used.

Performance data are determined from laboratory-prepared test laminates composed of 20–25 g/m² dry adhesive coated on 50 μm polyester (PET) protected with siliconized release paper. Different types of adhesives (thermoplastic, thermosetting, self-cross-linking, and emulsion PSAs) are dried at different temperatures and oven times. Hence, an exposure to ambient air for 15 min and 2 min oven time at 200–250°C are suggested for solvent-based adhesives; 1 min ambient air exposure followed by 3 min oven (212°F) exposure is suggested for emulsion polymers. A tackified CSBR coated on paper is tested by coating it on release paper, drying for 5 min at 70°C and then 2 min at 100°C, and then transfer-coating into paper (1 ± 0.1 mil dry). Samples on PET are prepared by direct coating of adhesives on chemically treated films and drying for 5 min at 158°F and 2 min at 212°F. The adhesive is then laminated onto a release liner and conditioned for a minimum of 2 h at standard temperature and humidity before testing [22]. Drying at 90°C for 3 min followed by conditioning at 23 ± 2°C and 50% relative humidity (RH) for 16 h is used in practice. The adhesive is coated on the face stock material with a Meyer bar to yield a 20–25 g/m² dry adhesive after 2 min in a ventilated oven at 105°C. Then a protective liner is placed on the adhesive film and conditioned at 20°C and 50% RH before testing [1]. Coatings on PE should be dried at 60°C; on soft PVC coatings should be dried at 45–50°C. Different recommendations exist concerning the coating/drying conditions of laboratory samples (Table 8.3). Sample preparation and drying are described in Ref. [1].

8.2.1.1.4   Conditioning

A specification for materials testing always must start with a definition of environmental conditions. Methods will normally specify 23°C at 50% RH for conditioning, but there are considerable differences in accepted tolerances [1]. If the laminate to be tested contains paper, broad tolerances in the testing environment will raise considerable problems (see also Section 8.1.2.2). One should therefore aim to achieve environmental conditions according to DIN ISO 137 (1982).

The PSA-coated films should be conditioned under different conditions; fresh films should be stored for 24 h at 23°C and 50% RH with no contact with the release paper. Aged films should be conditioned at 40°C for 14 days and then for 24 h at 23°C. For water-based PSA formulations the coalescence (film forming) occurs slowly compared with HMPSAs and solvent-based PSAs (interpenetration of resin and dispersion). Practical end-use conditions may strongly differ between test conditions; in these cases, tests under actual application conditions are required. Standard conditioning and testing

TABLE 8.3 Coating–Drying Conditions of Laboratory Samples of Water-Based Coatings

| | PSP | | Drying Conditions | | | | | |
| | PSA | | 1. Step | | 2. Step | | 3. Step | |
Carrier	Type	Thickness (μm) Dry	Temperature (°C)	Time (min)	Temperature (°C)	Time (min)	Temperature (°C)	Time (h)
PET	SB	25	RT	15	200–250	2		
3	—	—						
—	WB	—	RT	1	212*	2	—	—
—	WB	—	RT	15	90	5	—	—
Paper liner	WB	25	70	5	100	2	—	—
PET	—	—	158*	5	212*	2	23	2
—	—	—	90	3	—	—	23	16
—	—	20–25	150	2	—	—	20	
PET	—	25	115	3	—	—	23	24
PE	—	—	60	—	—	—	—	—
SPVC	—	—	40–45	—	—	—	—	—
—	WB	—	70	5	100	2	—	—
—	—	—	70	10	—	—	—	—
—	WB	—	115	3	—	—	23	24
—	WB	—	RT	30	105	5–10	—	—
—	WB	—	220	0.5	—	—	—	—
—	WB	—	105	2.0	—	—	—	—

Note: *Degrees Fahrenheit.

atmospheres for paper, board, pulp handsheets, and related products are described by 1994 TLMI Test II-A. Such standard atmospheres include a preconditioning atmosphere (10–35% RH and 22–40°C), a conditioning atmosphere (50.0 ± 2.0% RH and 23.0°C), and a testing atmosphere (the same as for conditioning).

The application climate for the main PSPs (labels, tapes, and protective films) is described in Ref. [1].

8.2.1.2 Test of Tack

According to a popular definition, a PSA or PSP possesses an aggressive tack. According to the practical definition, it is an adhesive that adheres instantaneously, without chemical interactions, as suggested by the German terminus technicus, Haftkleber, or by the more vague (because it is not time-dependent) French definition, autocollant (self-adhesive). The aggressive tack of the adhesive, required by the theoretical definition of a PSA, is necessary to ensure pressure sensitivity, that is, to adhere under slight application pressure.

The quantification of such a "slight pressure" is difficult and is correlated with the quantification of the required tack. Tack must be defined as well (this a problem; see later); on the other hand, it is a characteristic that can be measured. Numerous tack test methods exist in practice, but from the point of view of the theory, only the probe tack (see later) test is carried out relatively simply under mechanically reproducible and mathematically evaluable conditions. The test uses a quantified surface, with a quantified bonding and debonding rate (the force working perpendicularly) and a well-characterizable substrate quality.

Unfortunately, adopting the above definition of pressure sensitive, related to low application pressure and high tack (instantaneous adhesion), many products (e.g., protective films, separation films, masking tapes, etc.) based on PSAs cannot be considered pressure sensitive because they do not have instantaneous adhesion by slight pressure and a high tack. For the most part, such products must be laminated on the substrate by elevated pressure, like their plastomer-based competitors. However, different from their plastomer-based competitors (see also Chapter 7), such products possess a measurable tack, the so-called application tack (see later). Some products possess a low tack because they have a low coating weight (thickness) of the adhesive. Thus, from the technological point of view, the theoretical definition of tack requires the quantification of the adhesive thickness as a parameter (see later). Considering the industrial practice of self-adhesives as a complex case and the (idealized) case of a PSA with a high, instantaneous tack as a common case, and taking into account the influence of the coating weight on the tack (i.e., working with a high, upper critical coating weight), we will examine how the classic definition of tack can be used for PSPs.

Although general agreement exists that PSAs are more or less tacky materials, there is no general agreement on a definition of what tack actually is. Tack is an elusive property of PSAs; there are at least three common ways of measuring tack [4]. Tack and its measurement were described in detail in Refs. [1,2,17,41,54,55]. As accentuated in Ref. [54], although common test measurements are based on standard tack test methods, special applications require the control of so-called wet-tack, application-tack, etc., that is, of tack related to end-use practice (see later and *Technology of Pressure-Sensitive Adhesives and Products*, Chapter 8). The fundamentals of tack as an index for bonding are discussed in detail by Creton in *Fundamentals of Pressure Sensitivity*, Chapters 4 and 8.

Various standard and nonstandard tack test methods exist. The best known normed tack test methods are the rolling ball, the Polyken® tack, and the loop tack methods. Because of the different basic principles of the methods, the reflected adhesive characteristics greatly differ. A given pressure-sensitive material may be tacky or nontacky, depending on the tack measuring method used. In the application practice of PSPs the nature of the tack required may be different as well. For instance, different tack is necessary for a touch-blow labeled item (see also Chapter 4) and for a laminated item. Therefore, the usability of a given tack test method depends on the PSP to be characterized. The easy applicability of a test method also influences its choice.

Tack itself is an intrinsic property of the adhesive that depends on the PSA and PSP construction. Because of the application of the liquid PSA by joint build-up, together with a solid-state carrier material, the stiffness and plasticity/elasticity of the carrier component always influence the contact build-up that is, the tack of the PSP. Therefore, the intrinsic tack of the PSP (T_{iPSP}) is a function of the tack of the PSA (T_a), the nature of the solid-state components (N_s), and the construction of the PSP (C_{PSP}).

$$T_{iPSP} = f(T_a, N_s, C_{PSP}) \tag{8.3}$$

The construction of the PSP influences the choice of tack test method as well. The choice of tack test method depends on the versatility of the method for a given product under the given experimental conditions. The measured tack of a PSP (T_{mPSP}) depends on the intrinsic tack and the experimental test conditions (C_T), which include the test method and its application conditions.

$$T_{mPSP} = f(T_{iPSP}, C_T) \tag{8.4}$$

Tack is the ability of a viscoelastic macromolecular compound to build up an instantaneous bond by pronounced viscous flow, which behaves elastically by debonding. Depending on the relaxation phenomena in the polymer, the elastic debonding response may be different for the same polymer tested using different methods. As a supplemental variable parameter, the chemical and macromolecular build-up of the adhesive must be taken into account. In practice that means that for a given class of macromolecular compounds the correspondence between the different tack measuring methods may be different.

The test procedure to measure tack consists of two steps: bond formation and bond separation. Some test methods (quick stick or loop tack and Polyken tack) simulate real bond formation conditions; others, like RBT, measure the tack under very different experimental conditions. All of these tests measure the ability of the adhesive to form a bond in a very short time with minimal contact force. The RBT, loop tack, and probe tack tests do not produce results that correlate well with the others [4]. Table 8.4 summarizes the different test methods for the characterization of the tack, as suggested by suppliers of PSAs or raw materials for PSAs.

Some are normed, standard tack methods, whereas others serve mainly as scientific tools. The first and least sophisticated test for tack is to put a thumb on an adhesive coated onto a sheet of paper (see also *Polyken Tack*). The next level of complexity involves sticking a loop of adhesive-coated paper to different substrates. Each time the loop is pulled from the surface to which it sticks, the tester forms an opinion about how sticky

TABLE 8.4   Test Methods for Tack Measurement

| Method | Stress Type | | Test Principle | | | Material Characteristic | Norm | Unit |
	High Speed	Low Speed	Rolling Friction	Peel Resistance	Tensile Force	Tensile Force		
Rolling ball	—	—	•	—	—	—	PSTC 6	in. cm
Rolling cylinder	•	—	•	—	—	—	—	in. cm
Bull tack	•	—	•	—	—	—	—	g
Pitched wheel	•	—	•	—	•	•	—	g
Loop tack	—	•	—	•	—	•	PSTC 5	g/in²
Quick stick	—	•	—	•	—	•	TLMI L-IB1	lb/in.; oz/in²
							TLMI L-IB2	oz/in.; N/25 mm
							FTM 9	N/in.; N/25 mm gm; N, gm/cm²
Polyken tack	—	—	—	—	•	•	—	—
Probe tack	—	—	—	—	•	•	—	—

the adhesive is. The normalized test methods are used to compare different adhesives. Such data are not absolute values, but may be used to compare various products.

8.2.1.2.1 Standard Tack Test Methods

The standard tack test methods include the RBT, the loop tack test, and the probe tack test. There are modified variances of these methods as well. Different tack methods are adequate for different PSPs. For instance, using the probe tack test the load is regulated, as in the lamination of protective films; loop tack simulates real label application conditions, where labels are blown to a surface using air pressure. The RBT is more complex; it is friction related, but the correlations between polymer friction and its viscoelatic properties by bonding/debonding are not clear.

RBT Test The RBT test measures tack as a function of the distance traveled by a steel ball on an adhesive-coated substrate. In a different manner from other tack-testing methods, the rolling ball method does not measure the debonding force. Measurement of the tack as coefficient of friction, that is, by the RBT, was discussed in Ref. [54]. Examination of the friction force allows a correlation between tack and shear resistance. The friction force is the sum of an adhesion component and a deformation component, which depends on the cohesion. Therefore, tack will be a function of the cohesion as well. Hamed and Hsieh [58] measured tack as cohesion (see later). The probe tack test (see also *Polyken Tack*) can be viewed as a constant deformation rate test on the behavior of the fibrillated materials, but the RBT cannot.

The RBT, according to PSTC Method 6 [59], is measured as follows: a stainless-steel ball is allowed to run down a slope from a point down the inclined plane onto the adhesive tape; the distance traveled along the sample (25 × 250 mm, 1 × 10 in. strip) is measured in centimeters. In this test an 11.1-mm (7/16 in.)-diameter steel ball rolls down a plane with a length of 18 cm and inclined at an angle of 21°30' with adhesive thickness of at least 25 μm. In a modified RBT, according to PSTC-6, an inclined angle of 21°30', a necessary length of 12.5 cm, and a ball weighing 7.59 g with a diameter of 1.229 cm (0.484 in.) is used. The apparatus for this test is relatively simple, but unfortunately, the test is influenced by adhesive residues on the ball or by the viscosity and thickness of the adhesive layer. As noted by Benedek in Ref. [54], delamination of the PSA layer from the carrier may occur during RBT as well. High tack levels correspond to low RBT distances. If the sample has an RBT greater than 30 cm (for a common label or tape), the adhesive tack was not enough, the ball was not clean or chosen correctly, or the coating weight was too low. Low-tack PSPs (e.g., protection films) cannot be characterized adequately using the RBT test.

The RBT (as method) depends on the formulation. As noted in Ref. [60], by PIB rubber the compositions tended to coat the ball and change the results of the RBT test. Tse [61] investigated the effect of oil on tack and determined that RBT is influenced more by surface properties than by bulk properties.

Rolling Cylinder Tack (RCT) Test The RCT method is a modified version of the RBT, known as the Douglas tack test. A stainless-steel cylinder is used with a diameter of 24.5 mm, a length of 19.05 mm, weight of 75 g, and a travel path 203 mm long with an angle of 5°.

The reproducibility of the RCT values is better than the reproducibility of the RBT values; the running path is shorter, and less material is necessary (compared with

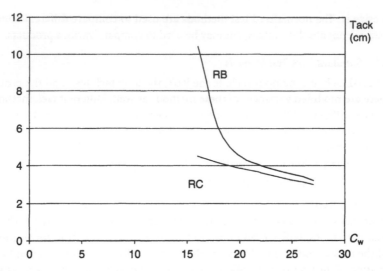

FIGURE 8.2   Dependence of the rolling ball (RB) and rolling cylinder (RC) tack on the coating weight (C_W).

competitive samples). For paper laminates values of 1.5–2.5 are very good and values of 7.5–8.0 are fair. No more than 15 cm can be accepted. For film laminates values of 2.5–3.5 are very good, values of 4.5–5 are fair, and values above 7–8 cm are not acceptable. As illustrated in Figure 8.2, RBT is more sensitive to the coating weight than RCT; that is, for very low coating weights (and tack) RCT is preferred.

Loop Tack/Quick Stick   Loop tack is a measure of the force required to remove a standard adhesive-coated film loop from a standard stainless-steel plate after short contact of the test strip with the steel plate in the absence of pressure.

According to the FINAT definition (FINAT Test Method 9 or FINAT FTM 9), the quick-stick tack value is the force required to separate at a specific rate a loop of material that was brought into contact with a standard surface using no external pressure to secure contact. According to FINAT, loop tack is actually quick stick, because of the special loop form of the specimen; quick stick is also called loop tack.

The FINAT FTM 9 quick-stick method differs from that of AFERA (4015) and the PSTC 5 quick-stick method because the latter measures peel at 90° without making a loop.

The quick-stick method is relatively simple and may be carried out using common tensile strength test machines. Unfortunately, the contact time is long and depends on the area, the contact area differs, and, when using the FINAT method, the peeling angle is not constant. The method has the advantage of allowing tack to be measured on a wide range of substrates such as stainless steel, glass, PE, and paper. It can, however, result in high tack readings on smooth substrates such as glass for adhesives with a low finger tack on rough surfaces. The EPSMA method for double-side tapes varies in that it specifies that a loop of PET film is brought into contact with the tape on a panel; this procedure discounts the effect of the material rigidity.

A 0.5 × 4 in. strip of 1-mil Mylar (polyester) film coated with the sample adhesive is formed into a loop with the adhesive on the outside; the loop is applied to the test plate until the PSA loop contacts 0.5 in.^{2} of the surface area on the plate. The loop is then removed from the plate at a rate of 12 in./min and the loop tack is the maximum force observed. According to another method, a loop formed from a 2.54-cm-wide strip of coated paper is lowered by a tensile testing machine onto a glass plate. The force required to remove the tape measured at a rate of 300 mm/min at 20°C is the measured loop tack. Tack (measured as loop tack) depends on the substrate. The following values were obtained for different substrates (in N/in.): stainless steel, 2.8; glass, 3.5; PE, 2.7; and PVC, 2.5.

In a modified PSTC 5 loop-tack (quick-stick) test, polished stainless steel is used as a substrate. When a contact surface of about 1 in.^{2} is achieved, the loop is separated from the panel at an 8 in./min separation speed. Quick stick or loop tack is the average of five test values. In some cases a 2.54 × 12.7 cm adhesive-coated strip is used. According to Ref. [62], tack should be tested in N/20 mm on a chromed steel plate. Quick stick to stainless steel and to kraft paper is measured; peel build-up and restick are also tested. In Refs. [4,63–65], loop tack was performed with a loop tack tester using a cross-head displacement rate of 5 mm/s (12 in./min) and a 25 × 125 mm (1 × 5 in.) loop of tape in accordance with PSTC-16 [59]. The free loop was 75 mm (3 in.) long. The maximum force per width of the specimen was recorded. The initial height above the substrate of the top of the free loop (at the bottom of the grips) was 50 mm (2 in.). The maximum displacement was 44 mm (1.75 in.) and the dwell time at maximum displacement was 1 s. Loop tack also can be tested through the test method ATM 136-84 [66]. Dunckley [67] discussed soft loop tack of such tackified dispersions as CSBR with very high tackifier loading. Table 8.5 lists some variations of the loop tack method. As Table 8.5 demonstrates, normed test rate and substrate surface are normally used, but both the type and dimensions of the carrier material employed and the contact area may strongly differ; thus, a comparative examination of loop tack data is difficult. The influence of the carrier on the loop tack was discussed by Lim and Kim in Ref. [2].

TABLE 8.5 Variances of Loop Tack Test Method

Sample			Test Conditions				
Carrier Material	Thickness (µm)	Dimensions (mm × mm)	Substrate	Contact Surface (cm²)	Test Rate (mm/min)	Norm	Note
PET	50	—	SS	2.54 × 1.27	—	—	—
PET	25	0.5 × 4 in.	SS	0.5 in.²	300	—	—
Paper	—	2.54	Glass	—	300	—	—
PET	—	—	SS	1.0 in.²	200	PSTC 5	Modified
PET	25	—	SS	2.5 × 1.25	300	PSTC-16	—
Paper	—	2.54	SS	2.54 × 2.54	200	—	—
PET	25	—	SS	1.27 × 1.27	300	—	—

A finite element analysis of the loop-tack test is developed in Ref. [68]. In this case the loop backing is modeled as a flexible, inextensible, elastic strip whose curvature is proportional to the bending moment.

Polyken Tack   The Polyken probe method is the most well known among the probe tack tests. The fundamentals of the probe tack test are discussed by Creton in *Fundamentals of Pressure Sensitivity*, Chapter 7. The Polyken probe tack testing machine provides a means of bringing the tip of a flat probe into contact with PSA materials at controlled rates, contact pressures, and dwell times and subsequently measuring the force required to break the adhesive bond in grams per centimeter. Polyken tack is less strongly affected by the resin softening point (it is applied under load). Because this method is based theoretically on the tensile test, numerous other tests are related to the probe tack test.

Tack test methods have been required for products other than PSAs as well. For instance, Ref. [69] describes the apparatus used to test the tack of contact adhesives. Such adhesives ensure joint build up through the contact of two adhesive-coated surfaces (see also Chapter 1). The equipment proposed does not measure the bonding force or tack. It tests the build up of the force only, as a function of the contact (and pressing) time and variable (50, 100, 150, and 200 g) pressing forces for a given adhesive-coated sample (with standard coating weight, sample dimensions) and contact area ($\phi = 20$ mm). However, it can be considered a first attempt to measure adhesion with the aid of an indenter.

A Polyken probe tack tester (Kendall) according to ASTM was described in Ref. [70]. A 120-g force and a 0.5-cm sample diameter were used. According to Ref. [71], Polyken tack is tested with a 100 g/cm contact pressure, 1-s dwell time, and a 1 cm/s test rate. Because this method implies a very small contact area, small imperfections in the adhesive film can lead to disparate values.

Historically, the probe tack test equipment was developed from the common industrial apparatus used in the field of other non-PSAs for the measurement of the bond strength as a function of time. In principle, the use of a probe test tack is an extension of thumb tack testing. Probe test methods were first used in the 1940s. Kamagata [72] tried to control conventional thumb tack testing using a small balance. In the 1950s, Wetzel [73] proposed a probe testing device that could be fitted in a standard extensiometer. Later, this method was redesigned into the Polyken probe tack tester. The probe is attached to a force gauge; the sample to be tested is attached to another weight to control the applied pressure, which then is lowered onto the probe and pulled off. The equipment applies a 100 g/cm^2 pressure with a 5-mm-diameter stainless-steel probe, a contact time of 1 s, and a rate of removal of 1 cm/s. The Kendall probe tack tester also exists for research investigation purposes. The Polyken probe tack tester and the test method have been suggested for the evaluation of readhering adhesives.

The probe tack method proposed by Wetzel [73] and developed by Hammond [70] has the disadvantage that a special apparatus is necessary and the contact area is too low (<0.2 cm^2). The main disadvantages of the probe tack are given by the extreme sensitivity of the method concerning the quality of the contact area. A more precise version of the probe tack method is that developed by Druschke [74], enabling the use of contact times as short as 0.01 s.

The industrial renaissance of this test method is due to its ability to be quantified exactly (normed contact force, contact time, and contact and decontact speed). Probe tack measures tensile strength and by means of the fibrillation theory it was possible to correlate the macroscopic mechanical properties of a material to its molecular mechanics, that is, modulus values. The use of probe tack as a characterization method for PSPs was forced, especially after demonstrating that this method allows the microscopic examination of debonding. Thus, the phenomena of cavitation and fibrillation were discovered.

Both phenomena were known since the beginning of pressure-sensitive formulation, but did not have real importance. Cavitation is observed in the debonding of plastic films, and the build-up of fibrils is a phenomenon known as legging from the formulation of removable adhesives. Such fibrils can be observed using the RBT test as well. The main question related to both phenomena (cavitation and fibrillation) relates to whether they are "simply" possibilities of energy transfer during debonding or whether they can be used as a "supplemental" criterion for pressure sensitivity as well. We use the term "supplemental" because, as accepted in common practice, pressure sensitivity is characterized by rheologic parameters, and the so-called Dahlquist's criterion (see also *Fundamentals of Pressure Sensitivity*, Chapters 4 and 5) notes that for a pressure-sensitive behavior a PSA must possess a modulus of about 0.1 MPa. A maximum tack implies a modulus of 1×10^5 to 2×10^4 Pa. Is fibrillation of PSAs a special case of patterned contact? Admitting the strong influence of the asperities on the fibrillation results and the discrepancies of the probe tack tests for "hard" adhesives, *ad absurdum* it can be supposed that patterning and fibrillation are correlated.

Probe tack also offers the advantage in comparison with other tack test methods that the properties of the carrier material do not affect the test results. However, probe tack possesses some theoretical and practical disadvantages as well. Concerning the theory, fibrillation by probe tack, as a phenomenon that is correlated with the build-up of new surfaces (i.e., energy consumption), depends on the number of fibrils, which is strongly dependent on surface roughness. From the point of view of hydrodynamics, fibrillation is a flow. Thus, various common parameters that influence flow, depend on its speed, and allow the use of invariants (like the *Re* number, etc.) in (macroscopic) industrial practice must be considered in this case also. It would be useful to compare this phenomenon with fibrillation of fiber-forming materials in polymer processing (e.g., spinning). From the practical point of view the main disadvantage of probe tack arises from its main advantage, mentioned above. PSA converters do not need the properties of a pure adhesive, but those of an adhesive product (carrier material plus PSA). Compared with peel resistance tests, probe tack supplies instantaneous values. The exactness of such values strongly depends on the apparatus and surface area. Peel resistance supplies statistical values of continuously repeated debonding. The experimental conditions used for probe tack tests are listed in Ref. [1].

The scientific rebirth of this procedure is forced by the possibility of a theoretical interpretation of the test results, which allows the correlation of contact mechanics, rheology, and macromolecular basis of the adhesive [29].

Due to a complicated stress distribution, peel test values cannot describe the molecular origins of the adhesion (see later). For this reason, axis-symmetric adhesion tests

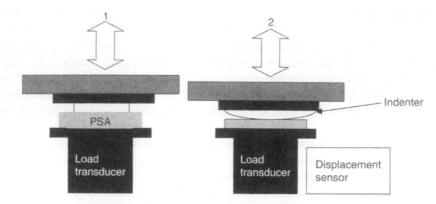

FIGURE 8.3 Schematic presentation of a probe tack tester with a flat (1) and spherical (2) indenter.

are favored in studying the interfacial and bulk contributions to adhesion. With this method, accurate measurements of the relationship between the energy release (G) and the crack tip velocity (v) can be made. The probe tack test is a variance of the axis-symmetric adhesion test (indenter test). A schematic presentation of the axis-symmetric test is given in Figure 8.3.

A rigid flat (1) or spherical (2) punch is brought in contact with the adhesive at a fixed rate. The layer is then separated from the punch, and the energy required for the pull-out process is measured by finding the area under the load versus the displacement curve.

For probe tack, the effective work of adhesion is given by the correlation between the area under the tensile load versus the displacement curve normalized by the area of contact at the maximum compressive load [75]. For strong adhesives, failure during the pull-out phase occurs by cavitation and fibrillation of the adhesive. Therefore, the tack test can be used to determine a phenomenological model for fibril deformation and to predict the peel rate and the peel force [76]. According to Kendall [77], the sphere adhesion (i.e., probe tack with spherical punch) result is similar to the peel test. (The geometry of a spherical punch pressing against a flat substrate has been used in fundamental studies of the tackiness of cross-linked elastomers.) PSA samples may be in the form of cylinders as well.

Although they present alignment difficulty and fibrillation upon pull-off, flat punch tests have often been the choice of industrial research laboratories and quality assurance and are the basis of commonly used ASTM standards [78]. The softness of the adhesive and the roughness strongly affect the measurement and may produce discrepancies in the test results. Such discrepancies are accentuated in case of SBCs (Figure 8.4).

Classic tack test methods (Polyken and RBT) yield results that differ from the loop-tack data. They deviate from the parallelism generally observed for changes in loop-tack values and peel resistance values. As illustrated by the data in Figure 8.4, such a deviation is a function of the adhesive system and the test method. This means that the tackiest formulation tested by a given tack test method may not be the best one tested by using another method and vice versa; the parallelism in tack found by two methods may

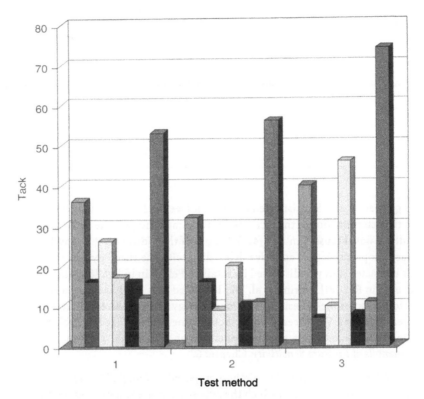

FIGURE 8.4 Discrepancies among rolling ball (1), loop tack (2), and Polyken tack values (3) measured for various natural rubber-based, styrene–butadiene rubber-based, and SIS block copolymer-based PSA formulations.

indicate a discrepancy using a third tack test method as a function of the base elastomer used for the formulation. The greatest deviations are observed for pressure-sensitive formulations based on SBC. Until recently, there was no explanation for such behavior. The investigations made by Hooker et al. [79] suggest that such deviations are due to the increased rigidity of SBC. For SIS-based PSA at normal temperature, surface roughness is detrimental for good contact. To avoid such discrepancies, a tack index was suggested. Such an index is based on the use of quick-stick, loop-tack, and RBT test methods [1].

Probe tack is the optimal technique for studying adhesion properties of soft polymers for short contact time conditions. Probe tack tests of PSAs on silicone release coatings were carried out as well.

A custom-designed probe tack apparatus allows the simultaneous acquisition of a nominal stress and strain curve and the observation of the adhesive film from underneath the transparent substrate. Probe tack tests demonstrate the interdependence between tack and cohesion. Experimental results indicate that the beginning of the probe test (where the cavitation process takes place) quantitatively corresponds to the small strain shear properties of the adhesives, whereas the end of the test (fibrillation, strain hardening) can be quantitatively related to the large strain elongation processes

TABLE 8.6   Variances of Probe Tack Test Method

Sample		Test Conditions			
PSA		Thickness (μm)	Contact Pressure (KPa)	Contact Time (s)	Debonding Rate (μm/s)
SIS	Tri-di-block	70	—	<1	—
	Tri-di-block	—	1	1	—
	Triblock	10	1	1	—
PIB	—	70–80	5 N	<1	100
PEHA	—	100	10	1	30

[50] (e.g., peeling off). Sophisticated equipment (see the Mechano-Optical Tack Tester) allows the determination of the tack strength and tack energy. (The tack/cohesion interdependence was discussed in Ref. [41].) The use of the indenter test method to correlate the macroscopic aspects of adhesive debonding with the rheology and the macromolecular characteristics was discussed in detail by Feldstein in Ref. [80] and by Feldstein and Creton in Ref. [81]. Doring et al. [82] indicated that AFM was an excellent tool for characterization of the adhesive phase morphology and correlated the results to probe tack measurements. Variances of probe tack test method are listed in Table 8.6.

A detailed discussion of the fundamentals of probe tack is given by Creton in *Fundamentals of Pressure Sensitivity*, Chapter 6.

Other Tack Test Methods   Another method proposed by Bull [83] (in a manner different from the rolling ball method in which the adhesive surface is static and the adhesiveless contact surface rotates) has a polished rotating adhesiveless cylinder in contact with a rotating adhesive-coated cylinder. This method measures the force (resistance) necessary to move the cylinders. Unfortunately, a special apparatus is needed, and like in the RBT, debonding is influenced by the compression of the adhesive by the cylinder.

According to Druschke [74], the following factors influence adhesion (and the test methods of adhesion): the surface tension of the adhesive, face stock, and adherent; the mechanical (viscoelastic) properties of the adhesive, face stock, and adherent; the surface structure of the adhesive, release liner, and adherent; the thickness of the adhesive, face stock, release liner, and adherent; and the test conditions. Taking these parameters into account, Druschke improved the rotating drum method. A further development of the rotating drum method is the pitch-wheel method, or the toothed wheel method, which permits very short contact times, as does the rotating drum principle [84]. In this method the polished wheel used in the rotating drum method is replaced by a pitched wheel, rotating freely around its axis, fixed in the upper clamp of a tensile tester with a liner. A rotating drum is pushed against the wheel until almost pressureless contact is achieved. The tensile tester measures the force acting against the rotating drum with the adhesive layer.

The mechano-optical tack tester (developed by Elf-Atochem) [85] is a variance of the probe tack test apparatus. It allows determination of the tack strength, tack energy, and the actual contact area for tests at controlled contact force, contact time, and release rate. It simultaneously measures the contact area between the probe and the PSA, together with the tack force and energy during the contact and separation phase. This apparatus is made up of a quartz prism probe linked to a force transducer. The whole system can be moved downward and upward at constant speed in the bonding and debonding phases. During the contact phase the actual thickness of the adhesive is measured with a micrometer fixed to the prison support. An optical device that moves with the probe provides a measure of the contact area at any time. The ratio of the transmitted intensity and the initial intensity of the light gives the percentage of contact area. A constant contact area is applied for a very short contact time (less than 1 sec) and the probe is pulled up. The force and the contact area are monitored versus time. The product of the area subtended by the debonding force versus time by the pulling rate yields the tack energy.

Tack was measured as plasticity as well [41]. The viscosity of HMPSA as a function of the aromatic content of tackifier resin follows a similar trend to RBT values and holding power [4]. The resins with a middle range of aromaticity have the most tack and lowest viscosity. Therefore, tack and melt viscosity seem to be correlated. This is not surprising given that RBT and melt viscosity are likely controlled by the fastest relaxation mechanisms. Furthermore, the higher viscosity samples have the most holding power. This is also not surprising given that holding power and viscosity are often directly correlated and that the properties of tack and holding power are inversely correlated and increasing tack is usually at the expense of cohesion.

The influence of various test parameters on the tack (e.g., adhesive nature, face stock material, and coating weight) was discussed by Benedek in Ref. [41].

8.2.1.2.2 Tack Test Methods for Special Products

Special PSPs may require that the common adhesive properties work under special application conditions characterized by a humid atmosphere, leading to condensation of water on the substrate surface such as bottle labeling (see also Chapter 4), medical tapes (see also Chapter 4), etc., or low temperature (which may cause condensation), such as freezer labels (see also Chapter 4). In other cases the PSP is applied under high laminating pressure to compensate the low coating weight used and advanced cross-linking employed to achieve removability, for example, protection films (see also Chapter 4).

Special adhesives can even stick to slightly moist materials [86]. Wet-tack is the tack measured on substrate surfaces covered with water. Generally, such tack is tested under standard test conditions with the exception of the substrate surface used, which is covered or saturated with water.

Low-temperature tack is tested under standard test conditions, with the exception of the substrate surface used, which is cooled to the application temperature required.

Application tack is measured under standard test conditions, but must be correlated with the industrial laminating conditions and peel resistance test.

8.2.1.3 Test of Peel Resistance

For a technologist, the peel resistance, that is, the resistance to the debonding of a PSP, is more important than its tack. Peel resistance is important as a value per se, but in the actual applications of PSPs, the time–temperature dependence and environment dependence of the peel gained a special importance. The removability [13,15] and water resistance [12,74] of PSPs play an especially decisive role in their common use.

The best known test of adhesion is the method used for lacquers according to DIN 53151 and ISO 2409. Adhesive strength (peel) also may be determined according to ASTM D 1000. Here the peel or tensile strength measurement should be used for the evaluation of the bond strength. For a long time tensile strength measurements and peel measurements were carried out in parallel [87]. Kemmenater and Bader [88] proposed a tension-measuring device for the peel. FIPAGO accepted a pendulum device [89,90]; in this test, the adhesive strength was tested via peel work in millimeters per kilopound. Today, the most common method for testing the adhesive strength of PSAs is the peel test. Different versions of this method using different peel angles, peel rates, and substrates are used. The history of peel testers was described by Liese [89]. FIPAGO recommended the adhesive strength tester of PKL used for gummed papers. The Werle tack tester measures adhesive strength in a similar manner based on peel adhesion [90].

Peel resistance is determined in accordance with ASTM D-3330-78, PSTC-1, and FINAT FTM 1 and 2 and is a measure of the force required to remove a coated flexible sheet material from a test panel at a specified angle and rate of removal. According to American norms, a 1-in.-wide coated sheet is applied to a horizontal surface of the clean, stainless-steel test plate with at least 5 in. of the coated sheet material in firm contact with the steel plate. A rubber roller is used to firmly apply the strip and remove all discontinuous and entrapped air. The peel adhesion of the sample is measured in a tensile tester. In Ref. [91], peel was measured using ASTM method D3330/D3330 M-99 Standard Test (Methods for Peel Adhesion of Pressure-Sensitive Tape). Adhesion of the latex was tested using ASTM D897 [92] and ASTM D330-90 [93]. Although there are standard methods for measuring peel, different "homemade" methods are used to measure different adhesive performance characteristics. Some test equipment allow the measurement not only at a constant peel rate but also with a constant force. From the point of view of application technology, this mode of measurement is often of greater interest.

The most important parameters of peel resistance were discussed in our previous works [2,41,54]. Test methods for peel resistance were described in Ref. [1]. Fundamentals of peel resistance are discussed in *Fundamentals of Pressure Sensitivity*, Chapter 8, by Kim et al. Peel resistance is a function of the test conditions, such as peeling angle, peeling rate, temperature, and dwell time.

8.2.1.3.1 Angle-Dependent Test

For different end uses, for example, labels for different applications or tapes, varying peel test angles are used. For a flexible and a rigid adherent a 180° peel or 90° peel is used and for peel between two flexible adherents a 180°T peel is carried out. Common angles are 90° and 180°; for label stock the 180° geometry according to FINAT FTM 1 found wider application than the 90° geometry. Peel tests at a peel angle of 180° were performed in Ref. [41] according to the FINAT FTM 1 conditions. PSA strips of 25 × 90 mm were applied on stainless steel plates or HDPE plates with the aid of a 1-kg FINAT standard test roller. The peel tests were measured after a dwell time of 20 min and 24 h at a peeling rate of 300 mm/min. According to Ref. [94], the optimum resin concentration for the best peel value depends on the peel angle. For 90° peel the highest value is obtained at the highest resin loading (ca. 35%); for 180° peel best values are obtained for less resin (30 phr) on PE.

Test of 180° Peel Resistance   According to the AFERA 3001 method for test of 180° peel resistance, one 10-mm-wide specimen of PSA film (prepared at least 24 h before running the test) is applied on a stainless-steel plate. Ten minutes later the specimen is peeled off by means of a tensile tester at a speed of 304 mm/min. The measurement is carried out at 23 ± 2°C. In fact, tests according to PSTC 1 and AFERA 4001 PB are carried out with a 25-mm-wide sample peeled off at 300 mm/min speed, and the peel force is given in grams per 25 mm.

A modified PSTC 1 method [95] prescribes a polished stainless-steel panel as a substrate. After 1 min of dwell time the tape is peeled away from the panel at an angle of 180° and at an 8 in./min separation speed. The 180° peel adhesion is the average of five test values and is expressed in grams per inch width of tape. Another modified PSTC 1 method on a 2.5 × 25 × 125 mm glass plate uses a tape-applicator. The peeling force is registered at 12-mm intervals and given in N/25 mm. Alternatively, a strip of 2.54-cm-wide coated paper is bonded to a glass plate by applying a constant pressure [96]. A tensile testing machine is then used to measure the force required to peel the paper strip at an angle of 180° at a rate of 300 mm/min at 20°C. The test speed for film laminates should be 300 mm/min and 75 mm/min for paper. O'Brien et al. [4] measured the peel energy or peel force per unit width, in accordance with PSTC 101: *Peel Adhesion of Pressure Sensitive Tape Test Method A—Single-Coated Tapes, Peel Adhesion at 180° Angle* [59]. Rectangular strips of 25 × 250 mm (1 × 10 in.) dimensions were tested using a Universal Testing Machine at 5 mm/s (12-in./min) cross-head displacement rate.

Peel adhesion should be tested according to FINAT FTM 1 at a peeling rate of 200 mm/min; 180° peel adhesion should be checked. ISO 8510-2 1990 (peel test for a flexible-to-rigid bonded test specimen assembly, Part 2.180° peel) can also be used. The test method of the peel resistance and the apparatus used are described in detail by Benedek in Ref. [1].

Test of 90° Peel Resistance   The 90° peel resistance is measured as the force required to remove an adhesive-coated material from a test plate at a specific speed and at an angle of 90°. The 180° peel resistance is a combination of tensile strength and shear, whereas a 90° peel resistance is one of tensile only. Peel decreases with the angle; 90° peel resistance is about twice that of 180° peel [97]. Wilken [98] obtained the maximum peel value between 90° and 180° peeling angle and demonstrated that below 90° the peel resistance increases as a result of the increased shear stress.

$$P^o_{180} \cong P^o_{90} > P^o_{\rightarrow 0} \tag{8.5}$$

Our data [1] demonstrate that 180° peel resistance is higher than 90° peel resistance. For 180° peel measurements the results depend on the face stock material [97]. A higher peel value will be obtained for thicker and stiffer materials [99]. Especially for soft PVC, 90° peel gives more exact values [100]. The 90° peel resistance of PSA tapes was measured on a glass plate coated with release agent [101]. According to Ref. [102], for carrier materials with higher thickness and stiffness the 180° peel resistance increases proportionally with carrier thickness (PETP) and then more slightly.

$$P^o_{180} > P^o_{90} \tag{8.6}$$

The difference between the peel values at different peeling angles (180° or 90°) depends on the chemical nature of the PSA; soft, rubber–resin PSA gives much lower 90° peel.

For the 90° peel test the apparent strain energy release rate can be calculated from the equation $G = P/b$, where P is the PSA tape width.

For such a test, a sample of PSA-coated material that is 1 in. (2.54 mm) wide and about 4 in. (about 10 cm) long is bonded to a target substrate surface (glass, unless otherwise noted) [62]. One end of the adhesive-coated sheet is peeled off the target substrate and held in a plane perpendicular to the target substrate, which is restrained in its initial plane. The coated sheet is pulled away at a constant linear rate of 300 mm/min and an angle of 90°.

A special case of 90° peel adhesion test is known as the butt tensile test. This test (90° adhesive test) restricts the evaluation to a single narrow area, in contrast with a standard peel test in which a continually new adhesive part is being examined [103].

Test of T-Peel Resistance   The T-peel adhesion test evaluates the "unwind" characteristics of the wound tape. The tape is bonded to another piece of the same film. The ends of the two films are inserted into the opposing jaws of a tensile machine, which then pulls the two films apart at a 180° angle. T-peel adhesion may be measured according to ASTM D-1976, DIN 53281, ISO 8510-1, or EN 1994D. This method is intended primarily to deter- mine the relative peel resistance of adhesive bonds between flexible adherents or when standard peel values are too low. The adherents shall have such dimensions and physical properties to permit them to bend through any angle up to 90° without breaking or crack- ing. Hamed and Preechatiwong [104] used T-peel measurement for the adhesion test of SBR pressed between PET and cotton fabric at 140°C. The T-peel test, according to ASTM D-1976-61T, is carried out a speed of 5 in./min; EN 1994D uses a speed of 300 mm/min.

A major attraction of peel tests is their apparent practical simplicity. However, in both testing and data interpretation, attention to detail is important if reliable and useful results are to be obtained. Kinloch et al. [105] proposed a peel test protocol. The peel test protocol [106] used for flexible laminates is intended to provide guidance on the measurement of peel strength of the laminate and then to demonstrate how the adhesive fracture energy, G_c, can be determined from the peel strength and other measurements. The protocol is divided into two parts: one for a "fixed arm peel" test and the other for a "*T*-peel" test. These geometries are different, but there is a common theme in converting the peel strength measurement into an adhesive fracture energy, G_c. The current challenge is to model accurately any exten- sive plastic deformation that may occur in the flexible peeling arm, because if the deforma- tion is not accurately modeled, then the deduced value of G_c may suffer a high degree of error (see also *Technology of Pressure-Sensitive Adhesives and Products*, Chapter 1).

For self-adhesive polyolefin films (see also Chapter 7) the so-called *peel cling* and *lap cling* are used to characterize the adhesive properties of the product. In this case, the peel cling is the T-peel of the SAF on itself; lap cling is the adhesion of SAF to itself, tested by shear measurement (see Figure 8.5). The cling of cling–stretch SAF is measured according to ASTM D-4649 at 1, 6, 14, and 40 days after manufacturing.

8.2.1.3.2   *Debonding Rate-Dependent Peel Resistance Test*

Peel resistance is proportional to the ratio of G'' and G' at the respective debonding and bonding frequencies [10,32]. At low peel rates the peel resistance depends mainly on

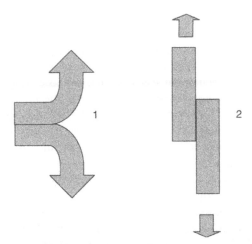

FIGURE 8.5 Principle of peel cling (1) and lap cling (2) test for self-adhesive films.

adhesion, but at high peel rates it is influenced by deformation. Therefore, factors that vary by peel test include peel rate, time after application, and substrate. Because of the viscoelastic properties of PSAs and the mechanical properties of the face material, the adhesive bond strength depends on the peel rate [1,2,41,54].

In most tests, single-point measurements at 5 mm/s are specified. In practice, test results demonstrate that the suitability of a range of products may vary with peel rate. Generally, the PSA is polymer based; consequently, it behaves like a viscoelastic medium. The bond strength is not constant for varying strain rates and temperatures. The rate of separation and temperature are related to the adhesive strength in a complicated manner, and these factors can influence the nature of the failure mode (adhesive or cohesive). This can lead to a situation in which an adhesive is selected on the basis of its performance at one set of strain rate/temperature conditions, but ultimately fails, because in use of the loading conditions are different. This should be considered when practical application conditions differ greatly from test conditions. For instance, diaper closure tapes (see also Chapter 4) must exhibit a maximum peel force at a peel rate between 10 and 400 cm/min and a log peel rate between 1.0 and 2.6 cm/min. Tapes exhibiting these values are strongly preferred by consumers. On the other hand, through careful design and formulation, PSAs can have a long modulus plateau over a wide frequency range, which ensures similar peel strength and failure mode at various removal rates.

In practice, the dependence of the failure mode on the peeling rate is used to delaminate removable labels without cohesive failure. Low-speed delamination ensures removability.

Universal testing of a roll of tape occurs at 12 in./min, whereas modern slitting and unwinding machines or tape application processes may run at several hundred feet per minute. With harder unwind, adhesive pick-off or transfer may occur. A similar statement holds true for release forces from siliconized release paper (see later). This phenomenon is like the change of adhesion/cohesion failure during the peel test as a function of the debonding speed. For industrial practice it is very important to know how stable the release value is with regard to the aging and peel rate conditions [107].

Peel resistance and tack increase with increasing test speed, but only up to a limit [108]; at very high measuring speeds the tack increases again.

The high-speed peel tester allows a 10–300 m/min testing speed for release force measurements [109]. With this method the separation force of the PSA-coated face stock from the release liner is evaluated at speeds similar to those typically used to convert and dispense the laminated material (see also Chapter 4). The method therefore provides a better characterization of the material than FINAT FTM 3. According to Ref. [110], the stripping speed was originally 12 in./min, but later the test speed increased to 400 in./min and higher. Label applications can run with 500 feet/min; thus, high-speed release testing equipment has been designed in this range. Very low peel values (from the release liner) may create label fly during conversion or application; high values may produce web break when skeleton stripping or dispensing failure during automatic application occurs. High-speed stripping of release (at 15 m/min) gives 25–30 g release force in comparison with low-speed stripping (at 3 m/min) 9–16 g. Release force generally increases over time. However, high-speed stripping values do not depend on storage time. As indicated in Refs. [111–114], the bulk viscoelastic properties of the adhesive are correlated to the release profile observed from peel force versus peel velocity measurements.

The test is carried out like a 180° peel test at jaw separation speeds between 10 and 300 m/min. To ensure good contact between the release paper and the adhesive, the sample is placed between two flat metal or glass plates and maintained for 20 h at 23 ± 2°C under pressure. The manner of stripping influences high-speed release test results (i.e., face from liner or liner from face). Generally, higher results are obtained when removing the face material from the release liner. According to TLMI Adhesion Test No. VIII-LD, the sample to be tested is secured to a rigid plate using double-side adhesive tape. Excess moisture in the carrier paper can be troublesome, and it is preferred that the papers have a moisture content per weight of about 4% or less. Predrying should be employed if necessary.

By increasing the thickness of the release backing the apparent high-speed release values were approximately doubled over the whole speed range when the backing was removed from the label. Such tests are important when accounting for the velocity-dependent influence of the release agents on the release liner. As indicated in Ref. [111], the largest impact of the HRA to release force typically occurs at low to intermediate peel velocities (0.005 to 0.17 m/s).

The debonding rate (stress rate) influences both test methods and industrial practice of PSPs. As illustrated in Table 8.7, the debonding rate, like the peeling rate, determines the removable/permanent character of an adhesive, affects the failure mode by peel test, and influences the value of peel resistance and peel resistance from release liner and the sensitivity of such tests to release age. In industrial practice the debonding rate also determines the failure nature (i.e., it produces inversion by unwinding a tape with excessive speed), affects skeleton stripping by labeling (i.e., the choice of the labeling machine), and influences the HRA concentration by formulation of the release agent.

8.2.1.3.3   Dwell-Time-Dependent Test

Dwell times (the contact time between adhesive and substrate) also influence the peel value. The dependence of the peel of industrial tapes on dwell time, application pressure,

TABLE 8.7   Debonding Rate as Test and End-Use Parameter

Laboratory Test		Industrial Application	
Test Method	Effect on	Process	Effect on

Peel test	Peel value
	Failure nature
	Removability
		Tape unwinding
		Labeling
		Release
		  formulation
		  (HRA level)
			Inversion
			Skeleton stripping

Release test	Peel value Peel dependence on release age		

and temperature was noted in the early stage of development. For common tapes an increased peel resistance of 1000–1500% occurred after a dwell time of 6 months. A rapid increase in peel (about 400–600%) occurred after 1 day. This is very important, especially for peel tests on difficult surfaces. In some cases (for single-coated tapes), the 180° angle PSTC 1 method is modified to allow 20-min or 24-h contact (i.e., increased dwell time) of the adhesive with the test panel [87]. The time/temperature influence on the peel test was discussed in Refs. [2,41,54]. Druschke [74] examined the different dwell times used by AFERA (10 min), PSTC (1 min), and FINAT (20 min and 24 h) for peel testing purposes.

8.2.1.3.4   *Carrier-Material-Dependent Test*

As discussed in Refs. [41,115], the carrier material strongly influences peel tests. Each carrier material allows a specific adhesion. Specific adhesion is the capability of the adhesive to develop a useful bond level on a particular surface to which it is applied. This is a surface-related characteristic, but the bulk properties of the carrier also influence adhesion. The mechanical characteristics of the carrier affect its dimensional stability. The decisive influence of the plastic deformation of the carrier was demonstrated in Ref. [116]. Deformation plays an important role in special products with a thin plastic carrier (e.g., tamper-evident labels, forms, protective films, etc.). For instance, elongation at break is about 200% (MD) and 500% (CD) for a common 50-μm PE carrier used for protection films. For thinner plastic films, 700–800% elongation was measured. Such pronounced longitudinal deformation causes diminution of the global cross-section (carrier + PSA). For PSPs with very thin carrier material, below the critical thickness, a common test of peel resistance cannot be used because of the carrier elongation, which also causes decreased coating weight to values lower than the critical value (see also *Technology of Pressure-Sensitive Adhesives and Products*, Chapter 1).

As discussed previously, the adhesive properties of PSPs depend on their components and construction. The main components of a classic PSP, the PSA, and the carrier material affect the adhesive properties through their intrinsic properties and their geometry. The main characteristic of PSA geometry is the coating weight; carrier geometry is characterized by thickness. Both must be above a critical value (see also *Technology of Pressure-Sensitive Adhesives and Products*, Chapter 1) to allow "functioning" of a PSP.

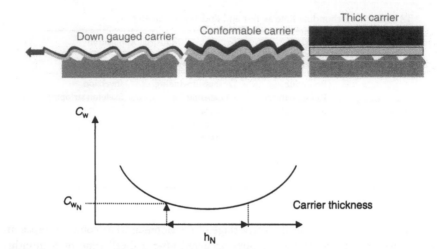

FIGURE 8.6 Schematic presentation of the correlation between carrier thickness (h) and coating weight (C_W).

The critical carrier thickness and the critical coating weight are interrelated, taking into account the role of the carrier conformability and deformability. Figure 8.6 presents a simplified correlation between carrier thickness (h) and coating weight (C_W). Above a nominal carrier thickness (h_N) the coating weight of PSPs increases to compensate for the lack of conformability. Below the nominal carrier thickness the coating weight must also increase to compensate for carrier (and adhesive) deformation.

The examination of commercial PSPs with various carrier thicknesses confirms the hypothesis of Figure 8.6. As illustrated in Figure 8.7, protection films with increased carrier thickness also possess increased coating weight. Increased coating weight imposes higher carrier dimensional stability due to the dependence between the coating weight and the peel resistance.

Figures 8.8 and 8.9 illustrate the dependence of peel resistance on carrier thickness for thin deformable plastic films. Below a critical carrier thickness, the PSP displays no peel resistance, although there is no carrier break.

Generally, a common test of the adhesive is carried out using standard, nondeformable (i.e., without plastic yielding), isotropic materials with controlled surface properties (e.g., PET). In practice, the elongation of the carrier film during de-bonding is the sum of the elongations (for the full carrier length) during and after peel off. Principally, the carrier elongation, which contributes to a supposed decrease in coating thickness, does not include "postelongation."

Peel tests with flexible carrier materials at large angle, with rigid carrier (disc sample) and with semiflexible carrier, were compared by Kendall [77].

8.2.1.3.5 Substrate-Dependent Test

Peel resistance should also be examined as a function of the substrate (adherent). The surface energy of the adherent, the compatibility of the surface with the adhesive, and its pH

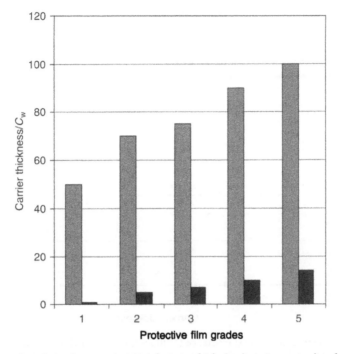

FIGURE 8.7 Correlation between increased carrier thickness (in micrometers) and higher coating weight (coating thickness, in micrometers) for protection films.

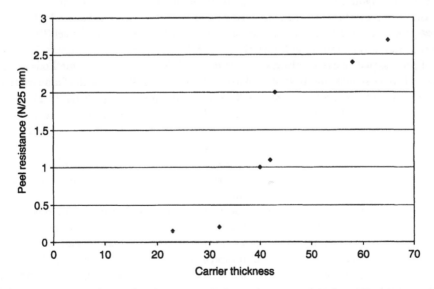

FIGURE 8.8 Dependence of peel resistance (N/25 mm) on carrier thickness (in micrometers). Cross-linked rubber–resin PSA on PE carrier material, 180° peel resistance.

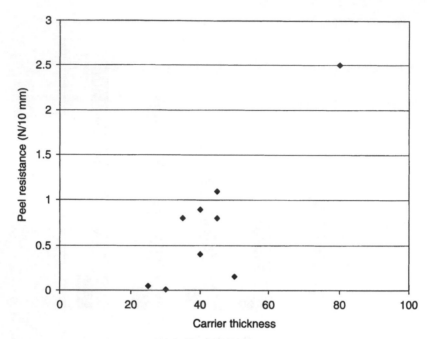

FIGURE 8.9 Dependence of peel resistance (N/10 mm) on carrier thickness (in micrometers). Cross-linked rubber–resin PSA on PE carrier material, 180° peel resistance.

influence adhesion. The strength of the adhesive bond for various substrates was ordered like the corresponding magnitudes of the thermodynamic work of adhesion [117].

The most important factor in measuring bond strength is the test plate, which can be float-process plate glass (FINAT), brushed steel (AFERA), or polished steel (PSTC and BWB). Glass is the most suitable material. The FINAT Test Method 1 proposes the use of float-process plate glass as substrate or test material. All other peel test methods except that of FINAT suggest stainless-steel plates as the adherent [48,74,97,118]. Standard tests of tack and peel are made on steel (PSTC-5 and 1), but measurements on PP, PP, cellulose, PET, PA, and glass are often suggested. Adhesion on different substrates [aluminum glass, polyamide, acrylic sheet, polycarbonate, rigid PE, rigid PP, and polystyrene (Pst)] was tested as well [119]. Peel tests at a peel angle of 180° were performed according to the FINAT FTM 1 conditions by Simal et al. [43] on stainless-steel or HDPE plates. Peel resistance should be evaluated with and without the use of a primer (see also *Technology of Pressure-Sensitive Adhesives and Products*, Chapter 10).

Paper used as a substrate displays a special behavior. Paper possesses an anisotropic structure, it is very resistant to in-plane stresses, and it is very sensitive of out-plane or z-directional stresses. Thus, paper is very susceptible to delamination. Paper delamination is characterized by a high initiation force and a relatively low propagation force. Experimental conditions strongly influence paper delamination [120]. Bikermann and Whitney [121] first reported that paper is much more likely to delaminate in peel if the tape overlaps the edge of the paper. In the case of interfacial failure, the peel curve is noisy but approximately constant, resembling peeling from stainless steel. The peel curve at higher peel rate is more complicated. The peel force initially rises until the peel force applied to the paper surface is high enough to remove fibers, after which the force drops to a low steady-state value, corresponding to catastrophic delamination (paper failure). Yamauchi and coworkers [122] reported peel forces and failure modes as a function of peel rates and paper types.

Peel resistance on cardboard is a special characteristic of PSAs. It is essential that PSAs are evaluated against the final intended surface, namely the National Bureau of Standards reference material 1810 [54]. Because cardboard has a grain that is dependent on the direction of the manufacturing process, this should be determined prior to any testing so that the same direction is used in all measurements. An adhesive tape can cause 100% delamination of the cardboard when stripped in one direction, but may only cause 5–10% fiber tear when stripped in the perpendicular direction. In the cardboard fiber tear test, a strip of tape is applied to the cardboard substrate, affixed with a 2-lb roller, and removed immediately by a pulling force extended on a tape at a 90° angle to the substrate surface. The adhesion properties of a product vary directly with the properties of its surface covered by fiber.

Peel resistance on plasticized PVC film is another characteristic of special PSAs. A PVC film/1 mil PSA layer/siliconized paper laminate is proposed for evaluation of the adhesion of plasticized PVC film [71]. After specified aging times, the siliconized paper is removed and the adhesive-coated PVC film then is adhered to a stainless-steel panel. After a 24-h dwell time the 180° peel resistance is determined according to PSTC Test Method 1.

PE peel tested on a PE plate is used for the high adhesion required on PE. This method tests the 180° peel resistance on PE using a rigid PE surface (plate). The test method is the same as for PE foil but with HDPE plates instead; the cleaning of the plates should be carried out with n-hexane. Adhesion to HDPE is lower than adhesion to LDPE [123]. For acrylics on film face stock material, adhesion performance was measured on stainless steel and HDPE as substrate, according to ASTM D330-90 [54].

There are different bond failure modes.

1. Face failure: a 100% transfer of adhesive to the substrate
2. Clean release failure: 100% adhesive release from the substrate
3. Cohesive failure: the adhesive splits between the face and the substrate

The nature of the failure is very important for removable PSAs. Papers with a different surface treatment yield different peel values for the same adhesive. Clay-coated papers yield paper tear if the superficial strength of the paper is not sufficient. Permanent adhesives should cause destruction of the paper (paper tear); partial destruction is noted as paper strain, which generally occurs at a force of 1500 g/in. [67]. Failure mode depends on the peel rate as well. Interfacial failure between the adhesive and the substrate is the common failure mode at moderate and high peeling rates (e.g., 300 mm/min) [14]. This type of failure has been addressed in a phenomenological way by several authors. The peeling energy (P) required to detach the adhesive from the substrate is proportional to the thermodynamic work of adhesion and depends on the rheologic characteristics of the adhesive (see later). Grammage of paper and paperboard (weight per unit area) used as a carrier material or substrate can be determined according to 1994 TLMI Test II-B. Moisture in pulp, paper, and paperboard can be determined according to 1994 TLMI Test II-C.

8.2.1.3.6 Zip Peel

For high-surface, film-based PSPs (e.g., protection film) instantaneous debonding above a critical debonding force plays an important role [1], taking into account the rapid manual delamination of large areas in industrial practice.

Theoretical peel studies work with constant peel force. However, in practice, in some applications a change of the peel force (i.e., of the peel resistance) is required during debonding. In such applications the low delamination peel allows the fast debonding of the protective layer from large surfaces (e.g., of protective films). Such a variable peel (zip-peel or zipper effect) can be obtained using various modalities. One consists of the use of an incompatible adhesive formulation containing components with low and very high glass transition temperatures. Another possibility is based on the use of a soft, elastic carrier based on a plastic film, which allows the deformation of the carrier material and, thus, energy dissipation. The so-called zipper effect can be due to the extreme deformation of the carrier (see later). Easy to delaminate protective tapes with acrylic adhesive were developed. This adhesive (or carrier) displays the zipper effect; that is, the higher the removal speed, the easier the peel. As a consequence, removing the protective tape will no longer be a time-consuming operation.

For such products, zip peel (sawtooth-like force/time plot) is desired. The test of zip-peel products requires the use of a modified tensile strength equipment with a spring balance (Prohaska device), which displays the resistance of the maximum peel force of the product.

8.2.1.3.7 Special Test Methods for Peel Resistance

Single-pull peel testing does not always predict the joint failure that occurs as a result of cyclic loading at subcritical loading levels [124–126]. Therefore, in Ref. [124] a combined peel test method was proposed. Samples were exposed to a single-pull, 180° peel test at the displacement rate of 250 µm/s. These data were analyzed to determine the load at first yield. Samples were then exposed to subcritical cyclic loading between 20 and 300 mN per cycle at frequencies of 1, 5, and 10 Hz to determine the rate of peel under the cyclic protocol. The results clearly indicate that these adhesive joints, when critically loaded at levels below the load at first yield experiment in the single-pull experiment, can open up at appreciable rates.

Subsequent adhesion is the value of adhesion after making contact with the release liner. From the adhesive performance point of view it is a measure of how much "loose" silicone will be transferred onto the laminated adhesive and what percentage of peel and tack performance is retained (subsequent adhesion) after delamination from the release liner. Therefore, special peel tests are carried out [10].

To evaluate the role of skin surfactants in affecting the mechanical properties of PSAs used in transdermal delivery systems, adhesion was tested using a drum peel configuration on a single-screw-driven mechanical test frame [127].

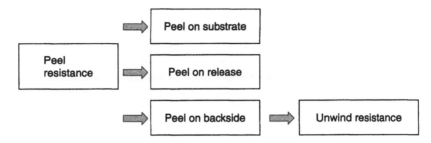

FIGURE 8.10   Various uses of the peel resistance test for PSP characterization.

Generally, metal–plastic film laminates are tested using the so-called roller peel, where the peeling angle is determined by the build-up of the mechanical construction. As stated by Dorn and Wahono [128], the stiffness of the film given by its material characteristics and geometry influences the usability of this method. For stiff plastic film (e.g., PP with a thickness greater than 150 µm) the results are denatured by the change of the peel angle due to the stiffness of the material.

Using the 90° peel test of (a cross-linked) adhesive tape, first the force required to initiate bond breaking (initial breakaway peel) and then the force needed to continue bond breaking, the initial continuing peel, is measured [129]. A pluck test (90° peel) with a slow peel rate (2.5 cm/min) is carried out as well. The initial continuing peel is lower (about 30%) than the breakaway peel.

Peeling off and the test of peel resistance are used for other than classic debonding tests, that is, debonding of the adhesive from a substrate. Figure 8.10 presents the use of peel test (peel from substrate, peel from release liner, and peel from carrier backside) for PSP characterization.

8.2.1.3.8   Special Adhesion Test Methods

The adhesion of a PSA can be tested using a drum-peel configuration on a single-screw-driven mechanical test frame (see above). This test allows a 90° peel resistance test. Such peel resistance testing was accomplished for medical PSAs using a peel test wheel (e.g., TA.XT2i, texture analyzer, by Texture Technologies Corp., Scarsdale, NY) [130].

In practice, the debonding of labels or protective films occurs mainly by peeling off. Tapes and packaging material applied using tapes may suffer other debonding operations as well, where delamination starts at an arbitrary place in the assembly that is not situated at the end of the laminate. Such debonding can be compared to blistering. As discussed above, the most common and convenient method to measure the fracture energy (G_c) is peel testing. A disadvantage of peel testing is the significant plastic deformation that the peel arm undergoes (see *Technology of Pressure-Sensitive Adhesives and Products*, Chapter 1). The extent of plastic deformation is a function of peel angle and peel rate. Therefore, the measured G_c values are said to be geometry dependent. An attractive alternative to peel testing is the pressurized blister. The small angle between the film and substrate minimizes plastic deformation, and the known stress distribution results in measured fracture energy closer to the true G_c. However, the experimental setup is very difficult, and the flow rate is not controlled. The shaft-loaded blister test (SLBT) may be a suitable alternative to either test in that it has both the ease and convenience of peel testing and the plastic deformation characteristic of pressurized blister testing.

The SLBT has a number of advantages over the more conventional peel test because of its easy experimental setup; also, the low angle of deflexion between the peel arm and the substrate reduces plastic deformation [131–134]. The theoretical framework for measuring the fracture energy using the SLBT was developed by Wan and Mai [135]. The SLBT test is experimentally similar to the pressurized blister test [136]. The two critical assumptions utilized for the SLBT are that the film undergoes pure elastic stretching (in other words, no bending occurs) and the load is approximated as a point load. Utilizing the SLBT, the apparent strain energy release rate was measured for a PSA tape bonded to either an aluminum or Teflon substrate [134]. Comparative measurements were carried out using the pull test and 90° peel test. Good agreement was found for the low-energy substrate only.

The probe tack test method using two crossed cylinders with equal radii of curvature was proposed for the test of adhesion [84]. Comparative studies of the changes in surface adhesion of PSAs were carried out with AFM and the spherical indenter test [137]. The forces probed with AFM are on the order of nanoNewtons, whereas the forces probed with the spherical indenter are on the order of milliNewtons. Penetration depths are only several to a few hundred nanometers for AFM, but several micrometers for the spherical indenter.

8.2.1.4 Test of Shear Resistance

The cohesion and the shear resistance, as well as their theoretical basis, were described in detail in Refs. [41,54]. The influence of the viscoelastic properties on the shear was described in Refs. [10,31]. A detailed discussion of the fundamentals of shear resistance is given by Antonov and Kulichikhin in *Fundamentals of Pressure Sensitivity*, Chapter 9. The cohesion of an adhesive may be estimated as its resistance to shear forces. Similar to tack and peel measurements, standard and special methods exist for the evaluation of shear resistance. Various methods based on different test parameters exist for shear testing purposes. These include shear resistance or holding power, shear resistance at a 20° angle, heat distortion temperature, heat resistance, 90° peel, shear resistance versus humidity, etc. A tensile strength measurement may also be used for the evaluation of cohesion [80]. Different methods for measuring shear resistance based on the examination of shear resistance, peel resistance, and tensile strength were described in Ref. [138]. On the other hand, shear measurements may characterize other adhesive properties. Shear tests were used to characterize the mechanical properties of adhesives [139]. The temperature resistance has been measured using 45° shear resistance at 70°C [80].

The most important methods (FINAT FTM 8 and PSTC 7) measure holding power or shear adhesion and record the time to failure at room temperature. Other methods are carried out at 65°C (Shell), 50°C (Ashland), or 70°C (Rohm). Some tests involve temperature measurement (shear adhesion failure temperature [SAFT] and heat distortion temperatures). The temperature is gradually increased and the temperature at bond failure is measured; for SAFT, a 40°F/h gradient is generally used. The slip distance is measured by Wacker Chemie GmbH (70°C, 5 min). Other methods measure the temperature at which a change from adhesive failure to cohesion failure occurs. The test conditions differ as well. Shear should be tested on 5-cm² samples at room temperature or 50°C with a 1-kg load and conditioned for 12 h [140]. Holding power is tested statically according to different norms such as AFERA 4012, TL7510-011, or U.S. Federal Test 20554 (PPT-T-60-D). These examples illustrate that many different shear measurement methods exist, although standardized methods have been defined. Various PSPs require different shear-resistance values. According to Dunckley [67], low-shear values are sufficient for office tapes, double-side tapes, and foam-mounting tapes (see Chapter 4).

8.2.1.4.1   *Static, Room Temperature Shear Resistance*

Shear resistance is determined in accordance with ASTM D 3654-78, ASTM D3654-88, PSTC 7, and FINAT Test Method 8 and is a measure of the cohesiveness (internal strength) of an adhesive. It is based on the time required for a statically loaded PSA sample to separate from a standard flat surface in a direction essentially parallel to the surface to which it was affixed with standard pressure.

Standard Test for Static, Room Temperature Shear   The standard test for static, room temperature shear test is conducted on an adhesive-coated strip applied to a standard stainless-steel panel in a manner such that a 0.5 × 0.5 in. portion of the strip is in fixed contact with the panel, with one end of the strip being free. The steel panel with the coated strip attached is held in a rack such that the panel forms an angle of 178–180° with the free-end tape extended; the latter then is loaded with a force of 500 g applied as a hanging weight from the free end of the test strip. In the test of shear resistance to a fixed load at 178° to the horizontal line, the 2° tilt is used to prevent peel-off. The elapsed time required for each test strip to separate from the test panel is recorded as shear strength. According to FINAT FTM 8 (a 178° shear adhesion test), a 25-mm-wide specimen of PSA film (prepared at least 24 h before running the test) is applied to one end of a stainless-steel plate to obtain a contact area of 25 × 25 mm. Five to 10 minutes after the specimen is applied to the plate, the assembly is suspended from a plate holder, which maintains the plate at an angle of 2° from vertical. A 1-kg weight is immediately attached to the clamp. The time recorder is automatically switched off by the weight as it falls. According to AFERA 4012-P1 (modified), a 25 × 12.5 rnm sample is used in combination with a 1000-g weight.

   As in the case of peel measurements, the influence of the contact time (dwell time) on shear test results should be considered. Shear adhesion testing according to PSTC 7 (1000 g/in.² load) implies a dwell time of 24 h. According to Ref. [141], shear resistance should be measured after a 20-min dwell time using a 25 × 25 mm or 25 × 12.5 mm area with a load of 5 or 10 N. Stainless steel or glass should be used as test substrates. Tests should be carried out at 25°C and 50% RH. Failure time or slip distance should be defined as shear resistance.

Modified Tests for Static Room Temperature Shear Resistance   Besides the standard, normalized shear test methods, there are several modified versions. For instance, in a test method according to Ref. [96], a strip of coated paper is fixed to the edge of a glass plate so that an area of 6.45 cm² (1 in.²) is in contact with the glass. The force required to remove the strip at an angle of 2° is measured at a separation rate of 1.0 mm/min and 20°C.

The influence of the sample dimensions should also be taken into account. Room temperature shear should be measured using an increased contact surface to avoid errors caused by coating defects. For a modified version of FINAT FTM 8, a load of 2 kg and a contact area of 12.5 cm² on aluminum test plates are used. A strip of PSA-coated material (20 × 150 mm) should be affixed with light pressure on an aluminum test plate (previously cleaned with toluene) so that the contact length is 62.5 mm. At least three measurements should be performed.

According to O'Brien et al. [4], holding power measurements were made with a modified PSTC-107, 180° shear adhesion of pressure-sensitive tapes. Instead of the 25 × 25 mm (1 × 1 in.) contact area described in PSTC-107, a 12.5 × 12.5 mm (1/2 × 1/2 in.) area of PSAT was utilized. The pressure-sensitive tape was adhered to stainless-steel coupons with a standard 4.5-lb (2-kg) roller, and a mass of 1.0 kg was suspended from the tape. The time (in hours) until the adhesive debonded entirely from the substrate was recorded up to a maximum of 167 h (10,000 min). Reducing the bonded area, and effectively increasing the shear load on the tape by a factor of 4 is a useful method to reduce the time to failure and to increase the sensitivity for detecting differences in tape performance.

The basic problem with the standard lap shear test is that the average measured shear strength is not a material property that characterizes the adhesive uniquely. Instead, it is a rather vague quantity that also is strongly dependent on the geometry of the joint being tested [142]. For instance, if the overlaps were doubled and all other variables were left unaltered, the strength of the joint would not be doubled. Similarly, if the joint geometry and adhesive are kept constant and the adherent material is changed, the apparent strength of the adhesive will change dramatically. From the point of view of the measurement efficiency, shear tests are more difficult and less efficient than tack or peel measurements.

For adhesive characterization purposes, not only the time but also the nature of the failure also is very important. The mode of bond failure is described by shear tests with different codes such as P, C, PS, and SPS, where P is panel failure, C is cohesive failure, PS is failure with panel staining, and SPS is slight panel staining. Modes of shear failure include the following.

1. Face stock failure: 100% transfer of the adhesive to the substrate (adherent)
2. Clear release failure: 100% adhesive release from the substrate (adherent)
3. Cohesive failure: (zip effect) adhesive splits between the face stock and the substrate; this results as a direct consequence of an imbalance between cohesive and adhesive properties and is an adhesive weakness

Static, normal temperature shear possesses low reproducibility and requires long measurement times. To obtain precise results in a short time, hot shear measurements were introduced (i.e., shear tests are carried out at elevated temperatures; see Section 8.2.1.5.2). Another possibility to accelerate shear tests is the use of the dynamic shear test (see Section 8.2.1.5.3).

8.2.1.4.2 Static Hot-Shear Resistance

This test is a modification of PSTC 7. A 1.27×2.54 cm specimen of the adhesive is mounted on a 7.5×20 cm stainless-steel panel; an aluminum foil is added as reinforcement for the face material. The panel then is positioned with its longitudinal dimension so that the back of the panel forms an angle of 178° with the extended piece of tape, with the 2.54-cm dimension of the adhesive extending in the vertical direction. The assembly then is placed in a 70°C oven and a 1-kg weight is attached to the free end of the tape. The time required for the adhesive to fail cohesively is reported in minutes.

There are many different variations of hot-shear resistance methods. According to Ref. [143], shear resistance at elevated temperatures is measured at 200 and 250°F (2.2 psi). High-temperature shear at 400°F, based on a 5 kg/in. bond and 0-min dwell time, was evaluated according to Ref. [107]. Shear at 70 and 50°C creep is determined as well. The method is similar to PSTC 7, but the weight is reduced from 500 to 250 g and the temperature is increased from 25 to 50°C. Room temperature shear tests (0.5×0.5 cm, 1000 g) and elevated temperature shear tests (with a 4.4-psi load) were also used [21]. To force the cohesive failure rather than substrate or panel failure, the polymer was tested using aluminum foil as substrate after a 72-h dwell time and a 20-psi load. Values of 1500 min to more than 3000 min were recorded. Hot-shear test and the hot-shear gradient allow the evaluation of guillotine cuttability, as noted by Benedek [1].

8.2.1.4.3 Shear Adhesion Failure Temperature: Dead Load Hot Strength Test (SAFT)

A variation of the hot-shear resistance test method can be achieved by mounting the laminate construction in an oven and suspending a weight from the sample in such a way that a vertical shear stress is applied to the bond. The oven temperature may be raised gradually to define the temperature at which the bond fails. This method, the SAFT (also called heat distortion temperature), determines the temperature at which a pressure-sensitive specimen delaminates under a static load in a shear mode. It is a method of determining the resistance to shear of a tape under constant load under rising temperature. When measuring the SAFT, the sample is subjected to a 30-min dwell time at room temperature. The load is attached, the sample conditioned 20 min at 100°F, and the temperature increased 1°F/min until shear failure occurs, up to a maximum of 350°F. The 178°C (modified PSTC 7) test is a measure of the ability of a PSA tape to withstand an elevated temperature rising at 40°F/h under constant force. For instance, for a rubber–resin-based adhesive tackified with a Wingtack resin, a temperature range to failure of 78–87°C was found. HMPSAs based on Kraton D display SAFT values (on Mylar) of 60–90°C [144].

The test assembly is supported by one end in a vertical position with a weight attached to the other end and heated in an oven for periods of 15 min; the temperature begins at 30°C and is raised progressively by 5°C increments after each 15-min period. The SAFT value is the temperature at which bond failure occurs. According to Ref. [145], the adhesive tape is applied to a glass plate (25×25 mm overlap) and a 2-kg roller is passed over the tape, once in each direction. The glass plate is placed in a temperature-programmed oven and clamped at an angle of 2° from vertical. A load of 500 g is hung from the tape and the oven temperature is increased by 4°C/min. The temperature at which the tape drops off the glass plate is recorded. According to Ref. [4], SAFT was measured using a shear tester. A 25 × 25 mm (1 × 1 in.) area of tape was adhered to stainless-steel coupons with a standard 2-kg (4.5-lb) roller. A mass of 1.0 kg was suspended from the tape. Samples were placed in an oven and equilibrated at 40°C. The oven temperature was then increased at 0.5°C/min. The temperature at which the adhesive failed cohesively was recorded.

The validity of the time/temperature superposition principle is demonstrated by the parallel use of dynamic shear resistance tests (controlled forces) and SAFT (controlled temperature). Practically, it is easier and faster to use the dynamic shear resistance test [96].

8.2.1.4.4 Dynamic Shear

This method tests the shear properties of the adhesive in a tensile tester under increasing load (force). Current static-shear test methods use a constant load at longer test times; they demonstrate (related to the nature of the test) poor reproducibility and require very long test times. As demonstrated in Ref. [146], SIS block copolymers display as high values of room temperature shear, and only highly plasticized formulations can be evaluated in a convenient period of time. Moser and Dillard [147] developed a dynamic test method correlated with the static test. For a current dynamic shear test following lamination as well as appropriate conditioning, the samples are placed in a typical tensile strength-testing machine and shear adhesion can be defined by applying shear stress at a given rate (dynamic shear). The test conditions (glass plate, 6.45-cm contact length, rate of 1 in./min at 20°C) are described in Ref. [1].

Other variants of the method exist [96]. Molecular disentanglement can take place at a testing rate of 0.01 in./0.25 mm per minute; thus, a dynamic shear test can be set up on an adhesion tester at this slow speed [148]. Whereas the width of the sample can be of any standard width, the height of the sample must be limited so that it does not become a tensile test of the backing. The maximum shear resistance achieved should be in the area of 2lb/1 kg. This would normally give a height around in 0.125 in./2 mm.

Dynamic lap shear is a variance of the dynamic shear test method. A 3 × 1 in. aluminum/aluminum assembly is used with an adhesive coating weight of 3 mil. The aluminum plates are pulled apart in a 180° configuration at a speed of 1 in./min [149]. A dynamic test method using a machine that has the ability to run extremely slowly is preferred (e.g., 0.02 in./min) [150]. A new dynamic method was developed with a testing speed of 2.5–100 mm/min (as a preliminary test) and 50 mm/min common testing speed, laminating pressure of 154 kPa, and temperatures of 23, 40, and 70°C (allowing a rapid assessment of holding power) [151]. A strip of PSA-coated material is adhered by its adhesive layer to a primed PE substrate with an adhesive contact bond area of 1 × 0.5 in. When testing is carried out at an elevated temperature, this substrate is first reinforced with aluminum foil to impart rigidity.

8.2.1.4.5 Special Test Methods for Shear Resistance

Some special test methods for shear resistance are based mostly on the modified standard test of shear resistance, such as the 20° hold test, the automotive shear resistance test, and the thick adherend shear resistance test.

The 20° hold test is similar to a standard shear resistance test except the test plate is inclined 20° from the vertical direction. This test measures a combined peel and shear

TABLE 8.8 Shear Resistance Test Conditions

Test Temperature (°C)		Substrate	Weight (g)	Sample Dimensions (mm × mm)	Norm	Note
RT		Glass	1000	25.0 × 12.5	AFERA 4012-P1	—
RT		—	1000	12.5 × 12.5	—	—
RT		Stainless steel	1000	25.0 × 25.0	—	—
RT		Stainless steel	500	12.5 × 12.5	PSTC 7	—
RT		—	500	12.5 × 12.5	PSTC 7	Modified, 24 h dwell time
RT		Glass	—	645 mm²	—	Dynamic, test rate 1 mm/min
RT		Stainless steel	500	12.5 × 12.5	ASTM D-3654-78	—
RT		Stainless steel	10N	25.0 × 25.0	FTM 8	—
RT		Stainless steel	1000	12.5 × 12.5	PSTC-107	—
RT		Aluminum	2000	20.0 × 62.5	—	—
—	40	Stainless steel	1000	25.0 × 25.0	FTM 8	—
	50	Stainless steel	250	12.5 × 12.5	PSTC 7	Modified
—	65	—	1000	—	—	—
—	70	Stainless steel	1000	25.4 × 12.7	PSTC 1	Modified
	70	Stainless steel	1000	12.7 × 25.4	PSTC 7	Modified
—	70–90	—	—	25.0 × 12.5	—	—

resistance of the adhesive mounted on a 1-mil Mylar film when applied under standard forces to a corrugated cardboard substrate. Shear resistance from cardboard, according to PSTC-7, is tested with a 200-lb bursting-strength cardboard. The tape is applied in the corrugation direction. With regard to carton-sealing properties, cardboard shear, cardboard adhesion (percentage of fiber tear), and high-humidity aging (80° RH/125°F, 3 days) are evaluated. The test method is described by Benedek in Ref. [1].

The automotive PSAs shear resistance test measures slip after a given time using a 500 g/in.² test sample surface at 158°F, according to the Fisher body procedure TM-45-134 [1].

The thick adherend shear test (TAST) uses a thick adhesive sample (6 mm, with an overlap length of 5 mm), and the shear test is conducted using a tensile testing machine. However, finite element analysis indicates that the stress distribution in the TAST is not pure shear. The values of maximum shear strain gained from the TAST are not comparable to values obtained from tests in pure shear [152]. Table 8.8 summarizes the experimental conditions used in shear measurements. The use of the measured shear values must be made in connection with the test conditions, the adhesive formulation, and the other adhesive characteristics.

Some years ago, tack was discussed as a cohesion-related property. Later, shear resistance was considered a cohesion- and cuttability-related characteristic. Recent developments in thermoplastic elastomers (TPEs) demonstrate that such generalized assumptions are not valid. The use of the shear resistance alone as an index of the cohesion or as an index of the nonelastic component of cohesion is incorrect. On the other hand, the use of cohesion as an index of tack is generally not possible. This statement is illustrated by recent advances in the use of the probe tack test method.

The shear test demonstrates the following disadvantages: the information obtained remains limited, the reproducibility is low, and the time required is too long. More reliable results are given by creep tests (measuring a creep time slope to determine the viscous and elastic components) or by measuring the creep (cold flow) directly (see also Section 8.1.3.1.3).

Shear creep tests are recommended for special soft products. The apparatus used for shear creep measurements suggested in Ref. [48] is similar to that described by van Holde and Williams [153]. The applied shear stress was 3.5×10^4 dyn/cm. The creep displacement was measured with a capacitance probe transducer for a period of 100 min following the application of the load [127]. According to Hu and Paul (see *Technology of Pressure-Sensitive Adhesives and Products*, Chapter 3), a controlled stress creep test was designed to mimic the flow behavior of adhesives under a representative shear stress. In this test, a constant stress of 2 psi was first applied to the adhesive for a period of time and the resulting strain was recorded. The stress was then removed and the elastic part of the adhesive began to remove the imposed strain. The strain recovery was recorded.

8.2.1.5   Global Evaluation of Adhesive Properties

A complete evaluation of the adhesive properties is possible after measuring all performance characteristics (i.e., tack, peel resistance, and shear resistance). Different graphical representations allow the comparison of the above properties as a function of the formulation and in relation to one another. The main types of diagrams used for evaluation of the adhesive properties are described by Benedek in Ref. [1]. The adhesive properties can be evaluated separately (one property) for different formulations or different properties can be evaluated for a given formulation. An overview of the performance characteristics for different chemical compositions or for a given formulation is also possible. The properties can be plotted separately as a function of the chemical composition and a base characteristic of the main components in a three- or two-dimensional graph. A combined method (an assembly of different test methods) was proposed for the evaluation of different adhesive characteristics of HMPSAs [9]. According to the proposal, peel adhesion should be tested according to PSTC 1, shear according to PSTC 7, and tack according to PSTC 6 (rolling ball), Polyken, or quick-stick tests.

The common test methods of adhesion (tack, peel resistance, and shear resistance measurement) are used for adhesive-coated and adhesiveless PSPs. The principles and practice for PSAs were described earlier. Such methods use standard carrier materials. The adhesive properties of the finished product may differ from the characteristics of standard samples due to the influence of the carrier material and to manufacturing conditions. On the other hand, in some PSPs the adhesive is not coated but imbedded in the finished product, or the carrier itself has an adhesive nature. For other products the adhesive is part of a composite structure with some of the characteristics of a carrier (from the point of view of mechanical resistance). For such products (see also Chapters 1 and 7) the general test methods will substantially differ from the common methods used for PSAs [1].

8.2.1.6 Product-Related Adhesion Tests

Special characteristics are tested for various special products. For instance, for splicing tapes (see also Chapter 4) the peel resistance and shear resistance are measured. Peel resistance for splicing tapes is tested as standard peel resistance on stainless steel and as peel resistance on siliconized liner [154].

The shear resistance is tested as dynamic, static, and practice-related shear. Both static and dynamic shear measurements are carried out at room and elevated temperatures. Static and dynamic shear measurements are carried out according to modified standard methods with paper as substrate. Practice-related shear testing uses a heated cylinder.

For packaging tapes based on HMPSA, adhesion on cardboard is of special importance [155]. The preferred adhesion value is 100 min (shear test). According to Ref. [156], for such tapes one of the most important properties is open-face aging. For packaging seal tapes, falling and flaptest (a combination of shear, peel adhesion, and tack) must be carried out [157].

For insulation tapes (see also Chapter 4), mechanical, electrical, and thermal properties are required. Elongation, shear resistance, and peel resistance are needed as well. For such tapes shear resistance should be measured after solvent exposure.

A bending test is used to evaluate UV light-cured PSAs [158].

A skin adhesion test is carried out for medical tapes. The skin underlying the tape sample is inspected visually to evaluate the amount of adhesive residue left on the surface of the skin. The samples are assigned a numeric rating from 0 to 5.

The following evaluation criteria are used in removability testing: A, percentage of adhesive transfer (less than 5%); L, legging; SL, slight legging. The examples illustrate the specific product-related character of the practical tests. A detailed discussion is given in Ref. [1].

8.2.1.7 Other Test Methods for Adhesive Properties

Other test methods exist for the characterization of adhesive properties. Some are classic procedures from the adhesive-limitrophe technology (e.g., Williams plasticity) and others use the new methods of rheologic characterization (e.g., DMA).

8.2.1.7.1 Test of Williams Plasticity

The Williams plasticity number (PN) indicates the deformability of the adhesive mass under a static load. Williams plasticity recently regained favor for PSA testing purposes. It was used extensively in the 1950s to follow molecular weight changes in reactions during polymer production, but was replaced by Mooney viscosity, which is normally faster and more convenient to measure. The PN for rubber is determined according to ISO 7323 [159].

In this test method the thickness of the sample (in millimeters) is measured while the sample is subjected to a constant load (5 kg) at 100°F (38–37.8°C) [62]. The Williams plasticity test is used for screening HMPSAs [160]; it is run at 100°F for 15 min [161]. For this test the (ball-shape) sample coated on a release liner is weighed (to 2.0 g) and conditioned for 15 min at test temperature, the load is applied, and the plasticity is measured 10 min

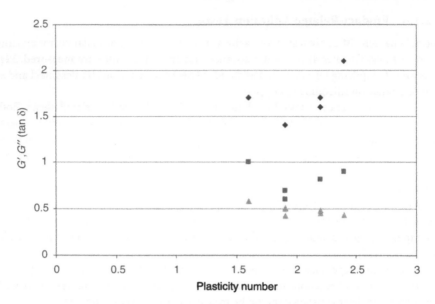

FIGURE 8.11　Interdependence between Williams plasticity number (PN) and G', G'' and tan δ (♦, G'; ■, G''; ▲, tan δ).

after the application of the load [26]. In some cases Williams plasticity is evaluated at different temperatures (38–200°F). The interdependence of Williams PN and modulus values (i.e., storage modulus, loss modulus, and tan δ) is illustrated in Figure 8.11.

The apparatus and the test method are described in Ref. [1]. The Williams PN is the thickness of the pellet in millimeters after 14 min of compression at 37.8°C in the plastometer under a 5-kg load.

This method yields valuable information about dried (coated) materials. Some companies use this method; others consider it inadequate for water-based PSA dispersions.

8.2.1.7.2　Rheologic Evaluation of Adhesive Properties

The interdependence of the adhesive properties with rheologic characteristics was discussed in detail by Benedek in Refs. [31,32]. Interfacial and rheologic processes as the main features of pressure sensitivity are discussed in *Fundamentals of Pressure Sensitivity*, Chapter 2. The viscoelastic behavior of PSAs in the course of the bonding and debonding processes is described in *Fundamentals of Pressure Sensitivity*, Chapter 5. The viscoelastic windows of PSAs are discussed in *Fundamentals of Pressure Sensitivity*, Chapter 6. A complex cross-linking system for an optically clear PSA used as a mounting aid in electronics that includes various mechanisms is described in Chapter 3. Using different chemistries and curing processes, an optically clear adhesive was developed for the required complex transition from a removable PSA to a permanent PSA and finally a structural adhesive for this application. The first stage of product manufacture utilized ionic cross-linking of the acrylic polymer during the coating and drying phases. The second stage used UV curing for cross-linking the urethane oligomer, which forms an interpenetrating network, and the third stage used thermal curing to completely cross-link the acrylic polymer. Such progressive transformation to different types of adhesives is monitored and confirmed by the shift in viscoelastic window.

Developments in the equipment available for the testing of rheologic properties allowed the use of rheologic characteristics for the screening of PSA formulations. The main parameters investigated are the viscosity, the T_g, and the modulus. For their study viscosimetry, DSC, and DMA are used.

DMA is a method used to study the mechanical characteristics of polymers (and other materials) and was applied for rheologic studies (i.e., for the flow behavior of adhesives) [162–164]. For over a decade DMA was used to measure, explain, and forecast pressure sensitivity on the basis of the rheologic characteristics of the polymers, taking into account the time–temperature dependence of the material characteristics (modulus). A number of parameters were obtained from the rheology experiments to predict structure–property relationships and, ultimately, PSA and PSP performance. The parameters include the first tan δ maximum temperature, which is a measure of the T_g of the rubbery matrix. In addition, the tan δ maximum peak height indicates how much energy the adhesive can dissipate [165]. The storage modulus G' at room temperature (25°C) was also noted to quantify how compliant the adhesive was. The tan δ minimum (above the T_g of the rubbery phase) was also measured to characterize the cohesive strength of the adhesive. Sample geometry introduces an error in both G' and G'', but the error is divided out of tan δ. As a consequence, the tan δ minimum may be better than either G' or G'' in some cases to measure cohesive strength. Finally, the third cross-over temperature was measured. For SBCs the third cross-over temperature (tan δ $= 1_{max}$) is the high temperature near the T_g of the glassy Pst domains at which the storage and loss moduli are equal. This is the temperature at which the adhesive begins to flow and loses cohesive strength.

According to Ref. [166], the most useful signals in PSA analysis are G' and G''. G' is a direct measurement of the tack strength, and G'' is a direct measurement of the peel strength. The controlled stress rheometer used for such measurements is a cone-and-plate parallel plate design, in which the material is sheared between a rotating geometry and a stationary base plate (see also *Fundamentals of Pressure Sensitivity*, Chapter 4). The main limitation of most commercial rheometers lies in the frequency range. Frequencies above 100 Hz are often hard to achieve and frequencies below 0.1 Ha require a significant time investment. Such tests use time–temperature superpositioning. The underlying basis for time–temperature superpositioning is that the process involved in molecular relaxation or rearrangements in viscoelastic materials occurs at accelerated rates and higher temperatures and there is a direct equivalency between time (the frequency of the measurement) and temperature. Once the master curve for the PSA has been generated, it is possible to analyze the data by creation of a performance grid. Several parameters for quantitative significance must be assigned to the performance grid. It is assumed that the application of the adhesive (tack test) occurs at an angular frequency near 0.1 rad/s. It is further assumed that the peel test approximates the higher frequency of 10 rad/s. The G' range of 50,000 to 200,000 Pa is accepted as ideal PSA performance.

This method presents the following main disadvantages: it is limited to the adhesive (see also *Technology of Pressure-Sensitive Adhesives and Products*, Chapter 1), it describes the behavior of "pure" polymers without micromolecular additives, and it is limited to the phenomena that occur in the bulky polymer. In the classic domain of adhesive-based PSPs the actual limits of a pure rheologic approach are illustrated by the test and application of cross-linked products. Natural rubber (NR) is a cross-linked product; however, its elastic properties are due mainly to chain entangling, and its cross-linked network is very elastic. Advances in macromolecular chemistry allowed the synthesis of other cross-linked polymers with rubbery elasticity. The compounds with physical cross-linking (thermoplastic elastomers) are hot processible (thermoplastic elastomers) and 100% solid materials. The compounds with chemical cross-linking (e.g., carboxylated styrene–diene copolymers) are supplied as water-based dispersions. Both are relatively hard materials; the SBC are elastomers without self-adhesivity and the carboxylated styrene–diene copolymers are products with a low level of self-adhesivity (they require a high concentration of tackifier) due to the bulky styrene monomer and the tightly cross-linked network. Such polymers behave like filled systems (see also *Technology of Pressure-Sensitive Adhesives and Products*, Chapter 1). They cannot be characterized by common test methods of adhesive properties or rheologic parameters. As discussed previously (Section 8.2.1.2.1), discrepancies appear through the use of the probe tack test method for SBCs because of the stiffness (rigidity) of the adhesive, which is the result of the special build-up of the network. Principally, the same problem, that is, the insufficient control of the length and distribution of the network bridges, can hinder the characterization and use of carboxylated styrene–diene copolymers. The T_g and sol/gel content alone do not describe the practical behavior (peel resistance and shear resistance) of the adhesive.

For formulations of pure macromolecular adhesives, DMA helps in the screening of possible recipes. DMA has numerous other uses as well. For instance, the mobility of the mid-block in TPEs can be evaluated by DMA. It is related to the softness, viscous flow (i.e., the loss modulus), the situation of the rubbery plateau modulus, and the ratio of the two moduli (storage and loss modulus), that is, tan δ. As discussed previously, the situation of tan δ related to temperature, the tan δ minimum temperature ($T_{\delta min}$), and the tan δ peak temperature ($T_{\delta max}$) characterize the rheologic behavior. The melting and disappearing of the end-block domains is indicated by the $T_{\delta min}$ and the cross-over temperature of the loss and storage modulus (T_{cross}). Between $T_{\delta min}$ and T_{cross}, the end domains soften and disappear. The difference between T_{cross} gives information about the melt viscosity of the formulation. In tests of release properties, the general shape of the release profile is consistent with the shape of the adhesive's viscoelastic function (specifically the loss tangent (tan δ) as a function of the dynamic frequency in the peel resistance window [111]. The ideal application range of HMPSA can be well estimated from the DMA and tack tests [167]. Ionic bonding in carboxylated SBR with ZnO is detectable by means of measuring the temperature dependence of tan δ [168].

Advances in DMA examining the whole product (adhesive and carrier) have been made as well. Willenbacher [169] used torsional resonance oscillation to achieve the moduli G' and G'' of the sample (PSA tape); two maxima are observed corresponding to the adhesive and the carrier. As discussed previously (see also Section 8.7.1), Williams plasticity is related to the modulus values as well.

A complementary characterization for PSAs is allowed by melt rheology. The study of viscosity and the storage modulus allows better characterization of HMPSAs.

A detailed discussion of the DMA data for adhesive formulation is given in Ref. [170]. As discussed by Muny [171], the cost of this equipment is still quite high, which limits its use to larger organizations. Vogt and Geiß [172] investigated various methods such as relaxation and creep tests to examine the material properties, creating basic data for design and engineering of adhesive-bonded joints. They suggested that dimensioning based on relaxation tests instead of creep tests in some cases may be a faster way to obtain sufficient long-term design. A combination of relaxation and creep data from DMA is proposed. The use of DMA for the study of high release agents is discussed in *Technology of Pressure-Sensitive Adhesives and Products*, Chapter 9.

A considerable amount of research has been devoted to the correlation of the linear viscoelastic properties obtained by DMA with the usual PSA parameters such as tack, peel, and shear strength. Much discussion exists in the literature regarding the storage modulus, G', and loss modulus, G'', which describe the viscoelastic behavior of PSAs. The two parameters are particularly useful for describing whether an adhesive meets the Dahlquist criteria ($G' \leq 10^5$ Pa) and whether the compound lies inside the viscoelastic window for PSAs. From DMA the empirical criterion of Dahlquist can be verified, which provides the requirement for proper bond formation. The tensile elastic modulus E' ($\approx 3\ G'$) should be lower than 0.1 MPa in a timescale of 1 s, which is typical for the bonding process.

However, O'Brien et al. [4] caution the reader that rheology should only be used as a screening tool to determine whether the adhesive components are compatible and whether the adhesive is soft enough to be a possible PSA (see Dahlquist's criterion). O'Brien et al. recommend that the reader exercise caution when making conclusions from DMA, because numerous examples exist in which an adhesive with "good" rheologic properties exhibits poor adhesive performance, often due to surface migration of an immiscible component or because the coated adhesive layer is too thin. As discussed in *Technology of Pressure-Sensitive Adhesives and Products*, Chapter 3, the maximum strain measured by most temperature sweep tests for hot-melt adhesives is generally <30%. In many real applications, the deformation of the adhesive may be well beyond the maximum strain limit. Normal peeling of an adhesive is one such situation. Measurement of the deformation at the peel front demonstrated that deformations are common several hundred percent and often up to 1000% (10 times). When cohesive failure occurs, deformations can be even larger. In such cases, controlled strain rheology tests are not sufficient to capture the full profile of the adhesive characteristics.

Deviations from the "rule" of Dahlquist's criterion exist as well. For instance, as discussed in Ref. [173], polyisobutene (PIB), a base polymer for PSAs, does not fulfill the Dahlquist criterion and does not demonstrate the fibrillation phenomenon typical of PSAs. Tackified radial acrylic copolymers synthesized by Simal et al. [43] do not fulfill Dahlquist's criterion. Such untackified copolymers detach from the substrate by lateral crack propagation and coalescence, similar to a highly cross-linked adhesive. There is no formation of foam and fibrils.

Generally, the storage and loss moduli are determined by DMA using a rheometer with parallel plate geometry for the oscillation measurements in the angular frequency range of 0.1–100 s^{-1}. Rheology data are very repeatable and the uncertainty in the glass transition is approximately ±0.5°C [4].

8.2.2 Test of End-Use Performance Characteristics

The adhesive properties of PSAs and PSPs are their most important end-use characteristics. Their manufacture (see also *Technology of Pressure-Sensitive Adhesives and Products*, Chapter 10) and application (see also Chapter 4) require adequate converting properties. For PSAs the converting properties are their end-use characteristics. Numerous tests are used for the evaluation of special products. The main special products are described in Chapter 4. They include removable PSPs, water-soluble/water-resistant, high- or low-temperature-resistant, flame-proof, conductive, biocompatible products, etc. Their end use strongly depends on their common and special properties. This chapter discusses the most important end-use properties of PSPs.

As demonstrated in Chapters 1 and 4, PSAs and PSPs can be classified as permanent or removable products. The principles of removability and formulation for removability are discussed in *Technology of Pressure-Sensitive Adhesives and Products*, Chapter 8.

8.2.2.1 Test of Removability

Removability was discussed in detail in Refs. [32,41]. As noted in Ref. [174], the most important special features of removable PSA labels are the removability, the migration/bleeding behavior, and the edge-lifting properties (see also Section 8.1.2.2). Removability is a complex PSP characteristic that includes the nature of the bond rupture and the force level required for debonding. Both are time–temperature functions because of their dependence on the polymer rheology, which means that instantaneous removability as well as time-dependent removability must be examined. Removable and repositionable products are required.

Various grades of removable products exist. In the automotive industry special removable tapes must display very high removability. Such selected tapes were removed cleanly from enameled, lacquered, or melamine test surfaces after 1 h at 150°C or 30 min at 121°C. For common PSA labels the experimental conditions are not so severe, but generally instantaneous removability and removability after aging (room temperature and high-temperature aging) are required. Removable adhesives can be formulated so as not to display adhesion build-up (i.e., the adhesion to the substrate does not increase to the point at which the label cannot be removed clearly even after exposure to heat; see also *Technology of Pressure-Sensitive Adhesives and Products*, Chapter 8). The most important tests covering removability are as follows.

- Removability from different surfaces after aging at elevated temperatures over a period of time, evaluated subjectively
- Removability from glass after aging at elevated temperatures, evaluated as peel resistance from glass
- Mirror test, residue-free removability from mirror glass, evaluated subjectively

Removability depends on the physical state of the adhesive as well. The various additives in a formulation may negatively influence its removability.

Principally, the test of removability is carried out by testing the peel resistance. Ideally, such a test leads to a low level of peel resistance and bond rupture occurs at the interface adhesive/substrate, that is, no adhesive residue is left on the substrate surface.

In practice, empirical peel resistance level is given as a standard value, and optical examination of the substrate surface ensures the evaluation of removability. Although in a first approximation such methods are simple, their efficacy depends on the rate of the test force and the adequate choice of dwell conditions.

Czech [174] described a very subjective test for removability. According to this test, a 2.5-cm-wide strip of the sheet coated with the PSA to be tested is applied to the horizontal surface of polyester-treated polyester or other test substrate, with at least 25-cm linear in firm contact. Three passes in each direction with a 2-kg hard rubber roller are used to apply the strip. After 1 day or 1 week of dwell time, the free end of the coated strip is doubled back, nearly touching itself, so the angle of removal will be about 135 grades. The free end is then pulled by hand at a variety of peel rates. The removability and peel force are judged according to the following ratings and recorded. *Good samples* are removed from the test substrate without damaging or leaving residue on the test substrate. They also exhibit high peel force and yet do not damage the paper backing over a range or peel rates. *Aggressive samples* are removed from the test substrate without damaging or leaving residue on the test substrate, but can only be removed from the test substrate at a slow peel rate without damaging the substrate backing. *Raspy samples* are removed from the substrate without damaging or leaving residue on the test substrate, but are too stiff to be removed smoothly. *Tear samples* display peel adhesion to the test substrate that is too high, causing the test substrate and other substrate backings to tear or delaminate at any peel rate. *Ghost samples* leave a very thin, adhesive residue on the test substrate when removed from the polyester and treated polyester samples. *Weak samples* have low tack and low peel adhesion. *Cohesive failure* samples leave adhesive residue on both the substrate backing and the test substrate.

8.2.2.1.1   Test of Removability of HMPSAs

According to Ref. [175], removability of hot melts is tested by aging stainless-steel panels with 1×6 in. strips of label stock at room temperature and elevated temperature (48°C) and cooling for 1 h at room temperature, followed by peel adhesion measurements of the label stock. Other test cycles include 24 h at room temperature, 24 h at 48°C, 1 week at room temperature, and 1 week at 48°C. For HMPSAs the initial peel from stainless steel should be below 1.0 lb or 16 oz/in. The resulting peel (after build-up) should be below 2.5 lb or 40 oz [176]. Tack values for removable HMPSAs (90° quick stick) are 0.5–1.3 lb/in., RBT values are 1–3 in., and peel resistance values are 0.9–2.6 lb/in.; the SAFT amounts to 128–156°F [166].

Not only the peel value but also the failure mode is important in the removability test of PSAs. Therefore, the amount of adhesive transfer for peelable tapes or labels is evaluated after storage at 40°C for 7 days [66].

8.2.2.1.2   Test of Removability of Water-Based PSAs

The evaluation of removable, water-based acrylic PSAs is carried out after 1 week of storage at 70°C on PET and HDPE [141].

8.2.2.1.3   Test of Adhesive Residue

When the removability test (described earlier) is performed, the surface (substrate) underneath the label sample is usually inspected to determine the amount of adhesive residue left on the surface of the adherent. Different subjective rating scales exist. In some cases each sample is assigned a numerical rating from 0 to 5 based on an arbitrary scale. In other cases, adhesive transfer is estimated with A, L, and SL ratings, where A is adhesive transfer, L corresponds to legging, and SL denotes slight legging or stringing. High-temperature resistance may be tested as the percentage of adhesive transfer after 30 min at 250°F [177]. Removability should be evaluated on glossy, stain-resistant acrylic enamel after aging at elevated temperatures and peeling at an angle of 45°. The test conditions are described in detail by Benedek in Ref. [1].

8.2.2.1.4　*Repositionability*

With repositionable adhesives the adhesive-coated paper may be readily lifted and removed from the contact surface [178].

8.2.2.1.5　*Readherability*

Readherability allows reapplication, that is, repeated bonding–debonding of a PSP. According to Ref. [138], in the case of readherability the adhesive-coated paper substrate may be readily lifted and removed from the contact surface and can be reapplied at least eight additional times to the paper adherend surface. Measurements performed with a 7 g/m² adhesive coating weight on a 70 g/m² paper as face stock material yield the following peel adhesion values on newspaper (test conditions: 180° peel, 300 mm/min): 110 g initially, 85 g after 50 repeated peel tests, and 75 g after 100 repeated peel tests. According to Ref. [179], the readherability of the PSA on paper has been tested by measuring the peel after repeated lamination and delamination (50–100 steps).

8.2.2.1.6　*Lifting (Mandrel Hold)*

Different lifting tests are carried out to determine whether the peel resistance/shear resistance of the adhesive balances the lifting forces of the face stock (deformation of the face stock material on a curved surface). In a test called the curved panel lifting test at 150°C, an aluminum panel with a radius of curvature of 23 cm and a length of 35.5 cm in the curved direction is used. The tapes to be tested are applied to the aluminum panel in its curved direction. The assembly-bearing panel is then put into an air-circulating oven at 150°C for 10 min, allowed to cool, and examined for failures. A rating of "pass" means no lifting has occurred; any lifting at either end of the strip is noted as "end failure" and the total length of tape that has lifted is rated.

This method should be used as a practical test for peel resistance of removable PSAs. Peel resistance is subjectively evaluated on the basis of lifting or flagging of the sample affixed to round, cylindrical items (e.g., H-PVC and PE cylinders with a diameter of 21 mm and glass cylinders with a diameter of 16 mm). The lifting (delaminated) length of the samples is measured at regular intervals (i.e., 1 day, 1 week). The test is described in Ref. [1]. Lifting (winging) is also evaluated after 5 h on vertical surfaces (see also Section 8.1.1.2.3).

8.2.2.2　Test of Water Resistance/Solubility

The formulation of water resistance/removability is discussed in *Technology of Pressure-Sensitive Adhesives and Products*, Chapter 8. Water absorbance as an important parameter of water resistance is described in *Technology of Pressure-Sensitive Adhesives and Products*, Chapter 7. Water-resistant/soluble PSPs possess a wide range of various applications (see also Chapter 4.). Water resistance is generally tested to characterize the water resistance/insolubility of a PSA laminate. Water resistance/insolubility is required when using labels under humid or wet conditions (see also Chapter 4). Classic specifications for pressure-sensitive raw materials contained the value of water absorbancy after storage in water (e.g., 24-h storage, according to DIN 53495, for Acronal 500 D, BASF, gives ca. 70% water absorbancy). This test can be carried out for storage in a humid atmosphere according to DIN 53473. Water resistance can be examined as the change in adhesive properties under the influence of water (or water vapor) or as the change in optical and dimensional characteristics of the label (tape) under the influence of water. With regard to water resistance, paper and film labels should be included in the evaluation; film-based products require special water resistance as well. In applications such as solar and safety window films, graphics, point-of-purchase advertising, and clear labels, the films may be applied to the glass by first wetting the adhesive and glass surface to allow positioning. The water is then removed from between the adhesive and the glass with pressure. During this wet-application step the adhesive must resist hazing and rapidly become optically transparent after removal of the water.

Table 8.9 summarizes the various grades of water-resistant/water-soluble products as a function of the test methods used to evaluate their water resistance.

Water solubility is required for wash-off labels, mainly those based on paper face stock. Water-releasable labels and water-dispersible labels exist. The interdependence of the parameters characterizing the water-resistant performance of a PSA or PSA laminate was discussed by Benedek in Ref. [180]. For certain applications a high water resistance (i.e., no influence of water on the PSA laminate) is required. For others, a strong influence of water on the laminate is desired. Water resistance must be differentiated from humidity resistance. Water-resistant products are special laminates; humidity resistance is generally required for most laminates. The evaluation of water resistance, measured as the stability (or change) of the adhesive properties under the influence of humidity, is mainly based on the examination of peel adhesion, tack, or shear after storage in a humid environment (see also Section 8.2.2.2.2).

Water resistance may be proved using different test methods. Generally, water resistance is measured indirectly by optical examination of the product and by evaluation of the bond strength after storage under wet conditions. The water resistance of conventional adhesive films is measured by the elution rate and transparency of the film when immersed in water. Water resistance can be evaluated by storing the laminates under water and removing them 24 h prior to testing to allow recovery of the adhesion. According to Ref. [27], resistance to water or solvents is tested by immersion in liquid for periods up to 1 week. Water resistance was evaluated immediately after 24-h water immersion or after a 7-day water immersion, followed by a 24-h recovery period [181]. Another test measured water resistance after a 24-h immersion; the film was bonded to stainless steel for 30 min and then placed in room temperature tap water, followed by a 180° peel measurement 24 h later [26]. In the Weyerhouser (shear) test, a laminated kraft paper strip (coated with PSAs) is immersed in water with a 350-g load attached; the time to failure is measured. Water-soluble HMPSAs for labels are formulated with 35–60% vinyl pyrrolidone or VA–vinyl

TABLE 8.9   Test Methods for Water Resistance/Solubility

Method	Test for	Evaluation Criterion	Test Parameters* Temperature	Test Parameters* Time	Formulation Parameters	PSP
Water whitening	Water resistance	Opacity	—	●	Nonpolymeric additives	Label, decal
Water absorbtion	Water resistance	Weight difference	—	●	Base polymer and nonpolymeric additives	PSA, carrier
Wet anchorage	Water resistance	Wet delamination	—	●	Base polymer and tackifier, carrier	Bottle label, medical tape
Wet adhesion on glass	Water resistance	Wet tack, wet peel	—	●	Base polymer and tackifier	Medical tape
Wet adhesion on polyethylene	Water resistance	Wet tack, wet peel	—	●	Base polymer and tackifier	Medical tape
Humidity resistance	Water resistance	Adhesive properties	●	●	Base polymer and nonpolymeric additives	Common PSPs
Washing machine resistance	Water resistance	Integrity	●	●	Base polymer and nonpolymeric additives	Special PSPs
Solubility in water	Water-solubility	Dissolution rate	●	●	Base polymer and nonpolymeric additives (surfactants, thickener, filler solubilizer)	Special PSPs
Dispersibility in water	Water-solubility	Dispersing rate	●	●	Base polymer and nonpolymeric additives (surfactants, filler, thickener, solubilizer); carrier	Splicing tape Medical tape
Water-removability	Water-solubility	Delamination rate (wet)	●	●	Nonpolymeric additives	Splicing tape Bottle label
Wash-off test	Water-solubility	Dissolution rate	●	●	Base polymer and nonpolymeric additives	Bottle label
						Bottle label

* For wet conditions.

pyrrolidone copolymer together with free fatty acids. Such labels can be peeled off in 5 min by subjecting the edge of the label to a stream of water at 35°C. Water dispersibility, water solubility, water activation, and solubility with alkalies can also be tested.

For plastic films, water absorbancy is related to moisture vapor transmission (MVT). As discussed in Chapter 4, MVT is required for medical applications. According to ASTM E 96-66 T Proc. E or DIN 53122, MVT can be tested gravimetrically. The water vapor transmission in 24 h at 20, 23, 25, 38, and 40°C is measured.

8.2.2.2.1 *Test of Water Whitening*

A PSA-coated film can suffer water whitening upon immersion in water. Water whitening is characterized by the speed of the loss of transparency and by the absolute value of the loss of transparency compared to the transparency of a dry coating. The test of water whitening is imposed by the application conditions of certain large-surface, film-based PSPs. These films may be applied to the glass by first wetting the adhesive and glass surface to allow positioning. The water is then removed from between the adhesive and the glass using pressure. During this wet-application step the adhesive must resist hazing and rapidly become optically transparent after removal of the water.

Common water-whitening tests measure the change in optical appearance [180]. According to this test, a clear Mylar label attached to stainless steel and immersed in water at 74°F for 48 h demonstrates neither whitening nor blushing. There is an obvious objection to water whitening as a measure of water resistance. Highly water-sensitive materials like gelatin or soluble cellulosic films are not whitened at all. However, in water-insoluble and slightly swollen latex films, the degree and speed of whitening are a good index of water sensitivity. In this method, a coated (partially uncoated) strip of the film is immersed in distilled water at room temperature and note is made of the time at which color changes appear (whitening of the adhesive layer). The longer the transparency persists, the better the water-whitening resistance. Target values are rated as follows: 10 s, fair; 15 s, good; and 17 s, very good. In a similar manner, comparison of water sensitivity was carried out by submerging latex-covered slides in water and measuring the percentage of transmittance over time. Water whitening was tested as the percentage of water absorbed in 24 h and the percentage of transmittance at 450 nm after being soaked in water for 24 h. Water absorbtion and whitening for a 99/1 BA/MAA latex was compared before and after dialysis (percentage of water absorbed in 24 h; percentage of transmittance at 450 nm after being soaked in water for 24 h) [182]. In another test, transmittance (transparency) of both samples (film-dry adhesive-film and film-wet adhesive-film) was measured comparatively with a colorimeter. The differences in transparency (between dry and wet laminate) are measured; target values are as follows: <8%, fair; and <3%, good [1]. According to Schultz and Kofira [183], comparison of water sensitivity was carried out by submerging slides covered in latex into water and measuring the percentage of transmittance over time.

8.2.2.2.2 *Test of Wet Anchorage*

Film coatings should not display rub-off of the adhesive layer (handmade test) after immersion in water (7 min).

8.2.2.2.3   Test of Wet Adhesion on Glass

A release-free film coating is immersed in water (room temperature, distilled water) for 7 min. Then the wet film coating is affixed to glass (bubble free and not floating). The adhesion of the wet adhesive to glass (the recovery of adhesive properties over time) should be tested by trying to remove the sample after different storage times (i.e., 1, 2, 5, 10, 15, 30, and 60 min). The shorter the time for adhesion recovery, the better the wet adhesion. Adequate values are less than 10 min; at least one test value at less than 30 min is required [1,180].

8.2.2.2.4   Test of Wet Adhesion

According to Ref. [184], water resistance should be measured using the peel resistance test of a sample after immersion in water. The PSA-coated polyamide film (which is hygroscopic) should be immersed in water after a dry dwell time of 10 min. After 3 h the peel resistance should be tested.

8.2.2.2.5   Test of Wet Adhesion on PE

Wet adhesion on PE is important for plastic bottles (see also Chapter 4). High-speed label application should be used for plastic bottles based on PE copolymers. Affixed labels should resist immersion in water or surface-active agent solutions. According to this test, labeled bottles are immersed in a diluted solution of surface-active agents and labels are peeled off after wet storage for another 24 h; peel forces are evaluated subjectively. No edge lifting is allowed. The method is described in Ref. [1]. Wet samples are tested the same way as dry samples, but wet labels (labels immersed for 7 min in water) should be used.

8.2.2.2.6   Test of Humidity Resistance

As discussed in Ref. [180], taking into account the strong influence of environmental conditions on the adhesive properties of PSA laminates, test conditions are standardized and a standard value of 50% RH is proposed. Testing humidity resistance simulates real application environments with increased humidity. Humidity resistance is tested for 7 days at 100°F, 85% relative humidity, under condensing humidity conditions [181]. The humidity resistance was measured for PSAs coated on a 2-mil PET film bonded to stainless steel for 30 min and then placed in a humidity cabinet at 100°F and 95% RH. After 7 days a 180° peel test is carried out [26]. In other tests, wet aging under 100% RH at 40°C for 240 h was carried out. Measurement of the hydrophobicity of special latexes was performed by measuring the contact angle of a drop of water placed on paper impregnated with latexes. The contact angle of the latex film produced with the nonmigrating surfactant not only had a higher original contact angle (125 vs. 80°) but also did not drop, even after several minutes [183].

8.2.2.2.7   Test of Water Removability

Labels must adhere in the presence of atmospheric moisture up to about 30°C, but should be easily removed by washing at temperatures of 45–80°C. The label should easily detach from the glass, with the PSA carried on the label and nontacky in the presence of water. When submerged in tap water at 60°C, the label removed itself in under 1 min [185]. Water dispersability requires that the paper used as a carrier material must be dispersable and the adhesive water soluble, but resistant to organic solvents. Water dispersibility is tested at different pH values.

8.2.2.2.8   Wash-Off Test

Wash-off adhesives are designed to easily dissolve in water. The test method can be described as follows: water removability is measured by applying a coated paper (8.2 × 3.7 cm) with 20 g/m^2 dry adhesive to a glass bottle and then agitating the bottle gently in water at 60°C. The label should easily detach from the glass with the adhesive film still attached to the label and be nontacky in the presence of water. Water solubility should be tested for affixed labels as old as 3–6 months stored at room or elevated (60–70°C) temperature.

8.2.2.2.9   Test of Washing Machine Resistance

Special labels of fabric (textile) used on clothing must be tested for washing machine resistance. Such labels should resist more than 50 wash cycles and up to 10 chemical (dry) cleaning process cycles [186]. Czech [31] described a test of water solubility for adhesive for splicing tapes that relates to water-soluble PSA layers at three pH levels (pH 3, pH 7, and pH 12). A 1 × 1 m linen drape is applied with 1 × 10 cm samples of PSA prepared and autoclaved according to the above procedures. The drape is folded such that the tape is stuck to two sides of the drape. The drape is then folded twice and placed in a Miele washing machine with other test drapes until the washing machine is filled appropriately with garments. Prior to beginning a laundry cycle, each of the filters in the washing machine was checked and cleaned of any residue. The washing machine was set for a cycle of 60 min. The laundry cycle include filling with water heated to 80°C and mixing with a commercially available detergent to create an aqueous alkali solution, followed by agitation in the soapy water, draining, rinsing, and spinning. After being laundered, each of the drapes and the filters of the washing machine were examined for any adhesive or fibrous residue.

8.2.2.2.10   Test of Repulpability

The test of repulpability is imposed for splicing tapes (see also Chapter 4). In Ref. [31], Czech described a method for a test of repulpability. According to this method, a 20 × 2.5 cm strip of double-side splicing tape is sandwiched between two strips of tissue paper and cut into approximately 1.5-cm squares. To these squares are added a sufficient number of 1.5-cm squares of tissue paper to equal a total of 15 g, after which all squares are placed in a Waring blender with 500 mL of water. After the blender is run for 20 s, the stock that splattered up the sides and into the cover is washed back into the bottom with a water bottle. The blender is again run for 20 s, washed as before, and run for a final 20 s. The stock is then removed from the blender, rinsed twice with water, and make into a hand sheet using a large sheet mold. The sheet is couched off the mold, pressed between blotters for about 1 min in a hydraulic press, removed, dried, and examined for any particles of unrepulped splicing tape. If no such particles are present, the tapes are considered satisfactory. Further details can be found in under TAPPI Test UM-213.

8.2.2.2.11   Test of Solvent Resistance

Resistance to water or solvents of cross-linked acrylics is tested by immersion in the liquid for up to 1 week [179]. According to Ref. [186], the British Association for Paper and Board, Printing and Packaging developed a test method for the chemical resistance of carrier materials against aggressive adhesive and ink components. The plastic film is tested after 48-h storage in a saturated solvent atmosphere.

8.2.2.3 Test of Temperature-Dependent Adhesion

The influence of environmental and experimental conditions (time, temperature, and shear rate) on the viscoelastic properties of PSAs was discussed in Ref. [32]. A great deal of attention was paid to the performance of PSAs at low temperatures because adhesive failure at those temperatures may occur. The surface temperature of the adherent is another factor. Most general-purpose adhesives are formulated for tack when applied to surfaces with temperatures as low as 38–40°F. If the temperature of the adherent is lower, a higher degree of adhesive cold flow is required to provide proper wet-out. Little concern is required for products that will be labeled at room temperature and subjected to lower temperatures later. In general, low-temperature properties are tested for function in the application field, but all of the adhesive properties must be checked to make a meaningful evaluation of low-temperature properties. For comparative purposes, static shear, peel resistance, dynamic lap shear, and low-temperature lamination were evaluated. Measurement of peel adhesion and tack at −5 and +5°C, as well as at −20°C, gives a reasonable indication of the end-use performance, both for low-temperature and deep-freeze applications (see also Chapter 4).

8.2.2.3.1 Static Shear Load Test at Low Temperatures

In the static shear load test at low temperatures a modified version of the PSTC test (1-in. contact surface on stainless steel and 1000-g weight) was suggested in Ref. [141]. Slippage and delamination of the PSA-coated strips after 5 days of storage in a constant-temperature cold chamber were observed. For dynamic lap shear testing at low temperatures, 3 × 1 in. samples are coated with adhesive, equilibrated in an environmental chamber at the desired temperature, and pulled apart in a 180° configuration at a speed of 1 in./min.

8.2.2.3.2 Test of Low-Temperature Adhesion (Peel Resistance)

For the evaluation of cold peel resistance, samples are applied when enclosed in an environmental chamber and allowed to equilibrate at the desired temperature for 10 min prior to being peeled in a 180° configuration at 12 in./min [151]. The temperature at which a PSA label is laminated to a substrate can be as critical as the usage temperature. Therefore, in low-temperature lamination tests, adhesive-coated strips and stainless-steel panels are conditioned at −20°C for 30 min and then laminated together at this temperature. After the samples are warmed to 23° (room temperature), 180° peel adhesion values are measured. According to common tests, a 60-lb paper label and an adherent surface are put in a test chamber at 35°F for 30 min. Labels are applied to various surfaces, including a HDPE panel, a polyester container, and a corrugated board. After a dwell time of 1 h at 35°F, the labels are removed by hand from the surfaces at approximately a 90° angle and a rate of 10–12 in./min. Labels that tear upon removal are considered satisfactory [71].

8.2.2.3.3 Test of Cold Tack

Cold tack is measured by attaching a label strip at the given temperature to a cardboard surface adjusted to the same temperature. The strip is applied with a 100-g roller, and the laminate is stored for 30 min at the required temperature in a deep freezer. Afterward, the label strip is removed slowly by hand under a defined angle, and the percentage of cardboard fiber tear is estimated. FINAT FTM 13 describes the standard method for low-temperature performance.

8.2.2.3.4 Test of High-Temperature Resistance

Temperature resistance is measured by testing peel resistance or shear at elevated temperatures. Hot-shear tests and SAFT characterize the high-temperature resistance of PSAs and PSPs (see Section 8.2.1.5). Temperature resistance has also been measured using the 45° shear resistance test at 70°C. For acrylic PSA-based aluminum tapes, shear resistance on stainless steel (400 g, 12.5 × 25 mm) at 155°C was measured.

8.2.2.4 Test of Aging

Formulation for time/temperature resistance of PSAs and PSPs is discussed in *Technology of Pressure-Sensitive Adhesives and Products*, Chapter 8, including aging and storage resistance. Particular needs exist for PSAs with good aging characteristics. Adhesion, cohesion, and tack must not change after long aging periods at high temperatures. The aging stability of the base elastomers can be very different. Industrial elastomers are classified according to ASTM-D 2000, SAE-J 200, and IN 78078 into different classes concerning their short-time aging at high temperatures and in the presence of oil. Antioxidants used for rubber–resin PSAs and tests of aging are discussed in detail in *Technology of Pressure-Sensitive Adhesives and Products*, Chapter 2. Filler effects on aging of NR are discussed in *Technology of Pressure-Sensitive Adhesives and Products*, Chapter 2, as well.

8.2.2.4.1 General Considerations

Aging tests are carried out on liquid and dried PSAs. Liquid PSAs are mostly limited to testing the viscosity of molten HMPSAs. For dried PSAs or laminates, aging tests are carried out to check the solid-state components of the laminate or the PSAs (shrinkage, migration, etc.; see also Section 8.1.2.2). Generally, aging resistance is tested indirectly as the stability of the adhesive performance characteristics after storage under forced conditions (i.e., high temperature, humidity, and light; Table 8.10).

TABLE 8.10 Test Conditions for Aging of Various PSPs

		Adhesive		Aging Conditions		
PSP	Application	Base	Cross-Linked	Time (Days)	Temperature (°C)	Test Criterion
Label	Bottle	Acrylic, WB	—	10	40	Tack, peel, shear
	—	Acrylic, WB	—	7–28	54/71	Tack, peel, shear
		Acrylic, WB	—	14	40	Tack, peel, shear
Tape	—	Acrylic, HM	• UV	20	70	Tack, peel, shear
	—	Acrylic, SB	•	20	80	Tack, peel, shear
	—	Acrylic, WB	—	60	40	Peel
	Masking	Acrylic, WB	—	30	80	Adhesive transfer
	Insulation	Rubber/resin	—	30	(−5, 3, 40, 70)	Tack, peel, shear
	Packaging	Rubber/resin	—	14	40	Tack, peel, shear
	Packaging	SBC	—	7	120/150F	Peel, exposed film tack
	Packaging	SBC	—	7	70	Tack, peel, shear
Protection film	—	Acrylic, WB	•	3	30	Peel, adhesive transfer
	—	Acrylic, SB	•	4	60	Peel, adhesive transfer
Release liner	—	Silicone	•	30	70	

TABLE 8.11   Test Conditions for Aging of HMPSAs

Adhesive	Aging Conditions		
Base Polymer	Time (Days)	Temperature (°C)	Test Criterion
SIS	4	177	Melt viscosity
SIS	1	177	Melt viscosity
SIS	3	150	Color, tack, peel resistance, shear resistance
SIS	7–35	70	Tack, peel resistance, shear resistance
SIS	1, 2, 3, 4	150	Tack, peel resistance, shear resistance
Acrylic block copolymer	4	177	Melt viscosity

Tack, peel resistance, and shear resistance are tested. These aging tests use a standard methods for the characterization of adhesive properties (or their combination) after storage of the adhesive (laminate) under well-defined conditions (temperature, light, humidity) for a defined time. The most important adhesive characteristics evaluated after aging are peel resistance and tack; generally, peel resistance increases with time and tack decreases. The loss of adhesive properties caused by aging depends on the chemical nature of the PSA. Solvent-based acrylics considered very aging-resistant products exhibit a loss of peel resistance of about 10% after accelerated aging. A more complex test concerns the aging of removable PSAs, where "clean" removability should be examined. In some cases, aging tests are carried out to determine the adequate adhesive.

Aging tests for HMPSAs are carried out (for melt-viscosity stability) at 177°C for 4 days [187]. Table 8.11 summarizes the aging conditions for HMPSAs.

Table 3.13 in *Technology of Pressure-Sensitive Adhesives and Products*, Chapter 3, compares melt-viscosity after aging at 177°C between SIS-based hot-melt and block acrylic-based formulas after 24-h aging at 177°C. According to Ref. [188], for HMPSA two different test series are carried out [color and adhesivity changes at elevated temperature (150°C, 24–100 h)] and the standard adhesive properties after aging at elevated temperature (70°C, 1–5 weeks) are tested. Aging specimens are very important for HMPSAs given that the adhesive properties may change significantly with time because the formulation is composed of low-molecular-weight tackifiers and oils, which can diffuse readily in the rubbery matrix. Because the rubber phase of these adhesives is well above the glass transition temperature, molecular movement and diffusion are not prevented and the system will always move toward thermodynamic equilibrium. The morphology of tackified SBCs can also be significantly altered by annealing or aging [189]. Scanning probe microscopy has been used to study surface morphology [190] and the ensuing changes that occur during the aging of PSAs [191,192]. Paiva et al. [192] studied poly(ethylene propylene) with a rosin ester tackifier and determined that tackifier-enriched domains increased in size and the polymer-enriched matrix increased in stiffness as the adhesive was aged.

TABLE 8.12 Test Conditions for Aging in a Humid Atmosphere

RH	Time (w)	Temperature (°F)		Test Criterion (°C)
50	8, 16, 27, 40	65	—	Tack, peel resistance, and shear
50	8, 16, 27, 40	73	—	resistance
50	8, 16, 27, 40	150	—	—
95	8, 16, 27, 40	73	—	—
80	1	—	66	—
50	4 h	—	29	—

In various applications such as solar and safety window films, graphics, point-of-purchase advertising, and clear labels, the adhesive bonds to the inside surface of the glass. The adhesive must also resist yellowing to maintain its crystal clear appearance during extended service. Acrylics do not contain light-sensitive unsaturation and are well suited for such applications.

The tests of aging characteristics of plastics have been summarized in Refs. [187,188,193]. Adhesive-coated laminates are exposed to weathering tests and their adhesive properties are tested. The weaknesses of special classes of adhesives are well known; hence, aging test methods have been developed to examine these shortcomings after storage. Consequently, the conditions proposed for accelerated aging (storage) can vary significantly. Examination of the adhesive properties after aging (240 h) under humid conditions (40°C,100% RH) can be carried out using standard peel test methods (see Section 8.2.1.3). Table 8.12 summarizes the aging conditions in a humid environment.

The following environments were proposed for aging tests of PSA laminates [188]: −65°F and 50% RH, +73°F and 50% RH, +150°F and 50% RH, and +73°F and 95% RH. Temperature and humidity cycling (MIL STD 331) and diurnal cycling from −65 to +165°F at 95% RH were performed. Test specimens are examined after 0, 8, 16, 27, and 40 weeks of exposure. Aging at 70°C for 1 day corresponds to 0.5 year of natural aging [194]. According to Ref. [195], aging tests are carried out at room temperature, 54°C, or 71°C for 1 to 4 weeks.

Aging behavior may be estimated by measuring changes in peel resistance or tack. Peel resistance should be tested initially and after 14 days at 40°C, as well as shear strength and RBT. For aging of tapes, storage at 80% humidity and 66°C for 6 days and at 50% humidity at 29°C for 4 h according to PSTC-9 was proposed in Ref. [196]. Exposed film tack may be evaluated by leaving tapes in a laboratory environment and periodically checking the finger tack. The aging of SIS-based HMPSAs was tested after storage at 150°C for 24, 48, 72, and 100 h [179]. Tests were carried out at 70°C after 1, 2, 3, and 4 weeks. Pure SIS films were aged at 95°C for 150 min [197]. For tackified SBC rheology, peel resistance, shear resistance, and tack properties were measured before and after aging at 40°C and ambient humidity for 2 weeks by O'Brien et al. [4]. Initial and aged adhesive performances were measured. Initial (unaged) tape specimens were conditioned at 25°C and 50% RH for 24 h prior to testing. Aged specimens were conditioned for an additional 2 weeks at 40°C and ambient humidity. Adhesive film aging may be tested at 150°F and 15% RH at 4- and 12-day intervals. Four-day exposure under these conditions, according to ASTM D-3611–77, represents 2 years of natural aging.

Adhesive build-up is tested for HMPSAs [198]. No significant increase in 180° peel adhesion to stainless steel was noted from initial testing through 24 h. For HMPSAs, peel resistance decreases after aging. An HMPSA with a 180° peel adhesion of 8.50 lb/in. (control) displays only 3.70 lb/in. after 7 days at 120°F and 4.70 lb/in. after 7 days at 150°F. Aging of SIS-based HMPSAs is performed at 95°C on polyester in a dark environment with dry air. According to Ref. [199], for a SIS formulation based on Kraton 1161 (1107) and (30%), C5 (49.5%) tackifier resin, probe tack (g/cm^2), initial/aged after 1 week, at 70°C; 180° peel, on stainless steel, initial/aged 1 week, at 70°C, and 178° shear initial/aged was tested.

As stated by Kauffmann [155], the effect of antioxidants can only partially be investigated by accelerated aging. The test procedures for the determination of an antioxidant are based on the artificial aging of the product and IR estimation of the carbonyl and hydroxyl absorbances (see also *Technology of Pressure-Sensitive Adhesives and Products*, Chapter 8).

The UV stability of HMPSAs is tested after 0, 1, 3, and 5 h using 180° peel, shear, tack, color, and density measurements [196]. Exposed film tack after days at 70°C may be measured as well. Aging for 1 day or 1 week at room temperature, 8 days at 88°C, and 8 days under a sunlamp is evaluated for chloroprene lattices [197]. According to Smit et al. [200], for UV stability tests coatings covered with PET and OPP were placed in a QUV chamber with UV 340 bulbs at 45°C. The samples were removed after 70, 208, 355, and 707 h. Tack and color (after 707 h) were assessed. Weatherometer tests should be carried out according to German Norm DIN 53387, according to Ref. [201].

An aging test is carried out by storing a laminate sample (100 × 100 mm) in an oven between two wooden plates, under a load of 1 kg/100 cm^2 at 70°C for 4 days [1]. In this test the degree of adhesive degradation during aging, as well as migration (see also Section 8.1.2.2.3), is examined. Another method subjectively tests adhesion of the coated PSA layer after aging under light and temperature exposure. For current industrial control tests, a 30-min aging time should be used [1]. Roll-storage aging is the resistance to adhesive property changes upon long-term contact with the face material (and liner).

8.2.2.4.2 Accelerated Weathering Test

One-mil films of adhesive coated on 1-mil PET films are exposed directly (adhesive side) to a light source in a UV device. A straight UV cycle is imposed (no moisture) [202]. According to Smit et al. [200], for UV stability testing samples were placed in a QUV chamber with UV 340 bulbs at 45°C (see also *Technology of Pressure-Sensitive Adhesives and Products*, Chapter 3).

8.2.2.4.3 Aging Test of the Release Force

Accelerated aging before a high-speed release test (see also Section 8.2.1.3.2) can be performed by placing a set of self-adhesive strips between two flat metal or glass plates and maintaining them for 20 h in an air-circulating oven at 70 ± 5°C. The strips then should be removed and conditioned for at least 4 h. The environment's effect on the release characteristics of the release paper is measured. Aged siliconized paper is applied to freshly prepared adhesives and the force required to remove the paper measured in the 180° peeling mode is measured [203].

8.2.2.4.4 Aging of PVC Face Stock

Aging tests on PVC (coated with PSAs) may be carried out after 1 week at 158°F [204]. This test appears equivalent or more severe than 7.5 days of exposure at room temperature.

8.2.2.4.5 Test of Migration for PSP

As mentioned previously (see Section 8.1.2), migration (bleeding) as a diffusion- and cold-flow-related phenomenon is time and temperature dependent, that is, aging accentuates migration. Different methods exist to test the bleed-through of an adhesive. According to Ref. [205], migration is tested by storing the label sample at increased temperatures and then examining it against a black cardboard. Suggested storage temperatures are 60 and 71°C for HM and solvent-based PSAs, respectively. Labels should be covered with siliconized paper, stored at 60°C, and examined subjectively after a determined time. The test can be run by placing samples of coated label stock, based on paper of a certain quality (e.g., litho paper), into an oven kept at elevated temperatures (minimum 70°C). The test samples are inspected for discoloration at weekly intervals [206]. According to Ref. [207], migration for splicing tapes is tested according to the well-known method, that is, by static load of multilayer samples at different temperatures and times. According to Krutzel [71], face stock penetration is tested using 60-lb white paper. The adhesive is coated on siliconized paper and laminated with Kromecoat litho paper. The laminate is stored in a 158°F oven for specified times (5 days). Bleed-through should be tested under a load of 1 lb/in.^{2}. According to Ref. [199], for an SIS formulation based on Kraton 1161 (1107) and (30%), C5 (49,5%) tackifier resin, penetration after 1 week at 70°C, was tested.

8.2.2.4.6 Edge Ooze Test for PSP

This property refers to the flow characteristics of the coated adhesive (see also Section 8.1.2.2) and is strongly dependent on the laminate build-up. Edge ooze affects cuttability (see also Section 8.2.2.7). The formulation for cuttability is discussed in *Technology of Pressure-Sensitive Adhesives and Products*, Chapter 8. Cuttability is discussed in *Technology of Pressure-Sensitive Adhesives and Products*, Chapter 10. The test method for edge ooze is described in Ref. [1].

8.2.2.5 Test of Conversion-Dependent Properties

Some product characteristics depend on the quality of the conversion on side and influence conversion on the other side (e.g., telescoping, cuttability, adhesive transfer, etc.).

8.2.2.5.1 Test of Telescoping

Telescoping of a slit roll is related to unwinding quality. For narrow webs it is more critical. The maximum acceptable value is generally given as a product specification. For instance, according to Ref. [207], for a 32.0-mm PVC tape, telescoping shall be limited to 6.35 mm.

8.2.2.5.2 Test of Cuttability

As discussed in *Technology of Pressure-Sensitive Adhesives and Products*, Chapter 10, web-like PSPs must be finished by conversion. Conversion includes confectioning, that is, the transformation of the web in other web-like (e.g., tapes or protection films) or discrete (e.g., labels, forms) items. In this process, cutting and die-cutting play a special role (see Chapter 10). Cuttability of the carrier material is required for labels during their manufacture. For tapes, cuttability is needed during their lamination. For protective films, cuttability is necessary after their application in a laminated state (e.g., laser cuttability). For all of these products cuttability is also required in the conversion phase of manufacture. Cuttability and its test were described in Refs. [32,180].

8.2.2.5.3 Test of Adhesive Transfer

Independent of the special end-use requirements the most important surface-related performance characteristics required for a face stock material is the anchorage of the adhesive on the carrier. Anchorage affects cold flow (see Section 8.1.3) and adhesive transfer. The role of the carrier as the basis for adhesive anchorage was illustrated by the data of Haddock [208]. Adhesive transfer is tested for medical products. The same medical adhesive coated on a cloth backing yields an average adhesive transfer index of 1.5 in comparison with a vinyl carrier, which displays a value of 3.5.

8.2.2.5.4 Test of Printability

Printability is discussed in *Technology of Pressure-Sensitive Adhesives and Products*, Chapter 10. Printability includes different performance characteristics such as the ability to lay flat, dimensional stability (under various mechanical and environmental stresses), good anchorage, and adequate machinability. Such requirements and their test methods were discussed in Section 8.1.2. When choosing a printing base, certain parameters such as required quality standards, performance ratio, and subjective assessment of the finished product must be taken into account. In this case quality includes printability, roughness on the front and back sides, gloss, opacity, density, thickness, tensile strength, and bending stiffness. Good printability requires low surface roughness, more or less surface absorbancy (depending on the printing ink), sufficient surface energy, resistance to solvents in printing inks, resistance to heat produced during ink drying or hot foil-stamping, and the absence of separating agents and other impurities from the imprintable side.

The test of printability is imposed for the carrier as well as for the finished PSP, taking into account preprocessing and postprocessing of the web (see *Technology of Pressure-Sensitive Adhesives and Products*, Chapter 10).

8.2.2.5.5 Test of Winding Properties

Winding properties influence the manufacturing of the carrier material [9], the coating, laminating, and converting, and the end use of PSPs. Winding properties play a special role for PSPs applied as webs (e.g., tapes and protective films). For tapes the winding and roll-down properties are very important. Such characteristics are the primary conversion parameters for tapes because of their high degree of confectioning (slitting). Table 8.13 illustrates the common values of adhesion on the back side of the carrier.

TABLE 8.13 Adhesion on Carrier Back Side

| PSP Grade | Carrier Material | | Adhesion on Carrier Back Side (N/25 mm) Tested As | | | Test Method |
			Peel Resistance	Peel Cling	Unwinding Resistance	
Tape	Packaging	PE	—	—	0.50	—
	Closure	PE	—	—	3.50	BDF A 07
		PE	2.50	—	—	—
	Flame resistant	PVC	—	—	8.70	OCT-906
		PVC	2.50	—	—	—
	Marker	PE	3.00	—	—	—
	Application		0.25	—	—	—
	Masking	PE	2.50	—	—	—
	Easy tear	PE	4.50	—	—	—
	Harness wrap	PE	2.50	—	—	—
	Special	PP	—	—	0.75	JIS/Z 0237
	Special	PVC	—	—	0.50	
Protection film	Common	PE	2.00	—	—	—
	Self-adhesive	PE	—	0.06	—	T-peel

Unwinding resistance is caused by such adhesion. Table 8.13 illustrates the unwinding force of common tapes situated as an order of magnitude in the debonding force (peel resistance) domain of removable labels or protective films.

The stresses that appear during winding depend on the running speed of the web. Their order of magnitude and their degree of influence differ during film manufacture, coating, and confectioning. In the case of film manufacture the maximum winding speed is 20–140 m/min for blown films, 120–400 m/min for cast film, and 280–350 m/min for biaxially oriented film [205]. Common coating speed values are 100–200 m/min. The rewinding speed for a narrow web is generally lower. For instance, one tape specification [207] states that when unwound at a rate of 7.3 m/min (i.e., at a rate that is an order of magnitude higher than common peel resistance test rates), the adhesive of the tape shall not transfer to the adjacent layer. During unwinding a tape may tear or split [175]. Both the unwinding resistance and the tape tear number depend on the temperature and unwinding speed.

It is known from the unwinding and slitting of tapes that the reel should be tempered before confectioning. The adhesion on the back of the tape (blocking) depends on the temperature. The resistance to unwinding can be considered equivalent to peel resistance. Therefore, the unwinding force is a function of unwinding speed and temperature. The unwinding force depends on a number of parameters,

$$F_w - f(F_a, E_a, E_f, h_a, h_f, w_t) \tag{8.7}$$

where F_a is the adhesion force; E_a and E_f are the moduli of elasticity of the adhesive and of the film, h_a and h_f are the thickness of the adhesive and film, and w_t is the width of the tape. The adhesive is a viscoelastic substance; therefore, the modulus of elasticity of the adhesive depends on the time and temperature, that is, the unwinding speed and temperature. The modulus of elasticity of the film carrier material also depends on the temperature. In this case, the adhesion force is the peel resistance of the PSA from the back of the tape. The peel resistance from the abhesive layer protecting the active face of the PSP is also measured for labels. For labels the release layer is a separate component, and delamination occurs during end use; it is an end-use characteristic. For tapes and protective films delamination occurs during rewinding (or end use, but the end-use delamination rate is very low). In this case it is both a converting and an end-use performance characteristic. Tape tear is due mainly to bubbles and bubble ranges ("chains") and small (diameter less than 2 mm) holes. Bubbles appear on the tape carrier on places where the film is too thin.

A test method is described in Ref. [209]. Tapes must be easily drawn from a roll. Strips of tape 25 mm wide are wound up on a roll. By means of balance, the force required to draw the adhesive tapes from the roll is measured. A force of 4.2 N/25mm is considered acceptable.

Unwinding resistance should be measured after production and aging. Shrinkage and elongation of the tape during unwinding/deapplication, peel force, fresh and aged, weathered, low temperature, peel from backside, and shear resistance at different temperatures are also measured. Winding hardness of the rolls is measured as well [210]. The high-speed unwinding force of pressure-sensitive tapes can be measure according to ASTM 3889, PSTC 13, and FINAT 4 methods.

8.2.2.6 Special Product-Related Tests

Various special PSPs require test methods according to their application field. Such methods were developed for the main classes of PSPs: labels, tapes, and protective webs. The next sections provide some typical examples.

8.2.2.6.1 *Test of Special Labels*

As discussed in Chapter 4, wine labels include permanent labels with cold-water resistance (ice water) and permanent labels that are removable with hot water. Items in all these categories must possess good adhesion on moist, condensation-covered surfaces [211].

The ability to lay flat in printing is completed by on-pack wrinkle-free squeezability [212]. Aging on the end-use substrate is a very important practical evaluation criterion. Therefore, shrinkage should be tested on different surfaces, such as 2-mil cast vinyl and 4-mil calendered vinyl [26]. Plasticizer resistance is important for exterior graphics on vinyl, semipermanent vinyl labels, and automotive decals; the target values are between 0.3 and 0.7%. Roll-storage aging is the resistance to adhesive property changes upon long-term contact with the face material and release liner. Migration of plasticizers, surface active agents, and antioxidants affect adhesives, printing inks, face stock, etc. Environmental resistance may be evaluated by retention of bond strength and other properties upon exposure to water, humidity, temperature changes, sunlight, etc.

8.2.2.6.2 *Test of Special Tapes*

Various tapes used in the automotive industry (see also Chapter 4) are often subjected to extremely high temperatures. In a closed automobile in the sun, temperature can reach

100–120°C. Car undercarriage protection and insulation tapes must resist 20 h at 150°C [213]. Their resistance to water or solvents is tested by immersion in the liquid for periods up to 1 week. For such foam tapes (with adhesive foam), both breakaway cleavage peel value and continuing cleavage peel value have been measured.

Tapes used in automobiles must fulfill special norms concerning volatiles and fogs (e.g., the PV3341 VW/Audi Norm and the PBVWL709 Daimler norm). Ref. [205] describes the test methods used. Kim et al. [214] developed a special volatile organic compound (VOC) analyzer. They completed an exhaustive investigation of the equipment used. Over the past decade, researchers have developed various techniques for measuring emissions of VOCs. An ASTM standard guide, a guideline from the Commission of the European Communities (ASTM, 1992; CEC, 1992), and European preliminary standard ENV 13419 parts 13 (CEN, 1998) have been published for such tests. Indoor air pollutants mainly include nitrogen oxides and VOCs, which can cause adverse health impacts on occupants. VOCs are primarily composed of BTEX (benzene, toluene, ethylbenzene, and o-xylene) and halogenated hydrocarbons. This article describes a method to determine VOC emission from various adhesives for building materials using a VOC analyzer. Such a method can be used for PSPs as well.

For application tapes (see also Chapter 4), peel on stainless steel, coating weight, rest-humidity content, and deposit (subjectively) after 24 h/RT on PVC are controlled.

The performances of EVA dispersions for flooring adhesives are tested as 180° peel resistance (after 14 min and 24 h at 23.3°C) and 178° shear resistance, according to Stratton [215].

Mechanical, electrical, and thermal properties are required for insulation tapes. Elongation, shear resistance, and peel resistance are required as end-use performances. Shear resistance should be measured after solvent exposure. Flame resistance is needed as well. Insulation tapes (see also Chapter 4) must resist temperatures between 5 and 60°C [216]. Thermally conductive pressure-sensitive insulation tapes are also manufactured as anticorrosion protection tapes [217]. Bilayered heat-shrinkable insulating tapes for anticorrosion protection of petroleum and gas pipelines have been manufactured from photochemically cured LDPE with an EVAc copolymer as an adhesive sublayer [218]. The liner dimensions of the two-layer insulating tape decrease by 10–50% depending on the degree of curing (application temperature 180°C). The application conditions and physicomechanical properties of the coating were determined for heat-shrinkable tapes with 5% shrinkage at a curing degree of 30% and shrinkage force of 0.07 MPa. Abrasion resistance of special tapes should be controlled using a sandpaper test.

Splicing tapes (see also Chapter 4) must be tested for the following properties: tack and adhesion on paper (on silicone raw paper also), dynamic and static shear strength at normal and high temperatures, resistance to greasing and bleed-through, and water solubility at pH 3, 7, and 12. According to Ref. [154], peel resistance for splicing tapes is tested as standard peel on stainless steel and as peel on siliconized liner. For splicing tapes, static and dynamic shear measurements are carried out according to modified standard methods with paper as substrate. Practice-related shear testing uses a heated cylinder. According to Ref. [219], for such tapes a water solubility between pH 3 and 9 should be given. As reported in Ref. [220], special acrylate-based, water-soluble adhesives for splicing tapes can change their adhesive properties as a function of atmospheric humidity.

For high-temperature splicing tapes, shear resistance at 250°C has been tested [26]. According to Blake [221], repulpability tests, tests of plasticizer migration, and carbonless paper deactivation are carried out for splicing tapes.

For medical tapes (see also Chapter 4), the conformability of the PSA, its initial skin adhesion value, the skin adhesion value after 24–48 h, and adhesive transfer [222] are measured as the main performance characteristics (in accordance with ASTM D-638 and D-882) [223]. The rate of removal is very low (15 cm/min). The preferred skin adhesive will generally exhibit an initial peel value of 50–100 g and final peel value (after 48 h) of 150–300 g (at a debonding speed of 254 cm/min). Adhesive residue is rated from 0 (no visible residue) to 5 (residue covering 75 to 100% of the tested area). Skin redness, degree of skin strippage, and adhesive left on the skin are also recorded. Adhesion is rated from 1 (tape off) to 7 (perfect adhesion). Adhesive transfer is rated from 0 (no residue) to 10 (heavy residue). Williams plasticity (see also Section 8.2.1.8.1) is used to evaluate adhesive conformability [208].

The preferred carrier materials are those that permit transpiration and perspiration or tissue or wound exudate to come through. They should have an MVT of at least 500 g/m² over 24 h at 38°C (according to ASTM E 96-80), with a humidity differential of at least 1000 g/m² [224]. The air porosity rate must fulfill a given value (50 s/100 cc per square inch). According to Smit [200], for medical PSA formulations the MVTR was measured using the inverted cup method.

Conformability of PSAs used for medical tapes is measured as creep compliance, according to Ref. [223], with a creep compliance rheometer. The value accepted is at least 1.2×10^{-5} cm²/dyn to about 2.3×10^{-5} cm²/dyn. The higher the creep compliance, the greater the adhesive residue left on the skin after removal. Good long-term wear, 4-day wear on a permeable backing, and 24-h adhesion without adhesive transfer are required. Adhesion to permeable medical backing is necessary as well.

Cytotoxicity of medical PSAs was tested according to the ISO elution method (Method 10993, p. 5).

High-shear, low-tack applications are suggested for certain mounting tapes (see also Chapter 4) and double-side-coated foam tapes. For these mounting tapes the target values of (180°, instantaneous) peel resistance, loop tack, and 72°F shear resistance are 34.5/6.5 N/25 mm and 400 min. A high coating weight is used. A thick, 88-μm adhesive layer was proposed for such tapes in Ref. [26]. Their peel resistance has been tested on steel and polycarbonate.

Electrical tapes (see also Chapter 4) must be approved by Underwriter Laboratories (UL) [224]. Underwriter Laboratory Classification 181, 5th edition, "Standard for Factory Made Air Duct Materials and Air Duct Connectors," contains specifications dealing with adhesive performances under conditions of heat, such as exposure to flame and moisture. This classification also refers to specifications for testing adhesives written by organizations other than UL.

Shear resistance should be measured after solvent exposure. Silicone PSAs used for electrical tapes can be contacted with silicone release if they are highly cross-linked [225]. The polymers are sufficiently cross-linked if they absorb less than 20% of their cured weight of heptane. Electrical and electronic tapes have been tested according to ASTM D-1000. Numerous other test have also been introduced. Discoloration of the substrate surface, water vapor permeation, electrolytical corrosion, influence on dielectrical properties, thermal curing, oil resistance, aging, etc., have been tested.

For packaging tapes based on HMPSA, adhesion on cardboard and open-face aging are of special importance [226,227]. A flap test, a combination of shear and peel resistance and tack, is carried out as well. One of the most important properties for such tapes is open-face aging. Shear adhesion to a carton at elevated temperature (40°C) is also tested for packaging tapes. The preferred value is 100 min. For packaging-seal tapes a falling test must be performed as well [161].

For transfer tapes used for structural bonding, plasticizer resistance is tested at elevated temperatures (70°C, 30% plasticizer) [28].

8.2.2.6.3 Test of Special Protection Films

According to automotive suppliers, protective films for cars should meet special requirements concerning outward appearance, adhesive properties, stainability, and paint protection. Outward appearance is controlled by visual evaluation and concerns color, thickness, and light permeability. Adhesive properties are measured initially and after aging on the paint, as well as by overlapping. Stainability is tested on automotive paint, but component stainability should also be tested, which means that the film should be tested on various automotive parts (glass, sunroof, rubber seal, emblems, stainless-steel moldings, headlight lens, etc.). Paint protection must include resistance to acids, alkalies, rail dust, gasoline, and oil. An initial peel of 4 N/25 mm, easy removability (hand peel-off), and no visible changes to the protected surface are required for automotive protective webs. The test methods used for automotive protective films are related to the practical transport, storage, and end-use conditions of vehicles and include mostly empirically stated conditions (see also Chapter 7).

Passive protective films for PC plates have a thickness of 50 μm; processible protective films have a thickness of 70 μm. Such films are tested for adhesion build-up at normal and elevated temperatures. Both short-term and long-term adhesion performance must be tested. Instantaneous adhesion at room temperature, adhesion after storage (2 days) at room temperature, and adhesion build-up after storage at 80°C (4 h) are evaluated.

PC plates are hygroscopic and are generally dried before use. Drying is carried out at 130°C for 0.5–48 h, depending on the plate thickness (0.75 to 12 mm). Masking should be applied at both 50 and 100°C. Therefore, for masking tapes short-term adhesion build-up is tested after storage at 130°C. Shrinkage at 130°C is also important. Long-term adhesion build-up is tested after storage at room temperature and 40°C after 2–8 weeks. The end adhesion should be situated at about 4 N/20 cm. If adhesiveless films are used, the strong influence of the laminating temperature should be taken into account. If adhesive-coated films are applied, attention should be paid to the dependence of the peel build-up on the coating weight.

Laminating can ensure sufficient instantaneous adhesion. If the product stiffens when it is cooled and during storage at room temperature, surface contact decreases and delaminating may occur. On the other hand, for the same product, high-temperature laminating (80°C) may lead to unacceptable adhesion build-up. For thick masking films used for PC or PMMA plates, shrinkage and the ability to lay flat are important quality criteria.

The lamination temperature of self-adhesive polyolefin protective film is 200–250°C. After being laminated, the film is cooled in a water bath. The 90° peel resistance of such films is about 180–250 g/30 cm (with tolerances of 50 g). Relatively thick films (120, 150, or 170 μm) are typically used. Such films are corona pr-treated at 44–50 dyn/cm.

The protection of aluminum, especially anodized aluminum surfaces, is a complex domain of coil coating. Standard tests of adhesion are generally made on stainless steel. Although the correlation between peel resistance on stainless steel and aluminum can be clear, the increase in adhesion with the dwell time on stainless steel and aluminum may differ significantly (see also Chapter 7). Other problems appear to be related to the use of protective films with "easy peel" and "cleavage peel." The dependence of cleavage peel on coating weight, as well as the interdependence between cleavage peel and standard peel, is important.

For antislide protection films tests of friction should be performed according to DIN51130 [205].

The current focus on extending the high temperature resistance of silicone PSAs is driven by higher soldering temperatures and a desire for clean removal in masking applications (i.e., masking tapes and protection films). The typical lap shear measurements are no longer adequate as an indicator of high-temperature performance. New applications and requirements have led to performance measurement by monitoring the propensity for bubble, residue, and ghost pattern formation, along with substrate discoloration at temperatures >300°C. A variety of ways to prepare a stainless-steel panel for testing have been introduced in the marketplace. In some cases, fine sandpaper is used to roughen the smooth stainless-steel plate prior to use, whereas in other cases, the stainless-steel plate is simply solvent rinsed and baked prior to use. It is not completely clear how the two methods of preparation impact performance, but some data support a greater sensitivity and discrimination toward ghosting with the use of preroughened panels [228].

Corona-treated protective films (applied at 40–60°C) used to protect plastic plates are tested via tensile strength testing machine as samples of 66 × 300 mm with a 50 mm/min peel rate.

The PB content of SAFs should be tested by IR, using a calibration plot from the supplier [229]. For adhesive, sealable copolymers (e.g., hot-laminated films) compression-bonded samples are tested on different surfaces [230]. Samples (23 × 15 cm) are preheated for 1 min under no load and then held for 1 min under a load of 44,500 N before being cooled.

References

1. Benedek I., Test Methods, in *Pressure-Sensitive Adhesives and Applications*, Marcel Dekker, New York, 2004, Chapt. 10.
2. Lim D.-H. and Kim H.-J., General Performance of Pressure-Sensitive Adhesives, in *Pressure-Sensitive Design, Theoretical Aspects*, Benedek I., Ed., VSP, Utrecht, 2006, Chapt. 5.
3. Benedek I., Design and Formulation Basis, in *Pressure-Sensitive Design and Formulation, Application*, VSP, Utrecht, 2006, Chapt. 1.

4. O'Brien E.P., Germinario L.T., Robe G.R., Williams T., Atkins D.G., Moroney D.A., and Peters M.A., Fundamentals of Hot-Melt Pressure-Sensitive Adhesive Tapes: The Effect of Tackifier Aromaticity, *J. Adhes. Sci. Technol.*, **21** (7) 637 2007.
5. Donker C., *Proc. of the Pressure Sensitive Tape Council*, 2001 pp. 149–164.
6. Tse M.F., *J. Adhesion*, **66**, 61–88 (1998).
7. Schlademan J.A., in *Proc. of the Pressure Sensitive Tape Council* 2003, in O'Brien E.P., Germinario L.T., Robe G.R., Williams T., Atkins D.G., Moroney D.A. and Peters M.A., Fundamentals of Hot-Melt Pressure-Sensitive Adhesive Tapes: The Effect of Tackifier Aromaticity, *J. Adhes. Sci. Technol.*, **21** (7) 637 2007.
8. Polymerdispersionen Polymerlösungen, in *Prüfmethoden*, PM-CDE 004d, Juli, 1979, BASF, Ludwigshafen, Germany.
9. Benedek I., Manufacture of Pressure-Sensitive Products, in *Developments in Pressure-Sensitive Products*, Benedek I., Ed., Taylor & Francis, Boca Raton, 2006, Chapt. 8.
10. Benedek I., Physical Basis for Pressure-Sensitive Products, in *Developments in Pressure-Sensitive Products*, Benedek I., Ed., Taylor & Francis, Boca Raton, 2006, Chapt. 3.
11. Maack Business Service, Trouble Shooting Guide for Processing Films, in *Conversion Industry Reference Report*, Report MBS No. 903, Zürich, Switzerland, 1994.
12. Hosseinpur D., Mohhammadi N., and Moradian S., A Simple Method for Characterizing the Surface Properties of Polymers, in *Proc. 25th Annual Meeting of Adh. Soc., and the Second World Congress on Adhesion and Related Phenomena*, Feb. 10–14, 2002, Orlando, FL, p. 446.
13. Benedek I., Converting Properties of Pressure-Sensitive Products, in *Pressure-Sensitive Adhesives and Applications*, Marcel Dekker, New York, 2004, Chapt. 7.
14. Benedek I., Converting Properties of Pressure-Sensitive Products, in *Developments in Pressure-Sensitive Products*, Benedek I., Ed., Taylor & Francis, Boca Raton, 2006, Chapt. 10.
15. Frutti F. and Chezzi F. (Boston S.p.a., Bollate, Italy), EP 0163761/04.05.84.
16. DIN 53279 T1/08.76, Prüfung von Klebstoffen für Bodenbeläge, Prüfung des Einflusses auf die Maßhaltigkeit, Klebstoffe auf Basis von Kunstkautschuklösungen, *Kaut. Gummi, Kunststoffe*, **39** (6) 557 1986.
17. Benedek I., The Role of Design and Formulation, in *Pressure-Sensitive Design, Theoretical Aspects*, Benedek I., Ed., VSP, Utrecht, 2006, Chapt. 3.
18. Benedek I., End-Use Properties of Pressure-Sensitive Products, in *Developments in Pressure-Sensitive Products*, Benedek I., Ed., Taylor & Francis, Boca Raton, 2006, Chapt. 11.
19. Benedek I., Chemical Composition, in *Pressure-Sensitive Adhesives and Applications*, Marcel Dekker, New York, 2004, Chapt. 5.
20. Czech Z., Developments in Crosslinking of Solvent-Based Acrylics, in *Developments in Pressure-Sensitive Products*, Benedek I., Ed., Taylor & Francis, Boca Raton, 2006, Chapt. 6.
21. Benedek I., *Development and Manufacture of Pressure-Sensitive Products*, Marcel Dekker, New York, 1999, p. 544.

22. Benedek I., *Pressure-Sensitive Formulation*, VSP, Utrecht, 2000, p. 81.
23. Goller K., *Adhäsion*, (4) 101 1974.
24. Goller K., *Adhäsion*, (4) 126 1973.
25. Goller K., *Adhäsion*, (7) 266 1973.
26. Pierson D.G. and Wilczynski J.J., *Adhes. Age*, (8) 52 1980.
27. Lombardi R., *Paper, Film and Foil Conv.*, (3) 74 1988.
28. Altenfeld F. and Breker D., *Semi-Structural Bonding with High Performance Pressure Sensitive Tapes*, 2nd ed., 3M Deutschland, Neuss, 1993, p. 278.
29. Benedek I., Physical Basis, in *Pressure-Sensitive Adhesives and Applications*, Marcel Dekker, New York, 2004, Chapt. 3.
30. Bleisch G. and Kuchler W., Maschinengängigkeit ist Maßstab, *Papier Kunst. Verarb.*, (6) 29 1995.
31. Czech Z., Synthesis, Properties and Application of Water-Soluble Acrylic Pressure-Sensitive Adhesives, in *Pressure-Sensitive Design, Theoretical Aspects*, Benedek I., Ed., VSP, Utrecht, 2006, Chapt. 6.
32. Benedek I., Rheology, in *Pressure-Sensitive Adhesives and Applications*, Marcel Dekker, New York, 2004, Chapt. 2.
33. Jordan R., *Coating*, (2) 37 1986.
34. Heying M.D., Lutz M.A., Moline P.K., and Watson M.J., US Patent 6,121,368., in Lin S.B., Durfee L.D., Ekeland R.A., McVie J., and Schalau II G.K., Recent Advances in Silicone Pressure-Sensitive Adhesives, *J. Adhes. Sci. Technol.*, **21** (7) 605 2007.
35. Aufderheide B.E. and Frank P.D., PCT Patent WO0205201 A1, in Lin S.B., Durfee L.D., Ekeland R.A., McVie J., and Schalau II G.K., Recent Advances in Silicone Pressure-Sensitive Adhesives, *J. Adhes. Sci. Technol.*, **21** (7) 605 2007.
36. Weilen W., Fink H.F., Klockner O., and Koerner G., *Coating*, (10) 376 1987.
37. ASTM D-2578-65; Standard Test Method for Wetting Tension of PE and PP films.
38. Templer K. and Wultsch F., *Wochenbl. f. Papierfabr.*, (11/12) 483 1980.
39. Pitzler G., *Coating*, (6) 218 1996.
40. Osterhold M., Breuchen M., and Armbruster K., *Adhäsion*, (3) 23 1992.
41. Benedek I., Adhesive Properties, in *Pressure-Sensitive Adhesives and Applications*, Marcel Dekker, New York, 2004, Chapt. 6.
42. Tobing S.D., Andrews O., Caraway S., Guo J., Chen A., and Anna Sh., Shear Stability of Tackified Acrylic Emulsion PSAs, in *Proc. 24th Annual Meeting of Adh. Soc.*, Feb. 25–28, 2001, Williamsburg, VA, p. 273.
43. Simal F., Jeusette M., Leclère Ph., Lazzaroni R., and Roose P., Adhesive Properties of a Radial Acrylic Block Copolymer with a Rosin-Ester Resin, *J. Adhes. Sci. Technol.*, **21** (7) 559 2007.
44. Leclère Ph., Lazzaroni R., Brédas J.-L., Yu J. M., Dubois Ph., and Jérôme R., *Langmuir*, **12**, 4317 1996.
45. Tong J.D., Leclère Ph., Rasmont A., Brédas J.-L., Lazzaroni R., and R. Jérôme, *Macromol. Chem. Phys.*, **201**, 1250 2000.
46. Rasmont A., Leclère Ph., Doneux C., Lambin G., Tong J.D., Jérôme R., Brédas J.-L., and Lazzaroni R., *Colloids Surfaces B Biointerfaces* **19**, 381 2000.
47. Tong J.D., Leclère Ph., Doneux C., Brédas J.-L., Lazzaroni R., and Jérôme R., *Polymer*, **42**, 3503 2001.

48. Yarusso D.J., Ma J., and Rivard R.J., Properties of Polyisoprene Based PSAs Crosslinked by Electron Beam Radiation, in *Proc. of the 22nd Annual Meeting of the Adhesion Society*, Feb. 21, 1999, Panama City Beach, FL, p. 72.
49. Clair D.J. St., *Adhes. Age*, (11) 23 1988.
50. Roos A. and Creton C., Adhesion of PSA based on styrenic block copolymers, in *Proc. 24th Annual Meeting of Adh. Soc.*, Feb. 25, 2001, Williamsburg, VA, p. 371.
51. Feldstein M.M., *Polym. Mat. Eng.*, **81**, 427 1999.
52. Derra R., Impact of the frame regulation on the use of adhesives in contact with Foodstuffs, in *Proc. 31st Munich Adhesive and Finishing Symposium 2006*, Oct. 22–24, 2006, p. 326.
53. Müller H., Release Liner, *ATP News-Letter*, (4) 6 2004.
54. Benedek I., Adhesive Properties of Pressure-Sensitive Products, in *Developments in Pressure-Sensitive Products*, Benedek I., Ed., Taylor & Francis, Boca Raton, 2006, Chapt. 7.
55. Park Y.-J., Hot-Melt PSAs Based on Styrenic Polymer, in *Pressure-Sensitive Design and Formulation, Application*, Benedek I., Ed., VSP, Utrecht, 2006, Chapt. 2.
56. MacDonald N.C., *Adhes. Age*, (2) 21 1972.
57. Symietz D., *Adhäsion*, (11) 28 1987.
58. Hamed G.R. and Hsieh C.H., *J. Polym. Phys.*, **21**, 1415 1983.
59. *Test Methods For Pressure Sensitive Adhesive Tapes* (Ed). Pressure Sensitive Tape Council, Northbrook, IL, 2000.
60. Hyvis/Napvis Polybutenes, BP Chemicals, Ref 208/1000.
61. Tse M.F., *J. Adhesion*, **70**, 95–118 1999.
62. Iovine C.P., Jer S.J., and Paulp F. (National Starch Chem. Co., Bridgewater, NJ), EPA 0212358/04.03 1987, p. 3.
63. Plaut R.H., Williams N.L., and Dillard D.A., *J. Adhesion*, **76**, 37–53 2001.
64. Woo Y., Plaut R.H., Dillard D.A., and Coulthard S.L., *J. Adhesion*, **80**, 203–221 2004.
65. Rivals I., Personnaz U., Creton C., Simai F., Roose P., and van Es. S., *Meas. Sci. Technol.*, **16**, 2020 2005.
66. *Surfynol Technical Bulletin*, Air Products and Chemicals Inc., 1991.
67. Dunckley P., *Adhäsion*, (11) 19 1989.
68. Williams N.L., Plaut R.H., and Dillard D.A., Analysis of the loop tack test for PSAs, in *Proc. of the 23th Annual Meeting of the Adhesion Society*, Myrtle Beach, SC, Feb. 20, 2000, p. 252.
69. *Coating*, (4) 88 1972.
70. Hammond P. Jr., *ASTM Bulletin*, **360** (5) 123–133.
71. Krutzel L., *Adhes. Age*, (9) 21 1987.
72. Kamagata K., Saito T., and Toyama M., *J. Adh. Soc. Japan.* (6) 309 1969.
73. Wetzel F., *ASTM Bulletin*, (221) 64 1957.
74. Druschke W., *Adhäsion*, (5) 30 1987.
75. Flanigan C.M., Crosby A.J., and Shull K.R., *Macromolecules*, **32**, 7251 1999.
76. Lin Y.Y. and Hui C.Y., Modeling the Failure of an Adhesive Layer in a Peel Test, in *Proc. 24th Annual Meeting of Adh. Soc.*, Feb. 25, 2001, Williamsburg, VA, p. 230.
77. Kendall K., Molecular Adhesion and Elastic Deformations, in *Proc. 24th Annual Meeting of Adh. Soc.*, Feb. 25, 2001, Williamsburg, VA, p. 11.

78. Lakrout H. and Creton C., Probe Tack Tests of PSAs With Flat and Spherical Punches, *Proc. of the 22nd Annual Meeting of the Adhesion Society*, Feb. 21, 1999, Panama City Beach, FL, p. 430.

79. Hooker J.C., Creton C., Tordjemann P., and Shull K.R., Surface Effects on the Microscopic Adhesion of Styrene-Isoprene-Styrene-Resin PSAs, in *Proc. of the 22nd Annual Meeting of the Adhesion Society*, Feb. 21, 1999, Panama City Beach, FL, p. 415.

80. Feldstein M.M., Molecular Fundamentals of Pressure-Sensitive Adhesion, in *Developments in Pressure-Sensitive Products*, Benedek I., Ed., Taylor & Francis, Boca Raton, 2006, Chapt. 4.

81. Feldstein M.M. and Creton C., Pressure-Sensitive Adhesion as a Material Property and as a Process, in *Pressure-Sensitive Design, Theoretical Aspects*, Benedek I., Ed., VSP, Utrecht, 2006, Chapt. 2.

82. Döring A., Stahr J-., and Zollner S., in *Proc. of the Pressure Sensitive Tape Council*, 2000, pp. 213–222.

83. *Allg. Papier-Rundschau*, (14) 340 1987.

84. Druschke W., Adhäsion und Tack von Haftklebstoffen, *Proc. AFERA Meeting*, 1986, Edinburgh, Scotland.

85. Tordjeman P., Papon E., and Villenave J.J., *J. Polymer Sci. Polym. Physics, Part B*, **38**, 1201 2000.

86. *Label, Labels* (2) 98 1997.

87. Mohle M., *Adhäsion*, (6) 260 1966.

88. Kemmenater C. and Bader G., *Adhasion*, (11) 487 1968.

89. H. Liese, *Adhäsion*, (3) 110 1966.

90. *Coating*, (6) 188 1969.

91. Brooks H.D., Kelly J.Y., Madison P.H., Thacher C.D., and Long T.E., Synthesis and Characterization of Acrylamide Containing Polymers for Adhesive Applications, in *Proc. 24th Annual Meeting of Adh. Soc.*, Feb. 28, 2001, Williamsburg, VA, p. 150.

92. Schultz A.K. and Kofira N., The Use of Non-Migrating Surfactants in Pressure-Sensitive Adhesive Application, in *Proc. 24th Annual Meeting of Adh. Soc.*, Feb. 25, 2001, Williamsburg, VA, p. 163.

93. Ouyang J., Jacobson S., Shen L., and Reedell S., Characterization of Acrylic PUR-based Waterborne PSAs, in *Proc. 24th Annual Meeting of Adh. Soc.*, Feb. 25, 2001, Williamsburg, VA, p. 236.

94. Bonneau G. and Baumassy M., New Tackifying Dispersions for Water Based PSA for Labels, in *Proc. 19th Munich Adhesive and Finishing Seminar*, 1994, Munich, Germany, p. 82.

95. Koch C.W. and Abbott A.N., *Rubber Age*, (82) 471 1957.

96. Kenneth A., Stockwell J.R., and Walker J. (Allied Colloids), EP 0.147.067/29.11.1983.

97. Fukuzawa K., and Uekita T., The Mechanism of Peel Adhesion, in *Proc. of the 22nd Annual Meeting of the Adhesion Society*, Feb. 21, 1999, Panama City Beach, FL, p. 69.

98. Wilken R., *Finat News*, (1) 53 1988.

99. Zorll U., *Adhäsion*, (3) 69 1976.

100. Wiest H., *Adhäsion*, (4) 146 1966.

101. Li L., Macosko C., Corba G.L., Pocius A., and Tirrell M., Interfacial Energy and Adhesion between Acrylic PSA and Release Coatings, in *Proc. 24th Annual Meeting of Adh. Soc.*, Feb. 28, 2001, Williamsburg, VA, p. 270.

102. Sehgal K.C., *Fundamental and Practical Aspects of Adhesive Testing*, SME Society of Manufacturing Engineers. Adhesives 1985 Conf. Papers, Sept. 10, 1985, Atlanta, GA.

103. Johnston J., *Adhes. Age*, (11) 30 1983.

104. Hamed G.R. and Preechatiwong W., Peel Adhesion of Uncrosslinked Styrene Butadiene Rubber Bonded to PET, in *Proc. of the 23th Annual Meeting of the Adhesion Society*, Feb. 20, 2000, Myrtle Beach, SC, p. 511.

105. Kinloch A.J., Black B.R.K., Hadavinia H., Paraschi M., and Williams J.G., in *Proc. 24th Annual Meeting of Adh. Soc.*, Feb. 25, 2001, Williamsburg, VA, p. 44.

106. Moore D.R. and Williams J.G., in *Fracture Mechanics Testing Methods Polymers, Adhesives and Composites*, Moore D.R., Pavan A. and Williams J.G. Ed., Elsevier Science Publishers, Amsterdam, 2000.

107. Chang E.P. and Wang I.F., Excimer Fluorescence Method for Determining Cure of Release Coating, in *Proc. of the 22nd Annual Meeting of the Adhesion Society*, Feb. 21, 1999, Panama City Beach, FL, p. 421.

108. Hamed G.R. and Hsieh C.H., *Rubber Chem. Technol.*, **59**, 883 1986.

109. *Coating*, (1) 8 1985.

110. Muny R., *Labels and Labelling*, (1) 36 1999.

111. Gordon V., Leay T.M., Owen M.J., Owen M.S., Perz S.V., Stasser J.L., Tonge J.S., Chaudhury M.K., and Vorvolakos K.A., Resin–Polymer interaction in silicone release coatings, in *Proc. of the 23th Annual Meeting of the Adhesion Society*, Feb. 20, 2000, Myrtle Beach, SC, p. 39.

112. Gordon G.V., Tabler R.L., Perz S.V., Stasser J.L., Owen M.J., and Tonge J.S., Rheology in the release of silicone coatings, in *Book of Abstracts*, 215th ACS National Meeting, Dallas, TX, March 29, 1998.

113. Gordon G.V., Tabler R.L., Perz S.V., Stasser J.L., Owen M.J., and Tonge J.S., Silicone Release Coatings: An Examination of the Release Mechanism, *Adhes. Age*, (11) 35 1998.

114. Gordon G.V., Tabler R.L., Perz S.V., Stasser J.L., Owen M.J., and Tonge J.S., Release Force control: The Bottom line®. Synergism between Adhesive and Liner; in *Proc. of the Pressure-Sensitive Tape Council*, May 5, 1999, Technical Seminar, Washington DC.

115. Benedek I., Build Up and Classification of Pressure-Sensitive Products, in *Developments in Pressure-Sensitive Products*, Benedek I., Ed., Taylor & Francis, Boca Raton, 2006, Chapt. 2.

116. Benedek I., Bond Failure in Pressure-Sensitive Removable Thin Plastic Film Laminates, in *Proc. of the 22nd Annual Meeting of the Adhesion Society*, Feb. 21, 1999, Panama City Beach, FL, p. 418.

117. Chalykh A.A., Chalykh A.E., Stepanenko V.Yu., and Feldstein M.M, Viscoelastic Deformations and the Strength of PSA Joints Under Peeling, in *Proc. of the 23th Annual Meeting of the Adhesion Society*, Myrtle Beach, SC, Feb. 20, 2000, p. 252.

118. Ullmann K.L. and Sweet R.P., Silicone PSAs and rheological testing, in *Proc. of the 22nd Annual Meeting of the Adhesion Society*, Feb. 21, 1999, Panama City Beach, FL, p. 410.

119. *Coating*, (11) 316 1984.
120. Zhao B., Miasek E., and Pelto R., Peeling Tapes from Paper, in *Proc. 24th Annual Meeting of Adh. Soc.* Feb. 25, 2001, Williamsburg, VA, p. 364.
121. Bikermann J.J. and Whitney W., *Tappi*, **46** (7) 420 1963.
122. Yamauchi T., Cho T., Imamura R., and Murakami K., Peeling Behavior of Dehesive Tape from Paper, *Nordic Pulp Paper J.*, (4) 128 1988.
123. *Adhäsion*, (10) 399 1965.
124. Conti J.C., Strope E.R., Gregory R.D., and Mills P.A., Cyclic Peel Evaluation of Sterilized Medical Packaging, in *Proc. of the 22nd Annual Meeting of the Adhesion Society*, Feb. 21, 1999, Panama City Beach, FL, p. 116.
125. Conti J.C., Strope E.R., and Jones E., in *Proc. of the 20th Annual Meeting of the Adhesion Society*, 1997, p. 425.
126. Conti J.C., Strope E.R., Jones E., and Rohde D., in *Proc. of the 21st Annual Meeting of the Adhesion Society*, 1998, p. 418.
127. Trenor S., Suggs A., and Love B., An Examination of How Skin Surfactants Influence a Model PIB PSA for Transdermal Drug Delivery, in *Proc. 24th Annual Meeting of Adh. Soc.*, Feb. 25, 2001, Williamsburg, VA, p. 144.
128. Dorn L. and Wahono W., *Adhäsion*, (7/8) 23 1988.
129. Fisher D.K. and Briddell B.J., (Adco Product Inc., Michigan Center, MI), EP 0426198 A2/08.05.91.
130. Shuman R.J. and Josephs B.D., (Dennison Manufacturing Co., Framingham, MA), PCT, WO 88/01636.
131. Lai Y. and Dillard D.A-., *Int. J. Solids Struct.*, **34**, 509 1997.
132. O'Brien E., Doyle K., Ward T.C., Gio S., and Dillard D., Adhesion of Model Epoxy Bonded to Glass Subjected to Chemical Stress Measured by the Shaft Loaded Blister Test, in *Proc. 24th Annual Meeting of Adh. Soc.*, Feb. 25, 2001, Williamsburg, VA, p. 475.
133. O'Brien E.P. and Ward T.C., Characterization of Thin Films: Shaft Loaded Blister Test, in *Proc. of the 23th Annual Meeting of the Adhesion Society*, Feb. 20, 2000, Myrtle Beach, SC, p. 202.
134. O'Brien E.P., Ward T.C., Guo S., and Dillard D., Characterizing the Adhesion of Pressure- Sensitive Tapes Using the Shaft Loaded Blister Test, in *Proc. 24th Annual Meeting of Adh. Soc.*, Feb. 25–28, 2001, Williamsburg, VA, p. 113.
135. Wan K.T. and Mai Y.W, *Int. J. Frac.*, **74**, 181 1995.
136. Gent A.N., *J. Adhes.*, **23**, 115 1987.
137. Paiva A., Foster M.D., Crosby A.J., and Shull K., Studying Changes in Surface Adhesion of PSAs with AFM and Spherical Indenter Test, in *Proc. of the 23th Annual Meeting of the Adhesion Society*, Feb. 20, 2000, Myrtle Beach, SC, p. 43.
138. *Glossary of Terms used in Pressure Sensitive Tapes Industry*, PSTC, Glenview, IL, 1974.
139. Hahn O., Schlimmer M., and Ruttert D., *Adhäsion*, (12) 9 1988.
140. *Prüfung von Haftkleber*, D-EDE/K, BASF, Ludwigshafen, Germany, Juni/Juli, 1981.
141. National Starch Chem. Co., Durotak, Pressure Sensitive Adhesives; Technical bulletin, Zutphen, The Netherlands, 4/1986.

142. Smith L.J., *Adhes. Age*, (4) 28 1987.
143. Chum C.M., Ling M.C., and Vargas R.R. (Avery Int. Co., USA), EPA 1225792/18.08.1987.
144. *Adhes. Age*, (12) 25 1977.
145. British Petrol, *Hyvis*, Technical bulletin, 1985.
146. Schlademan J.A., The Role of Tackifier Compatibility and Molecular Weight on Bulk Properties of PSAs, in *Proc. of the 22nd Annual Meeting of the Adhesion Society*, Feb. 21, 1999, Panama City Beach, FL, p. 75.
147. Moser A. and Dillard J.G., Private Communication; in *Adhes. Age*, (12) 25 1977.
148. Johnston J., Alternate Methods for the Basic Physical Testing for PSA Tapes and Proposals for the Future, in *Proc. of the 22nd Annual Meeting of the Adhesion Society*, Feb. 21, 1999, Panama City Beach, FL, p. 327.
149. Mudge R.P. (National Starch Chem. Co., Bridgewater, NJ), EPA 0225541/11.12.1985.
150. James D.J. and Holyoke H.C., *Adhes. Age*, (4) 23 1984.
151. Sobieski L.A. and Tangney T.J., *Adhes. Age*, (12) 23 1988.
152. Adams R.D., Thomas R., Guild F.J., and Vaughan L.F., Thick Adherend Shear Tests, in *Proc. 24th Annual Meeting of Adh. Soc.*, Feb. 25, 2001, Williamsburg, VA, p. 59.
153. van Holde K.E. and Williams J.W., *J. Polymer Sci.*, **11**, 243 1953.
154. Czech Z., *Adhäsion*, (11) 26 1994.
155. Kauffmann J.F., *Adhäsion*, (10) 163 1981.
156. Mitton S. and Mak C., *Adhes. Age*, (1) 42 1983.
157. *Allg. Papier-Rundschau*, (18) 572 1987.
158. Gerace M., *Adhes. Age*, (8) 85 1983.
159. Rubber, raw and unvulcanized, compounded, determination of plasticity number and recovery number; parallel plate method (1985-08-15), *Kaut. Gummi Kunststoffe*, **39**(1) 60 (86).
160. *Coating*, (3) 360 1985.
161. Ashland Oil Inc., Chemicals, *Bull.* No.1496-1, 1982.
162. Chu S.G., in *Handbook of Pressure Sensitive Adhesive Technology*, Satas D., Ed., Van Nostrand Reinhold, New York, 1989.
163. White C.C., Vanlandingham M.R., Drzal P.L., Chang N.K. and Chang S.H., *J. Polym. Sci. Part B - Polym. Phys.*, **43**, 1812–1824 2005.
164. Mazzeo F., *Proc. of the Pressure Sensitive Tape Council*, 2002, pp. 139–148.
165. McCrum N.G., Read B.E. and Williams G., in *Anelastic and Dielectric Effects in Polymeric Solids* (Ed). John Wiley & Sons, New York, 1967.
166. Aubuchon S.R., Ulbrich S., and Wadud S.B., Characterization of pressure Sensitive Adhesives and Thermosets by Controlled Stress Rheology and Thermal Analysis, in *Proc.24th Annual Meeting of Adh. Soc.*, Feb. 25–28, 2001, Williamsburg, VA, p. 427.
167. Hirdina Falk B., Characterisation of Pressure-Sensitive Adhesives. Correlations between Adhesive and Rheological Properties, in *Proc. 31st Munich Adhesive and Finishing Symposium 2006*, Oct. 22–24, 2006, p. 266.
168. Sato K., *Rubber Chem. Technol.*, **56**, 942 1984.

169. Willenbacher N., High Resolution Dynamic Mechanical Analysis (DMA) on Thin Films, in *Proc. 24th Annual Meeting of Adh. Soc.*, Feb. 25, 2001, Williamsburg, VA, p. 263.

170. Benedek I., Principles of Pressure-Sensitive Design and Formulation, in *Pressure-Sensitive Design, Theoretical Aspects*, Benedek I., Ed., VSP, Utrecht, The Netherlands, 2006, Chapt. 4.

171. Muny R.P., *Adhes. Age*, (12) 18 1986.

172. Vogt D. and Geiß P.L., Viscoelastic Material Behavior of Pressure-Sensitive Adhesives-Potentials and Drawbacks in Engineering Application, in *Proc. Munich Adhesive and Finishing Symposium 2006*, Oct. 22–24, 2006, p. 406.

173. Tobing S. and Klein D.S., *J. Appl. Polym. Sci.*, 79, 2230 2001.

174. Czech Z., Removable and Repositionable Pressure-Sensitive Materials, in *Pressure-Sensitive Design and Formulation, Application*, Benedek I., Ed., VSP, Utrecht, The Netherlands, 2006, Chapt. 4.

175. Davis I.J., (National Starch, Chem. Co., Bridgewater, NJ), US Pat., 4.728.572/ 01.03.1988.

176. *Mobil Plastics Technical Bulletin*, MA 657/06/90.

177. *Technical Information*, LHM, Celanese Resins Systems, 1991.

178. Schuman H. and Josephs B. (Dennison Manuf. Co., USA), PCT/US86/ 02304/25.08.1986.

179. Lehmann G.W.H. and Curts H.A.J., (Beiersdorf A.G., Hamburg, Germany), US Pat., 4038454/26.07. 1977.

180. Benedek I., Manufacture of Labels, in *Pressure-Sensitive Adhesives and Applications*, Marcel Dekker, New York, 2004, Chapt. 8.

181. Kenneth A., Stockwell J.R., and Walker J., (Allied Colloids Ltd., Low Moor, U.K.), EP 0147 067/28.11.1983.

182. Miller C.M. and Barnes H.W., Factors Affecting Water Resistance of Latex-Based PSAs, in *Proc. 24th Annual Meeting of Adh. Soc.*, Feb. 25, 2001, Williamsburg, VA, p. 164.

183. Schultz A.K. and Kofira N., The Use of Non-Migrating Surfactants in Pressure-Sensitive Adhesive Applications, in *Proc. 24th Annual Meeting of Adh. Soc.*, Feb. 25, 2001, Williamsburg, VA, p. 142.

184. Müller J. and Eisele D., Wasserbeständigkeit von Klebesystemen, in *ATP Newsletter*, (3) 6 2004.

185. US Pat., No., 3691140/12.09.1972; in Miyasaka H., Kitazaki Y., Matsuda T., and Kobayashi J., (Nichiban Co., Ltd. Tokyo, Japan), Offenlegungsschrift, DE 3544868 A1/18.12.1985.

186. *Coating*, (3) 65 1974.

187. Milton S. and Max Ch., *Adhes. Age*, (1) 12 1983.

188. Becker D. and Braun J., *Kunststoff Handbuch*, Bd.1, K. Hanser, München-Wien, 1990, p. 931.

189. Akiyama A., Kobori Y., Sugisaki A., Koyama T., and Akiba I., *Polymer*, 41, 4021 2000.

190. Dupont M.J.O. and Keddie J.L., *J. Adhes. Sci. Technol.*, 17, 243 2003.

191. Paiva A. and Foster M.D., *J. Adhesion*, 75, 145 2001.

192. Paiva A., Sheller N., Foster M.D., Crosby A.J., and Shull K.R., *Macromolecules*, **34**, 2269 2001.
193. Harrison D.J.P., Johnson J.F., and Yates J.F., *Polymer Eng. Sci.*, (14) 865 1982.
194. *Coating*, (7) 20 1984.
195. Hinterwaldner R., *Adhäsion*, (3) 14 1985.
196. Grossmann R.F., *Adhes. Age*, (12) 41 1969.
197. *Adhes. Age*, (3) 36 1986.
198. *Technical Service Report, 6110*, Firestone, August, 1986.
199. Merrill N. and Machielse J., Thermoplastic A-C® Polyolefin Additives for Hot-Melt Adhesives, in *Proc. 31st Munich Adhesive and Finishing Symposium 2006*, Oct. 22–24, 2006, p. 232.
200. Smit E., Paul C.W., and Meisner C.L., Acrylic Block-Copolymer Hot-Melt PSAs, in *31st Munich Adhesive and Finishing Symposium 2006*, Oct. 22–24, 2006, p. 295.
201. *ATP News-Letter*, (2) 3 2004.
202. *Tappi J.*, **67** (9) 104 1983.
203. Price S. and Nathan J.B. Jr., *Adhes. Age*, (9) 37 1974.
204. Mudge R., Ethylene-vinylacetate based, water-based PSA, in *TECH 12, Advances in Pressure. Sensitive Tape Technology, Technical Seminar Proc.*, Itasca, IL, USA, May 1989.
205. Dobman A. and Planje J., *Papier u. Kunststoffverarb.*, (1) 37 1986.
206. *Adhäsion*, (1) 11 1974.
207. Bull A.L., *Thermoplastic Rubbers, Technical Manual*, Shell Elastomers, TR 8.12, p. 13.
208. Haddock T.H. (Johnson & Johnson, USA), EPA 0130080 Bl/02.01.1985.
209. Becker H., *Adhäsion*, (3) 79 1971.
210. *Papier Kunst. Verarb.*, (6) 56 1995.
211. *Etiketten-Labels*, (5) 24 1995.
212. *Label/labels Internat.*, (5/6) 18 1997.
213. *Coating*, (1) 12 1994.
214. Kim S., Kim J.-A., and Kim H.-J, Analysis of VOCs Emission from Adhesives using VOC Analyzer, in *Proc. of the 29th Ann. Meeting of the Adhesion Society Inc.*, Feb. 19–22, 2006, Jacksonville, FL, USA, p. 119.
215. Stratton W.M., *Adhes. Age*, (6) 21 1985.
216. Wechsung A.B., *Coating*, (9) 268 1972.
217. Torigoe W., (Dainichi Nippon Cable Ltd., Amagasaki, Japan), U.S. Pat, 4,645,697,1987, *Adhes. Age*, (5) 26 1987.
218. Rybov V.M., Chemikov O.M., and Nosova M.F., *Plast. Massy*, (7) 58 1988.
219. U.S.Pat.3,096,202, in Gleichenhagen P. and Wesselkamp I., EP.0,058,382 B1, Beiersdorf A.G., Hamburg, Germany, 1982.
220. Czech Z., *European. Adhes. Seal*, (6) 4 1995.
221. US Pat., 3661874, in F. D. Blake., (Minnesota Mining and Manuf., Co., St. Paul, MN, USA), EP 0141 504 A1/15.05.85.
222. Sun R.L. and Kennedy J.F. (Johnson and Johnson Products Inc., USA), US. Pat., 4,862,888,1988, *CAS, Hot-Melt Adhesives*, 26, 1, 1988.

223. US Pat, 3321451, in Krampe S.E. and Moore C.L., (Minnesota Mining and Manuf. Co., St. Paul, MN), EP 0202831 A2/26.11/86.

224. *Adhes. Age*, (8) 62 1986.

225. Pennace J. and Kersey G.E, Flexcon Co. Inc., Spencer, MA, USA, PCT, WO 87/035537/18.06.87.

226. Donker C., Luth R., and van Rijn K., Hercules MBG 208 Hydrocarbon Resin, A New Resin for Hot-melt Pressure Sensitive (HMPSA) Tapes, in *Proc. of 19th Munich Adhes. Finishing Seminar*, Munich, Germany, 1994, p. 64.

227. Jacob L., New Developments for Tackifiers for SBS Copolymers, in *Proc. of 19th Munich Adhes. Finishing Seminar*, Munich, Germany, 1994, p. 107.

228. Lin S.B., Durfee L.D., Ekeland R.A., McVie J., and Schalau II G.K., Recent advances in silicone pressure-sensitive adhesives, *J. Adhes. Sci Technol.*, 21 (7) 605 2007.

229. Panagopoulos G., Pirtle S.E., and Khan W.A., *Film Extrusion*, 103 1991.

230. Ward R.M. and Kelley D.C., *Tappi J.*, 86, 140 1988.

Appendix: Abbreviations and Acronyms

1. Compounds

AA	acrylic acid
ABC	acrylic block copolymer
ABS	acrylonitrile–butadiene–styrene
AC	acrylic
AHM	acrylic block copolymer based hot-melt
AMPS	2-acrylamido-2-methylpropane sulfonic acid
ANQ	5-amino-1-4-naphthoquinone
APO	amorphous polyolefin
ATBC	acetyltributyl citrate
ATEC	acetyltriethyl citrate
BA/AA	butyl acrylate/acrylic acid
BAEMA	butyl aminoethyl methacrylate
BNR	butadiene–nitrile rubber
BPI	propylene isophthalamide
BPO	benzoyl peroxide
BR	butyl rubber
CLC	carcass-like cross-linker
CMC	carboxymethyl cellulose or critical micelle concentration
CPP	cast polypropylene
CR	polychloroprene
CSBR	carboxylated butadiene rubber
CTA	chain transfer agent
DACP	diacetone alcohol/xylene cloud point
DBP	dibutylphthalate
DCPD	dicyclopentadienyl
DDM	dodecyl mercaptan

DMAEMA	dimethylamino methacrylate
DNA	deoxyribonucleic acid
DOP	dioctyl phthalate
EAA	ethylene–acrylic acid
EBA	ethylene–butyl acrylate
EBAc	ethylene–butyl acrylate
EHA	ethylhexyl acrylate
EHA/AA	ethylhexyl acrylate/acrylic acid
EMAA	ethylene–maleic anhydride
EO	ethylene oxide
EPDM	ethylene–propylene–diene multipolymer
EPR	ethylene propylene rubber
EPVC	emulsion PVC
EVAc	ethylene–vinyl acetate copolymer
EVOH	ethylene–vinyl alcohol copolymer
GMA	glycidyl methacrylate
HC	hydrocarbon
HDDA	hexanediol diacrylate
HDI	hexamethylene diisocyanate
HDPE	high-density polyethylene
HEA	hydroxyethyl acrylate
HEC	hydroxyethyl cellulose
HEMA	2-hydroxyethyl methacrylate
HFBA	heptafluorobutyl acrylate
HLB	hydrophylic–lipophylic balance
HMDI	hexamethylene diisocyanate
$H_{12}MDI$	cycloaliphatic dicyclohexylmethane
HMHDPE	high-molecular-weight high-density polyethylene
HMPSA	hot-melt pressure-sensitive adhesive
HIPS	high-impact polystyrene
HPC	hydroxypropyl cellulose
HPVC	hard polyvinylchloride
HRA	high-release additive
HTPB	hydroxyterminated polybutadiene
IB	isobutene
IBMA	isobutylmethacrylamide
IOA	iso-octyl acrylate
IPDI	isophorone diisocyanate
IP	isoprene
LDPE	low-density polyethylene
LLDPE	linear low-density polyethylene
M14	sodium tetradecyl 3-sulfopropyl maleate
MAA	methacrylic acid
MAGME	methyl acrylamidoglycolate methyl ether
MAM	MMA/BA/MMA block copolymer

MDI	4,4'-diphenylomethane diisocyanate
MDPE	medium-density polyethylene
MeP	methyl pentene
MMA	methyl methacrylate
MMAP	methyl cyclohexanone/aniline
MMT	montmorrilonite
MQ	methylsiloxane resin
Na-CMC	Na-carboxymethyl cellulose
Na-MMT	sodium montmorillonite
NIPU	nonisocyanate polyurethanes
NMA	*N*-methylolacrylamide
NR	natural rubber
NRL	natural rubber latex
NVCL	*N*-vinyl caprolactam
NVP	*N* vinyl vinylpyrrolidone
OMS	odorless mineral spirit
OPP	oriented polypropylene
OPS	oriented polystyrene
PA	polyamide
PAA	poly(acrylic acid)
PACM	polyacrylamide
PANQ	poly(5-amino-1-4-naphthoquinone)
PB	polybutene
PC	polycarbonate
PDFA	pentadecafluoro-octyl acrylate
PDMAEMA-co- MMA-co-BMA	poly(*N*-dimethylaminoethyl methacrylate-co- methyl methacrylate-co-butylmethacrylate)
PDMS	polydimethylsiloxane
PE	polyethylene
PEG	polyethylene glycol
PEO	poly(ethylene oxide)
PET	polyethylene terephthalate
PETA	pentaerythrytol triacrylate
PETP	polyethylene terephthalate
PEU	polyester urethane
PFO	phenol formaldehyde oligomer
PI	polyisoprene or polyimide
PIB	polyisobutylene
PMAA-co-EA	poly(methacrylic acid-co-ethyl acrylate)
PMAA-co-MMA	poly(methacrylic acid-co-methyl methacrylate)
PMMA	poly(methyl methacrylate)
PMS	paramethylstyrene
PO	propylene oxide
polyHEMA	poly(2-hydroxyethyl methacrylate)
PP	polypropylene

PPE	poly (ethylene-propylene)
PS	polystyrene
PTFE	polytetrafluoroethylene
PU	polyurethane
PVA	polyvinyl alcohol
PVC	polyvinyl chloride
PTFE	polytetrafluoroethylene
PVDF	poly(vinylidene fluoride)
PVA	polyvinyl alcohol
PVE	polyvinyl ether
PVOH	poly(vinyl alcohol)
PVP	polyvinyl pyrrolidone
RR	rubber–resin
RSH	thiol
SAN	styrene–acrylnitryl
SB	star branched(polymer)
SBAC	solvent-based acrylic
SBC	styrene block copolymers
SBR	styrene–butadiene–rubber
SBS	styrene–butadiene–styrene
SDED	sodium dodecyl diphenyl ether disulfonate
SDPS	sodium dodecyl phenyl sulfonate
SDS	sodium dodecyl sulfate
SEB	styrene–ethylene–butene
SEP	styrene–ethylene–propylene
SF	sol fraction
SHM	styrenic hot-melt PSA
SI	styrene–isoprene
SIBS	isoprene/butadiene mid-block
SIS	styrene–isoprene–styrene
SPVC	suspension PVC
TBC	tributyl citrate
TDI	toluene diisocyanate
TEC	triethyl citrate
TGPTA	triethyleneglycoltriacrylate
TMAEMA	*N*-trimethylammonium ethyl methacrylate chloride
TMPTA	trimethylol propanetriacrylate
TPE	thermoplastic elastomers
TPU	thermoplastic polyurethane
TRIS	trimethylolpropane mercaptopropionate
ULDPE	ultra-low-density polyethylene
VAc	vinyl acetate
VC	vinyl chloride
VLDPE	very-low-density polyethylene
VP	4-vinylpyridine

2. Terms

AF	adhesive failure or adhesive formulation
AFERA	Association of European Tape Manufacturers
AFM	atomic force microscopy
ASE	alkali-soluble emulsions
ASTM	American Society for Testing and Materials
ATR-FTIR	attenuated total reflectance/Fourier transform infrared
ATRP	atom transfer radical polymerization
BGVV	Bundes Gesundheitsamt (German Sanitary Administration, formerly BGA)
BOPP	biaxially oriented polypropylene
bp	boiling point
BUR	blow-up ratio
CAS	Chemical Abstracts Selects
CB	cardboard
CD	cross-direction
CEN	European Committee for Standardization
COF	coefficient of friction
CR	constant shear rate
CRT	clean room technology
C_w	coating weight
C_{wcr}	critical coating weight
d	size of deffect
D_0	diffusivity of the drug molecule
Da	Dalton
DCI	direct charge imaging
DIN	Deutsche Industrie Norm (German Industrial Standard)
DSC	differential scanning calorimetry
DDI	dart drop impact
DMA	dynamic mechanical analysis
DMTA	differential mechanical thermal analysis
DPI	dots per inch
DZ	diffusion zone
δ	diffusion path length
E	modulus of elasticity
EB	electron beam
ECG	electrocardiogram
EKG	electrocardiogram
EN	European Norm
EPA	Environmental Protection Agency
FDA	Federal Drug Administration (USA)
FFP	film-forming polymer

FINAT	Fédération International des Fabricants Transformateurs d'Adhesif et
	Thermo-collants sur Papiers et Autres Supports
	(International Federation of Manufacturers and Converters of Adhesives and Heat-Induced Adhesives on Papers and Other Carrier Materials)
Fipago	Federation internationale des producteurs des papiers gommés (International Federation of Gummed Papers)
FTIR	Fourier transform infrared
G'	storage modulus
G''	loss modulus
GARField	gradient at right angles to the field
GI	gastrointestinal tract
GMP	good manufacturing practices
GPC	gel permeation chromatography
h	thickness of gel layer at dissolution
HS	hot shear
IR	infrared
ISO	International Organization for Standardization
k	empirical constant
K_p	skin permeability coefficient
KS	Krämer–Sarnow (method)
LBL	layer by layer (construction)
LC	liquid crystalline
LCST	lower critical solution temperatures
LLC	ladder-like cross-linker
LT	loop tack
λ	retardation time
λ_{12}	coefficient of spreading of the 1st phase over surface of 2nd phase
MD	machine direction
M_e	entanglement molecular weight
MFI	melt flow index
MFT	minimum film forming temperature
MI	melt index
M_n	number average molecular weight
M_w	weight average molecular weight
Mp	melting point
MOPP_c	mono-oriented polypropylene copolymer
MOTT	Mechano-optical tack tester
MPR	melt-processed rubber
MVTR	moisture vapor transmission rate
MW	molecular weight
M_w	weight average molecular weight
MWD	molecular weight distribution

n	constant
NIR	near infrared
NMR	nuclear magnetic resonance
p	bonding pressure
P	peel strength
P_0	specific adhesion
P_{cr}	critical peel strength
P_i	initial peel strength
P_{max}	maximum value of peel strength
P_{skin}	peel strength to human skin
P_{steel}	peel strength to steel substrate
PCB	printed circuit board
PN	plasticity number (Williams)
PSA	pressure-sensitive adhesive
PSMA	Pressure Sensitive Manufacturers Association
PSP	pressure-sensitive product
PSTC	Pressure-Sensitive Tape Council
Q	heat
QS	quick stick
R	gas constant
RB	rolling ball
R&B	ring and ball
RBSP	ring and ball softening point
RBT	rolling ball tack
RC	rolling cylinder
RCT	rolling cylinder tack
Re	Reynold's invariant
RFID	radiofrequency identification
RH	relative humidity
RT	room temperature
S	full surface of the substrate
SAF	self-adhesive film
SAFT	shear adhesion failure temperature
SB	solvent based
S_{def}	nonfilled area of a defect
SEC	size exclusion chromatography
SLBT	shaft-loaded blister test
SP	softening point
SR	swell ratio
SS	stainless steel
S_{true}	"true contact" area
t	time
t^*	expected wearing time
t_{cr}	critical wearing time
T_{iso}	isotropization point

T_g	glass transition temperature
T_m	melting point
tan δ	phase angle, $\equiv G''/G'$, dissipation factor
TAPPI	Tape Pulp and Paper Institute
TDD	transdermal drug delivery
TEM	transmission electron microscopy
TEWL	transepidermal water loss
TLMI	Tag and Label Manufacturers Institute
TM	tapping mode
TMA	thermomechanical analysis
TWT	toothed wheel tack
τ	relaxation time
T	shear stress
UCST	upper critical solution temperature
USP	United States Pharmacopaea
UTM	Universal Testing Machine
UV	ultraviolet
V	debonding rate
VIP	variable image printing
VOC	volatile organic compound
VTF	Vogel–Tamman–Fulcher
W	work
W_i	weight fractions
WB	water-based
WLF	William–Landel–Ferry
ZN	Ziegler–Natta
Θ	contact angle
γ	surface energy
γ	deformation
$γ_i$	surface tension of *i*-th type component
η	viscosity
σB	stress at break
εB	ultimate elongation

Index